URBAN PLATFORMS AND THE FUTURE CITY

This title takes the broadest possible scope to interrogate the emergence of "platform urbanism", examining how it transforms urban infrastructure, governance, knowledge production, and everyday life, and brings together leading scholars and early-career researchers from across five continents and multiple disciplines.

The volume advances theoretical debates at the leading edge of the intersection between urbanism, governance, and the digital economy, by drawing on a range of empirically detailed cases from which to theorize the multiplicity of forms that platform urbanism takes. It draws international comparisons between urban platforms across sites, with attention to the leading edges of theory and practice and explores the potential for a renewal of civic life, engagement, and participatory governance through "platform cooperativism" and related movements. A breadth of tangible and diverse examples of platform urbanism provides critical insights to scholars examining the interface of digital technologies and urban infrastructure, urban governance, urban knowledge production, and everyday urban life.

The book will be invaluable on a range of undergraduate and postgraduate courses, as well as for academics and researchers in these fields, including anthropology, geography, innovation studies, politics, public policy, science and technology studies, sociology, sustainable development, urban planning, and urban studies. It will also appeal to an engaged, academia-adjacent readership, including city and regional planners, policymakers, and third-sector researchers in the realms of citizen engagement, industrial strategy, regeneration, sustainable development, and transport.

Mike Hodson is senior research fellow in the Sustainable Consumption Institute, Alliance Manchester Business School, University of Manchester and a member of the Manchester Urban Institute, UK.

Julia Kasmire researches and teaches on how to use new forms of data for social scientists with the UK Data Service and the Cathie Marsh Institute at the University of Manchester, UK.

Andrew McMeekin is Professor of Innovation at the Alliance Manchester Business School and the Sustainable Consumption Institute (SCI), UK.

John G. Stehlin is Assistant Professor in the department of Geography, Environment, and Sustainability at the University of North Carolina at Greensboro, USA.

Kevin Ward is a Professor in the Department of Geography and Director of the Manchester Urban Institute at the University of Manchester, UK.

URBAN PLATFORMS AND THE FUTURE CITY

Transformations in Infrastructure, Governance, Knowledge and Everyday Life

Edited by Mike Hodson, Julia Kasmire, Andrew McMeekin, John G. Stehlin and Kevin Ward

Routledge
Taylor & Francis Group
LONDON AND NEW YORK

First published 2021
by Routledge
2 Park Square, Milton Park, Abingdon, Oxon OX14 4RN

and by Routledge
52 Vanderbilt Avenue, New York, NY 10017

Routledge is an imprint of the Taylor & Francis Group, an informa business

British Library Cataloguing-in-Publication Data
A catalogue record for this book is available from the British Library

Library of Congress Cataloging-in-Publication Data
Names: Hodson, Mike, editor. | Routledge (Firm)
Title: Urban platforms and the future city: transformations in infrastructure, governance, knowledge production and everyday life / edited by Mike Hodson, Julia Kasmire, Andrew McMeekin, John G. Stehlin and Kevin Ward.
Description: First Edition. | New York: Routledge, 2020. | Includes bibliographical references and index.
Identifiers: LCCN 2020024810 (print) | LCCN 2020024811 (ebook)
Subjects: LCSH: City planning—Environmental aspects. | Sustainable urban development. | Urban policy.
Classification: LCC HT241 .U7266 2020 (print) | LCC HT241 (ebook) | DDC 307.1/216—dc23
LC record available at https://lccn.loc.gov/2020024810
LC ebook record available at https://lccn.loc.gov/2020024811

ISBN: 978-0-367-33418-5 (hbk)
ISBN: 978-0-367-33419-2 (pbk)
ISBN: 978-0-429-31975-4 (ebk)

Typeset in Bembo
by codeMantra

CONTENTS

FIGURES

TABLES

CONTRIBUTORS

Sarah Barns is a digital strategist, researcher, and creative producer, whose work today draws on a twenty year career working across policy research and urban strategy. Between 2014 and 2017 Sarah held an Urban Studies Foundation Post-doctoral Research Fellowship for her project *Platform Urbanism: data infomediaries, city labs and open data experiments in urban governance*, based at Western Sydney University. Today Sarah develops strategic partnerships and projects in the field of public space media, and advises in the field of urban data policy. Her book *Platform Urbanism: Negotiating platform ecosystems in connected cities* is published by Palgrave in 2020, and she has published in journals such as *Urban Geography*, *Geography Compass*, *Urban Policy and Research* and *City, Culture and Society.*

Shauna Brail, PhD, is an Associate Professor, Institute for Management & Innovation at the University of Toronto Mississauga. As an urban planner and economic geographer, 'Shauna's research focuses on the transformation of cities as a result of economic, social, and cultural change.

Ayona Datta is a Professor in Urban Geography in University College London. Her research interests are in gender citizenship, urban futures, and smart cities in the global South. She is author of "*The Illegal City: Space, law and gender in a Delhi squatter settlement*" (2012), co-editor of "*Translocal Geographies: Spaces, places, connections*" (2011) and "*Mega-Urbanization in the Global South: Fast Cities and New Urban Utopias*" (2016). Ayona is co-editor of *Urban Geography* journal, on the Trustees Board of the IJURR Foundation and on the editorial boards of the journals *Antipode*; *Digital Geography and Society* and *EPD: Society and Space.*

Niels van Doorn is an Assistant Professor in New Media and Digital Culture at the University of Amsterdam. He is also the Principal Investigator of the *Platform*

Labor research project (2018–2023), funded by the European Research Council. This project examines how digital platforms are (1) impacting how people work and secure a livelihood, and (2) reorganizing relations between market, state, and civil society in post-welfare times.

Tharsila Maynardes Dallabona Fariniuk is a post-doctoral researcher at the Pontifícia Universidade Católica do Paraná, and a lecturer on architecture and urbanism at UNIFACEAR, Brasil. Her research focuses on smart cities and emerging urban spatialities permeated by ubiquitous technologies.

Rodrigo José Firmino is a Professor in Urban Management at the Pontifícia Universidade Católica do Paraná (PUCPR) in Curitiba, Brazil, and a CNPq (Brazil's National Council for Scientific and Technological Development) Research Fellow. He is editor in chief of the journal Urbe (ISSN 2175–3369), and one of the founding member of the Latin American Network of Surveillance, Technology and Society Studies (LAVITS). Rodrigo has been publishing articles and essays in major journals about surveillance and space, augmented technologies and cities, and digital urbanisation, cyberculture, smart urbanism, among other themes.

Richard Heeks is Chair in Development Informatics at the Global Development Institute, University of Manchester; Senior Research Fellow at the University's Sustainable Consumption Institute; and Director of the Centre for Development Informatics (http://www.cdi.manchester.ac.uk). He has been consulting and researching on informatics and development for more than 30 years. His book publications include India's Software Industry (1996), Reinventing Government in the Information Age (1999), Implementing and Managing eGovernment (2006), ICTs, Climate Change and Development (2012) and Information and Communication Technology for Development (2018). His research interests are data-intensive urban development, e-resilience and e-sustainability, digital development, and the digital economy in developing countries. He has a PhD in Indian IT industry development, directs the MSc programme in ICTs for Development, and runs the ICT for Development blog: http://ict4dblog.wordpress.com.

Mike Hodson is senior research fellow in the Sustainable Consumption Institute, Alliance Manchester Business School, University of Manchester and a member of the Manchester Urban Institute. His research focuses on the shape of future sustainable cities, how transitions in urban infrastructures may contribute to sustainable cities, and governing processes shaping such transitions. He has published and presented widely on this agenda in both books and in leading peer-reviewed journals within the areas of urban studies, science and technology studies, and environment and planning. His work on urban digital platforms represents an extension of this agenda.

Julia Kasmire researches and teaches on how to use new forms of data for social scientists with the UK Data Service and the Cathie Marsh Institute at the University of Manchester. She approaches this task as an interesting combination of thinking like a computer (essential for data sciences) and thinking like a human (essential for social sciences) in the context of complex adaptive systems. She is deeply committed to equality, diversity, and inclusivity and is currently dabbling with stand-up comedy as a form of science communication.

Maroš Krivý is an Associate Professor and Director of Urban Studies at the Faculty of Architecture, Estonian Academy of Arts, and was previously a Research Associate at the Department of Geography, University of Cambridge. His research focuses on genealogies of neoliberal urbanism, and bridges the fields of urban geography, architectural history, and planning theory. He has published in leading journals, including *Planning Theory*, *Political Geography*, *IJURR*, *Architectural Histories*, and *The Journal of Architecture*, and contributed to numerous edited collections, most recently *Neoliberalism on the Ground* (University of Pittsburgh Press, 2020).

Anthony M. Levenda is an Assistant Professor in the Department of Geography and Environmental Sustainability at the University of Oklahoma. His research focuses on the urban governance of climate change and the politics of renewable energy transitions. He has published on this agenda in peer-reviewed journals within the areas of urban studies, science and technology studies, and environment and planning.

Andrés Luque-Ayala is an Associate Professor in the Department of Geography at Durham University, UK. His research focuses on the politics of urban infrastructures, unpacking the ways in which "the digital" transforms urban ecological flows. His most recent book, *Urban Operating Systems: Producing the Computational City* (w/Simon Marvin; MIT Press), explores the modest potentials and serious contradictions of reconfiguring urban life, city services, and urban-networked infrastructure through computational operating systems.

Dillon Mahmoudi is an Assistant Professor in the Department of Geography and Environmental Systems at the University of Maryland, Baltimore County and runs the Mapping Capital lab. His economic geography-based research focuses on questions at the nexus of cites, technology, and uneven development, seeking to trace how the flows of capital relationally produce uneven geographies. Through this lens, he publishes on software labor, emerging digital geographies, geographies of urban inequality, and digital political ecology through the combination of advanced quantitative methods, critical GIS, and qualitative research methods.

Juliana Marques is a data scientist with data_labe, a data journalism laboratory in Maré (Rio de Janeiro), one of the biggest favelas in Brazil. Formed in statistics at the State University of Rio de Janeiro (UERJ), she has over 7 years of experience in the telecommunications industry modelling emerging telecoms markets. At data_labe she focuses on data journalism and the democratization of data analysis.

Michele Masucci is the Vice President for Research, Professor of Geography and Urban Studies, and Director of the Information Technology and Society Research Group at Temple University. Her research focuses on the innovation ecosystem, community geographic information systems, digital inclusion, and university-community partnerships. Her current work examines how barriers to accessing information and communication technologies are interrelated with community development, environmental quality, and access to health, education, and opportunities for civic engagement.

Anurag Mazumdar is a PhD student in the Department of Geography and GIS, University of Illinois, Urbana-Champaign. His research focuses on the datafied politics of ride-hailing platforms in cities of the Global South, and he is interested in questions of digital labour, citizenship, and urbanism. Prior to academia, he has worked with academic journals, news websites, and think tanks, and brings this experience to his academic work.

Andrew McMeekin is Professor of Innovation at the Alliance Manchester Business School and the Sustainable Consumption Institute (SCI). He was Research Director of the SCI between 2013 and 2008 and Co-Director of the ESRC funded Sustainable Practices Research Group between 2010 and 2013. Andrew's current research focuses on sustainability transitions and urban platform innovation.

Susan Moore is an Associate Professor in Urban Development and Planning within The Bartlett School Planning, University College London. Her research interests include the relational geographies of urban (and suburban) development and built form, with a particular interest in the international proliferation of New Urbanism. Her research focuses on urban development and governance models and the circulation of so called "best practices". Recent projects include an analysis of the London 2012 Learning Legacy Agenda and an examination of the use of social media platforms as communicative infrastructures in local urban transformations.

Padmini Ray Murray is an independent researcher based in Bengaluru, where she founded Design Beku in 2018: a collective that emerged from a desire to explore how technology and design can be decolonial, local, and ethical. While her academic research has appeared in several peer-reviewed publications, she also creates new media work which reflects her research and interests: her most recent projects are "Visualising Cybersecurity" – a handbook that aims to alter how cybersecurity is depicted and discussed in the media (with the Centre for Internet and Society and Paulanthony George, 2019) and "A is for AI: A Dictionary of AI" (with Pratyush Raman, 2020).

Lizzie Richardson has a background in cultural and economic geography, with much of her research focusing on workplaces, work practices, and identities, including the changing geographies of work with digital technologies in the UK. She is also examining the socio-spatial dynamics of digital platforms for cultural and economic activities, particularly in urban contexts. She held a Leverhulme Early Career Fellowship in the Department of Geography at Durham University, after working as a University Lecturer in the Department of Geography at the University of Cambridge. She is currently a Lecturer in Digital Media and Society in the Department of Sociological Studies at the University of Sheffield.

Scott Rodgers is Senior Lecturer in Media Theory in the Department of Film, Media and Cultural Studies at Birkbeck, University of London. His research specializes in the relationships of media and cities and the geographies of communication. He also has interests in media production, journalism, urban politics, media philosophy, and ethnographic methodologies. Scott is currently exploring the convergence between digital platforms and the making of contemporary urbanism, locality, and publicness.

Jathan Sadowski is a Research Fellow in the Emerging Technologies Research Lab at Monash University. His work focuses on the political economy of digital systems. By casting a critical eye on technology, he seeks to uncover the often hidden interests, imperatives, and ideologies that influence the design and use of "smart" cities, homes, things, etc. Jathan's new book – *Too Smart: How Digital Capitalism is Extracting Data, Controlling Our Lives, and Taking Over the World* (The MIT Press) –analyses how the pursuit of power and profit is being materialized through smart tech in various ways and places in society.

Aaron Shapiro is an assistant professor of technology studies in the Department of Communication at the University of North Carolina-Chapel Hill and a former research fellow at New York University's Information Law Institute. His research examines urban technologies' social, political, and economic implications for vulnerable or marginalized communities. Aaron's book *Design, Control, Predict: Logistical Governance in the Smart City* is forthcoming with the University of Minnesota Press and studies how logistical problematics of efficiency and optimization motivate the integration of technology within urban institutions and the built environment.

Satyarupa Shekhar leads the work on urban governance at the Citizen Consumer and Civic Action Group, India. Satyarupa works to overcome challenges to access to basic services posed by a lack of data and information, while improving transparency and accountability. Her areas of interest are public administration, role of elected representatives, open data, and the use of technology to improve governance. She has been a part of innovative action-research projects that re-look at common urban challenges, including those on informal waste workers, public toilets, and decision making for municipal services. She has worked with Chennai's city government and one Tamil Nadu state department to improve their data management practices to plan civic infrastructure better, and has supported the creation of data that has been used by the city government and courts to provide evidence for failures in service provision. Satyarupa is recognised as a commentator on transparency and accountability and (open) data, and the role of data intermediaries.

John G. Stehlin is an Assistant Professor in the department of Geography, Environment, and Sustainability at the University of North Carolina at Greensboro. His research focuses on the political economy of urban mobility infrastructure and the emergence of active transportation as a new urban development logic. He is the author of *Cyclescapes of the Unequal City: Bicycle Infrastructure and Uneven Development* (University of Minnesota Press, 2019).

Gilberto Vieira is co-founder of data_labe, a data journalism laboratory in Maré (Rio de Janeiro), one of the biggest favelas in Brazil. He holds a Master in Culture and Territories. Through his activism work, he studies the mechanisms and articulations of cultural production in vulnerable territories, linking citizen policy and empowerment through new technologies and digital tools.

Kevin Ward is a Professor in the Department of Geography, Director of the Manchester Urban Institute at the University of Manchester and the Editor-in-chief of *Urban Geography.*

Alan Wiig is an Assistant Professor of Urban Planning and Community Development at the University of Massachusetts, Boston. An urban geographer, his research examines global infrastructure, smart urbanization, and the form, function, and politics of urban revitalization agendas across the North Atlantic.

Clancy Wilmott is an Assistant Professor in Critical Cartography, Geo-Visualisation and Design with the Berkeley Centre for New Media and the Department of Geography at the University of California, Berkeley. Previously, she has been a Vice-Chancellor's Postdoctoral Research Fellow in Design and Creative Practice at RMIT University, Lecturer in Human Geography at the University of Manchester, and a Postdoctoral Researcher in the Centre for Interdisciplinary Methodologies at the University of Warwick. Her research bridges geography and media studies, focusing on the intersections between cartographies, digitalities, and spatial practices, with an emphasis on the politics of knowledge and representation. She is the author of *Mobile Mapping: Space, Cartography and the Digital* (2020) with Amsterdam University Press.

Luke Yates is a Lecturer in Sociology and Researcher at the Sustainable Consumption Institute at the University of Manchester. His research focuses on collective action, politics, and everyday consumption practices. He is particularly interested in how these play out in processes of change such as through protests and occupations, politicised forms of consumption and lifestyle, and in the organisation of daily life. His current work examines the political struggles and social movements in and around the platform economy, and he is also working on drawing together conceptual strands from social practice theory, social movement studies, and theories of socio-political transformation.

ACKNOWLEDGEMENTS

How we think about relationships between digital platforms and urban contexts has been the subject of a rapidly developing body of social science literature over recent years. This book seeks to engage with various strands of this emerging literature and to begin building some thematic coherence around what we understand as urban platforms and their implications for the cities of the future. In doing this, the concern of the book is to ask what we mean when we talk about urban platforms and what their implications are for the organization of urban infrastructure, existing modes of urban government and governance, how knowledge about urban contexts is generated and mobilized, and what this might mean for everyday urban life.

To do this, the volume brings together a group of established and early-career researchers who each have a unique contribution to make to debates around urban platforms and the future city. It includes contributions from those in development studies, human geography, information and communications technology, media studies, and sociology. We would like to thank each author for their contribution and the ways in which they have engaged with the process of producing this book.

This process began with an international workshop – "Urban platforms and the future city: Transformations in infrastructure, governance, knowledge, and everyday life" – in February-March 2019, organized and hosted in Manchester by the Sustainable Consumption Institute and the Manchester Urban Institute. Many of the authors spoke at this workshop, where early versions of their chapters were presented and discussed in a critically supportive environment. We would like to thank the Sustainable Consumption Institute, the Manchester Urban Institute, Sharing Cities Sweden and Formas, the Swedish research council for sustainable development, project title Advancing urban innovation: Living labs for sustainable building and planning, for the support they each provided for the workshop.

In addition to chapters that were presented at the workshop, we also invited a small number of chapter contributions from academics whose work resonates with the issues and themes in this book. A number of these chapters are revised versions of articles that were previously published as journal contributions. We would like to acknowledge the use of the following revised articles:

Shapiro, Aaron. "The Urban Stack. A Topology for Urban Data Infrastructures". *TECNOSCIENZA: Italian Journal of Science & Technology Studies* 8, no. 2 (2017): 61–80.

Mahmoudi, Dillon, and Anthony M. Levenda. "Beyond the Screen: Uneven Geographies, Digital Labour, and the City of Cognitive-Cultural Capitalism". *TripleC: Communication, Capitalism & Critique. Open Access Journal for a Global Sustainable Information Society* 14, no. 1 (February 17, 2016): 99–120. https://doi.org/10.31269/triplec.v14i1.699.

Sadowski, Jathan. "When Data Is Capital: Datafication, Accumulation, and Extraction". *Big Data & Society* 6, no. 1 (2019): 1–12. https://doi.org/10.1177/2053951718820549.

Manchester and Greensboro, March 2020

1

INTRODUCTION

Mike Hodson, Julia Kasmire, Andrew McMeekin, John G. Stehlin and Kevin Ward

We are witnessing a profusion of digital platforms that play an increasingly pervasive role in mediating many areas of life. By 2015, every one of the top ten trafficked US websites was a platform, while in China platform businesses held each of the top eight spots in Chinese web traffic rankings (Moazed and Johnson, 2016), and American and Chinese platforms made up nine out of the top ten most downloaded apps of the 2010s (Meisenzahl, 2019). Platforms represent an increasingly hegemonic business model, predicated on the capacity to realise exchange value not through direct production of goods or services but through building and mobilising *networks* of producers, distributors, and consumers, constructing and shaping the activities and practices of participants in such networks, and deriving rents from these activities (Stehlin, 2018; Sadowski, 2020).

Platforms can be categorised in a number of ways, but a common feature is their function as a mediator between various network actors (Srnicek, 2016; Langley and Leyshon, 2017). These mediators have come to condition how we shop, what we buy in terms of both products and services, how we pay for things, how we move, how we communicate and socialise with each other, and how we relate and manage personal information, as well as the organisation of industrial processes and the infrastructural architectures that facilitate all of these activities. Platforms increasingly mediate everyday consumer experience of digital life, relationships between firms, and the 'back end' of a vast range of digital processes, such that platforms like Amazon and Google, for example, are nearly inescapable (Hill, 2019). The 'winner take all' advantages conferred by strong network effects create a tendency towards monopoly, leading to fierce battles, mergers (both horizontal and vertical), and sudden failures in the platform economy. These dynamics are fuelled by huge inflows of venture capital into the platform sector, driving massive valuations and equally fragile bubbles.

Platforms are also increasingly prevalent as an urban phenomenon, in part because cities provide the largest and richest markets for their services. But their importance goes far beyond the general phenomenon of 'platform capitalism' occurring in an urban setting (Srnicek, 2016). Instead, digital platforms fundamentally, and unevenly, reconfigure urban space and life itself. Developers of digital platforms are increasingly, and rapidly, seeking to intervene in nearly all aspects of urban life, from built environments and infrastructure systems to environmental monitoring and civic engagement (Barns et al., 2017). This can be seen across a range of digital platforms that directly reshape urban space, including well-known platforms like Airbnb on home-sharing, CityMapper on journey planning, FreeNow on car sharing, and Uber on ride-hailing; rising giants like China's Didi Chuxing and Indonesia's Go-Jek app ecosystems; and less prominent platforms like Germany's Coup Mobility electric scooter sharing service to Uganda's SafeBoda moto taxi-hailing platform. These platforms form the infrastructure of an emerging digitally mediated urbanism that is reconfiguring cities across the globe.

Accordingly, this book interrogates the emergence of 'platform urbanism' (Barns, 2020), bringing together contributions from across a number of disciplines, including geography, innovation studies, urban planning, and media studies. It takes the broadest possible scope, examining how platform urbanism transforms urban infrastructure, governance, knowledge production, and everyday life (and, in turn, how these aspects challenge notions of platform urbanism). It brings together researchers from across five continents and has four main aims:

- To advance the leading edge of theoretical debates at the intersection of urbanism, governance, and the digital economy;
- To draw on a range of empirically detailed cases to contribute to theorising the multiplicity of forms platform urbanism takes;
- To illuminate international comparisons between urban platforms across sites, with attention to the leading edges of theory and practice through a rich array of cases;
- To explore the potential for a renewal of civic life, engagement, and participatory governance through 'platform cooperativism' and related movements.

What is platform urbanism, and why now?

Our intention in this book is to emphasise the processes of urban platform*isation*. This complements and extends existing literature and debates on platform capitalism (Srnicek, 2016; Langley and Leyshon, 2017) and also moves beyond a relatively static analysis of platform types derived from a few large firms and towards a sensitivity towards trajectories (Stehlin et al., 2020). The focus here is on both platforms as entities and platformisation as a process, where the latter complements understandings of the former.

Most interest to date has addressed platforms as disruptors of industries, understood through the lens of Schumpeterian processes of creative destruction (Geissinger et al., 2018; Kibum et al., 2018). Celebrated examples include Apple's iPhone and App Store ecosystem supplanting a generation of mobile handset manufacturers, Spotify reconfiguring the consumption of music from the tangible CD and record to a digital music service, and Uber's challenge to establish models of private taxi hire. Much academic focus has been on understanding how platform business models interact with incumbent firms, how value is created and captured for innovating firms, and how new jobs are produced and existing jobs are destroyed (Gawer, 2014).

Seen broadly through this lens, platforms are architectures that organise and control networks, providing a context and infrastructure for people and firms to create and exchange value in new ways, through matching them with each other and with content, goods and/or services created on the platform. At a general level we can say that 'platforms are digital infrastructures that enable two or more groups to interact. They therefore position themselves as intermediaries that bring together different users: customers, advertisers, service providers, producers, suppliers, and even physical objects. More often than not, these platforms also come with a series of tools that enable their users to build their own products, services, and marketplaces' (Srnicek, 2016, p. 43). Or, to put it in a slightly different way: platforms allow for the creation of new markets, where consumers and producers can connect with each other and exchange goods, services, and information (Moazed and Johnson, 2016). Crucially, the platform firm does not own the services, but only the underlying network infrastructure and the system architecture that controls it, which allows rapid deployment and 'viral' network growth at low marginal cost.

Platforms do not simply enable such activities in an open-ended way, but actively steer and shape them, raising the issue of their internal algorithmic structure. This consists of a 'stack' with three elements: (1) technological architecture; (2) service provision (or community, network, etc. of producers and consumers); and (3) data (Bratton, 2015). The primary function of this structure is to convert disparate pieces of user information into processed 'data capital' (Sadowski, 2019) by consistently expanding the network of users and developing new ways of measuring and monetising activity across these networks. This means that forms of urbanism generated through certain kinds of platform activities, even those with seemingly emancipatory potential, have definite tendencies towards entrenching platform power that are difficult to overcome.

Digital platforms form a specific element of a shift towards a digital economy, where the interweaving of corporate interests, new business models, emerging information technologies, and data capture is becoming a hegemonic model (Srnicek, 2016; Langley and Leyshon, 2017). Indeed, platformisation speaks to a wider societal transformation that scholars have grappled with in debates not only about the digital economy (Mosco, 2017) and the development of 'smart' governance responses (Kitchin, 2014; Shelton et al., 2014; Leszczynski, 2016) but also about the circular economy (Geissdoerfer et al., 2017), the sharing economy (Sundararajan, 2016), and platform 'cooperativism' (Scholz and Schneider, 2016).

Platforms and platform urbanism

This explosion of digital technologies and data has increasingly played out in urban contexts and through urban platforms. Debates have proliferated in policy, business, and academic circles regarding efforts to integrate digital technologies and urban territories as 'smart' cities (Kitchin, 2014). These debates fuse the disruptive potential of digital technologies with attempts to re-think how services and infrastructures that underpin urban life can be governed through 'smart' systems and technologies. Our focus on platform urbanism (Barns 2020), though, goes beyond more general questions of how ICTs are changing urban living and governance and is also distinct from analyses of 'smart cities' and other forms of data-informed urban governance. To a certain extent, platform urbanism has begun to supplant smart urbanism; where 'smart city' policies installed 'urban operating systems' (Marvin and Luque-Ayala, 2017), platform urbanism implies monitoring techniques and governance efforts that emerge from the proliferation of platforms that are given as much latitude for operation as possible. For example, instead of public-private partnership contracts for 'smart' traffic monitoring, we are more likely to see Uber's data 'dashboard', Uber Movement, an even more opaque technological interface, performing a similar governmental role. In the tension between the platform and the urban, a multiplicity of 'types' of urban platforms proliferate (Stehlin et al., 2020).

Thus, it is important to clarify what we mean by 'urban platforms' and what it is that is *urban* about these platforms. At issue is not simply whether the bulk of platform activity occurs in cities, but the extent to which the activities they enable are constitutive of and/or parasitical on urban space: how they create a platform city (Anttiroiko, 2016). As Rodgers and Moore (2018a) note, platforms can be parasitic on urban infrastructures, novel infrastructures in their own right, and co-generative with the urban, where at issue is 'how the urban shows up in, through and as platforms; and at the same time, how platforms show up in, through and as urban' (2018b). At the same time, a distinction can be made between how single platforms intersect with other urban dynamics and how wider, dynamic ecosystems of platforms and urban information infrastructures rework entire dimensions of urban life (Andersson, 2017). As digital technologies and data become central to new, 'disruptive' business models that promise the breakup of established monopolies, settled labour relations, and existing technologies, it is also urgently necessary to understand the social interests that promote platform development and deployment, the kinds of resources they marshal, and to what ends.

Why, though, the turn now towards research on platform urbanism? Long-term processes of capitalist development, punctuated by crises and responses that inform the search for new forms of capital accumulation, have begun to synthesise with the increasingly important role of information, data, and knowledge in economic life (Srnicek, 2016). These developments have had a long gestation, arguably back to previous capitalist crises of accumulation in the 1970s and the weakening of post-War models of nationally based Fordist industrial economies amid a secular decline in manufacturing profitability in the global North (Brenner, 2006). The increasingly explicit role of information and knowledge in the organisation of economic life has been understood over this period through the lens of a shift to post-industrial society (Bell, 1976) and as a phase of disorganised capitalism (Lash and Urry, 1987) predicated on economies that increasingly manipulated signs, symbols, and spaces (Lash and Urry, 1994). The issue is that 21st-century capitalism is increasingly not just about extracting data but also about utilising this in producing knowledge to generate value through cultural circuits (Thrift, 2005; Scott, 2012). This political economy prioritises financialisation, interweaves this with innovations in IT, regulation, and organisation to create new possibilities for profit (Thrift, 2005; Krippner, 2011). These dynamics were intensified by the financial and economic crisis of 2007–2008, which has fuelled the search for new investment horizons and accumulation strategies. It is within this confluence of political-economic change and the importance of data and knowledge, and technological and organisational innovation that the circumstances that have given rise to the platform can be understood.

Equally crucial in the post-crisis moment has been the intensification of urban development processes, as cities across the globe have been remade as part of a new 'spatial fix' for global capital (Harvey, 2007) that combines intensive growth in urban cores with an increasingly technocratic approach to urban management, particularly environmental governance (Hodson and Marvin, 2010). Long-standing questions about the right to the city and the production of urban space (Lefebvre, 1968, 2003) have been reinvigorated at this moment (Harvey, 2008) because competing visions of urban futures grapple implicitly and explicitly with the contradictions of the city as 'an engine for capital accumulation, on the one hand, and a site for social/class struggle, on the other' (Merrifield, 2014, p. 1). The current iteration of this struggle poses the view of the city as an innovation machine (Florida et al., 2017) against wide-ranging visions of a more cooperative, civically oriented, and/or socially just digital urbanism. At stake in this is the fundamental question of who has the right to shape the platforms that govern urban life.

This raises questions at both the level of how we understand the organisation of urban space in a context of increasing platformisation but how political struggles have begun to articulate alternative possibilities for platform-based urbanism (Aronoff et al., 2019). At the urban scale, these issues necessitate a response. Urban futures, including how they are shaped through platforms, are, like all futures, up for grabs and the subject of struggle. This can be seen in debates suggesting that there is a shift from ownership to access (Scholz and Schneider, 2016; Sundarajan, 2016) and between users (through sharing). Platforms, through these decisions, are also reconfiguring urban space, almost always in uneven ways.

In this sense, when platforms meet urban context a richer understanding is needed than seeing the platform as a business model or in terms of its technical possibilities. The purposes to which platforms are mobilised are many and competing (e.g. to generate profit, to support civic priorities, to denigrate or enhance public service provision, to contribute to reducing carbon emissions, and so on). How this struggle unfolds is a question of politics, participation, and governance. That is to say, it is necessary to understand who the social interests are that are shaping the ways in which platforms are introduced into urban contexts. It also becomes important to understand how the socio-technical configurations that support new urban platforms are organised. The challenge is that it is hard to know in advance the effects that multiple urban platforms will have on existing urban contexts.

The relationship between digital platforms and the urban has a socio-technical dimension, raising questions about the implications of digital platforms for existing urban *infrastructure* systems and in particular what forms of reconfiguration this produces across scales and contexts (Hodson et al., 2017). Understanding urban platforms as socio-technical focuses our attention on their *governance*, and how the capacity to govern is and should be organised and is being reconfigured in relation to platforms at an urban scale. Governance interests, whether policymakers, business, civil society groups, and so on, bring not only different priorities but also forms of knowledge and other resources to the process of configuring platforms in urban contexts, leading to significant power asymmetries, largely opaque functionality, and often outright hostility to democratic oversight.

Platform governance is not merely an administrative process of aligning social interests in a given city or region but also needs to be understood in response to multiple political-economic-ecological challenges across scales, the effects of which will be geographically variegated. These challenges are many, but include, for example, a rapid growth of urban areas, particularly in the Global South; processes of privatisation and liberalisation that call into question public modes of governing; the prevalence in some contexts of public sector austerity; and the challenges presented by climate change, resource 'scarcity', and 'carbon control' politics (Jonas et al., 2011), and so on. There is often a disjuncture between the scale of economic and ecological challenges facing urban contexts and the fragmentation and/or denigration of political and social capability to respond.

The fragmentation of urban *knowledge* across government, civil society, and private firms leads to uncertainty over how to respond to these challenges, and has, in turn, promoted something of an experimental turn in both policy circles and in academia in seeking to make and understand urban futures (Bulkeley and Castan Broto, 2013; Evans, 2016). This is not only about fragmented knowledge but also the privileging of particular forms of market-oriented urban expertise. There is a politics to this. The mobilisation of platforms as experimental knowledge, backed by particular social interests or coalitions of interests, often represents efforts to gain digital control of part or whole urban systems of provision from a particular point of view. Signifiers of this include the massive investments in urban platforms by venture capital (Stehlin et al., 2020); battles over regulation between platforms and place-based political interests (e.g. taxi drivers protesting Uber in London and various other cities, motorcycle and public transport drivers protesting Go-Jek in Jakarta, and so on); Airbnb's resistance to providing public data on the scale and scope of its 'home-sharing' service; and a vibrant landscape of civic alternatives to corporate visions of urban platforms (e.g. public sector and cooperative experiments in Barcelona and Helsinki). Platforms are also central to reconfiguring relations of production and consumption with implications for urban *living* and everyday urban life.

The question that follows all of this is: in what (variety of) ways are platforms configured in urban contexts, how does the urban challenge and shape platforms, with what implications, and through what kinds of political contestation? To address this, the contributions to the book are organised into four parts. This consists of chapters that deepen understanding of the socio-technical configurations that support urban platforms and their implications for existing urban infrastructure systems, modes of urban governance, urban knowledge, and urban life. These chapters are often sensitive to particular urban contexts but collectively provide possibilities for a more general analysis that is generated from multiple urban contexts.

The book and its structure

The book is organised into four parts. These are as follows:

1. What kind of urban infrastructure are platforms?

   Platforms are architectural innovations, in the sense that they change linkages and interdependencies between socio-technical elements of *infrastructure systems* (Henderson and Clark, 1990; McMeekin et al., 2019). In terms of the scale and social interests promoting platforms, we are seeing the emergence of both planetary platforms (initially led by big technology companies such as Amazon, Apple, Facebook, and Google but now including more specifically urban-oriented firms like Airbnb, FreeNow, Grab, and Uber) and urban/metropolitan-scale initiatives spearheaded by local governments and affiliated stakeholders. The latter include both situated expressions of corporate planetary platforms (e.g. Google's Sidewalk Labs or Didi Chuxing's smart traffic signals programme) and civic or co-operative responses (Stuttgart's Polygo citizen card or the Coopcycle app developed by food delivery workers). Platforms nevertheless display common features: a digital architecture that coordinates interactions; a dependence on physical data storage and circulation infrastructure; and a fundamental drive to expand network effects, in which each additional user increases the importance or value of the rest (Srnicek, 2016).

Infrastructure, therefore – what Star calls a 'system of substrates' (1999) – is a 'useful framework for analyzing the use and implications of online platforms that are effectively serving as shared public infrastructure...' (Nash et al., 2017, p. 369). A key point of this is the hidden politics where "'platform' above all appears to be deployed in ways that suggest neutrality, and perhaps immateriality'" (Rodgers and Moore, 2018a, 2018b)[1]; but where 'platforms are not utilities or conduits that simply channel circulations. Platforms actively induce, produce and programme circulations' (Langley and Leyshon, 2017, p. 17). This active quality is reflected in the use of 'infrastructure' as a noun, indicating physical, technical, and social networks, but also as a verb – 'infrastructuring' (Bowker and Star, 2006) – signifying its capacity to mediate and shape relationships and processes between things and people (see Easterling, 2014). In short, as infrastructure, platforms have an internal, largely 'black-boxed' architecture with definite affordances (Gibson, 1979) that shape how they can actually be used. But they also display emergent properties, as the accumulation of user data creates new potential avenues for growth on which platform owners and investors capitalise.

While venture capital and large corporations are the dominant social interests driving investment in platform infrastructures, platform capitalism is highly variegated, incorporating governments, business accelerators, civic associations, and activist groups. Nevertheless, market power over network organisation endows certain major firms like Amazon and Google with substantial control over a wide array of platform services. 'A key question', Andersson notes, 'is to what extent single corporations control entire panoplies of interconnected platforms. Many of the newer, smaller platforms are dependent on preexisting, larger ones, whose dominance is further solidified' (Andersson, 2017, pp. 380–381). This layering and interconnection of platforms has been characterised as 'platformisation' (Anttiroiko, 2016), suggesting an ongoing process of network formation, contestation, and consolidation. This means that the intersections between platform, infrastructure, and the urban take manifold forms. Platforms, as infrastructure, enable action at a distance, but build on and repurpose existing physical infrastructures (housing, roads, ICT infrastructure, GPS satellites, energy grids) without which they could not function. Thinking through the lens of infrastructure 'means thinking about the degree to which platforms are parasites of different kinds of urban infrastructure, but it also means taking things one step further, and thinking about platforms per se as new forms of urban infrastructure' (Rodgers and Moore, 2018b).

The centrality of infrastructure to thinking about urban platforms but also the multiplicity of issues this raises contribute to a research agenda that needs to be addressed. The chapters in this part of the book make a start in doing so by bringing together how urban platforms are constituted as infrastructure with the more mundane, far-reaching, and largely invisible communications and energy infrastructures that support platform urbanism.

Aaron Shapiro's chapter begins with a detailed examination of the urban platform 'stack': the interdependent linkages of the infrastructural base, control, and interface layers internal to a software platform that, in the case of urban platforms, directly organise everyday material urban objects and practices into a new set of relationships mediated by platform algorithms and controlled by their owners. Aaron draws on two case studies: the LinkNYC WiFi hotspot network launched in 2016 and on-demand service platforms like Uber, Postmates, and Caviar. He demonstrates how the opacity of the linkages between the distributed infrastructural base of smartphone users and the control and interface layers enables ostensible services – free public WiFi infrastructure or on-demand delivery – to function as data collection and commercialisation platforms.

The next three chapters explore the infrastructures that organise and sustain platform urbanism itself. Shauna Brail's chapter, for example, focuses on the new inter-urban relationships forged by planetary-scale ride-hailing platforms like Uber, Go-Jek, and Gett. Shauna argues that the explosion of the platform economy has disrupted the 'global' or 'world city' model of financial centres and corporate command-control nodes established by Saskia Sassen and her interlocutors (Sassen, 1991; Taylor, et al.., 2000). Cities like San Francisco, Tel Aviv, Jakarta, and Singapore are part of a new global infrastructure of software programming, data management, and firm governance that is crucial to the *production process* of ride-hailing. Shauna points to the importance of the intangible qualities of urban agglomeration and social life as keys to cities' functions in this dynamic network, but these factors are potentially undermined by the unequal division of labour between a small cohort of well compensated software engineers and the massive ranks of precarious workers whom the platform deploys.

Alan Wiig and Michele Masucci continue this theme, excavating the hidden layers of information technology (IT) infrastructure that immaterial metaphors of 'cloud' or 'fog' computing tend to occlude, as well as the human relationships and everyday practices that sustain the platform city. They focus on post-industrial North Philadelphia, noting how telecommunications and digital infrastructure have surged into gaps left by deindustrialisation, occupying ageing buildings with servers and cellular towers but doing little to restore employment prospects for nearby residents. Job training programs intended to prepare low-income residents of colour for highly paid digital labour in the platform economy have faded, enrolling them instead primarily as consumers of, or precarious workers within, digital platforms. This calls our attention to the way that, for cities, fostering an 'ecosystem' of urban platform infrastructures is entangled with promises of modernisation and improvement that may never materialise.

Finally, Dillon Mahmoudi and Anthony Levenda move beyond the confines of the city itself, examining the political ecologies that link urban platforms, firms like Amazon, Facebook, and Google that provide the basic infrastructure of data hosting, consumer tracking, and web mapping, and the proliferation of energy- and material-intensive rural servers through which data flow. In the case of the Pacific Northwest, the platform city rests on a hidden metabolic relationship to the vast landscape of hydroelectric power generation, which attracts data centres to rural areas with cheap electricity and permissive tax regimes. Counter to narratives of 'immaterial' labour that pervade analyses of 'cognitive-cultural capitalism' (Scott, 2012), they highlight the material linkages between high-wage 'symbolic analysts', everyday precarious gig economy workers, and the spatial division of labour between the platform city and its energy-producing hinterland.

2. Do platforms represent a new model of urban governance?

   Many platforms appear to represent, or even promise, a *new model of governance*; one that results in new configurations of public, private, and non-profit actors. Given that '[i]n a generic sense, a *platform* is any physical, technological or social base on which socio-technical processes are built' (Anttiroiko et al., 2017, p. 329, original emphasis) there is the potential for almost limitless configurations to constitute new forms of urban governance, with novel models of ownership and control and with diverse interactions with pre-existing urban governance arrangements. Whereas much of the economics and management literature seeks to elucidate the *essence* of platform governance as two-sided markets and technological architectures (Gawer, 2014), the current volume has an orientation for understanding shifting urban governance through social *processes* of platformisation.

   In relation to urban public governance a 'platform is not only a tool for managing information but also in a wider sense a framework within which to involve key stakeholders in governance processes and to seek solutions to complex social problems' (Anttiroiko et al., 2017, p. 329). Platforms that promise a 'turnkey' dashboard or smarter, 'algorithmic' governance relocate substantive decision-making into the platform software itself (Barns, 2016; Leszczynski, 2016). Others, like Uber and Airbnb, often actively resist established governance altogether. The plurality of actors and sectors involved in urban platforms, and growing concerns about the effects on democracy that these platforms provoke, may also be informing a significant shift towards more civic and cooperative forms of value capture (Scholz and Schneider, 2016).

   In terms of the shape of both urban governance configurations and their purpose, questions of ownership are key. Critical to this is ownership and control of the technological infrastructures and architectures that generate data. At an urban scale, the key struggle is not over individual platform technological arrangements but whether and how technological architectures are organised as an urban system and to what end. That is to say, how cloud platforms, sensors, algorithms, technological operating systems, and the standards on which they operate are organised at the scale of city or metropolitan area. Cities, for example, can be seen as sites from which to experiment with platform infrastructure, to cultivate the normalisation and institutionalisation of the corporate control of infrastructure and to *extract* and monetise data. From this perspective, Alphabet, Google's parent company, is seeking to develop a broad range of urban services, including city maps, real-time traffic information, free wifi, and self-driving cars. The critical point is that 'Alphabet essentially wants to be the *default platform* for other municipal services' (Morozov, 2017)[2]. They would set the rules which other platform services would have to adhere to.

Critics of corporately controlled urban platforms may highlight numerous issues: their potential to undermine democracy, their promotion of particular (uneven) geographies and, with that, the development of enclaves and the promotion of inequality, and the private appropriation of benefits generated from new platform architectures. The suggestion is that experimentation with alternative models of collective ownership can begin to confront these criticisms and conceive of forms of institutionalisation and ownership through which platforms may be able to strengthen democracy, to address insecurity and inequality, and to appropriate benefits generated for the public good rather than private gain. Experimentation with forms of civic, municipal, or cooperative urban platforms can be seen in Barcelona, Madrid, Helsinki, London, and also in less celebrated urban areas, such as Preston in the UK. By contrast to corporate urban platforms, municipal authorities in Barcelona are seeking to constitute an alternative, a 'test bed for developing a more citizen-focused form of participative democracy' based on a civic view of services and resources but where the view of the necessity of a common (default) platform is central (Tieman, 2017)[3].

A wider analytical point relates to how pre-existing modes of urban governance and service provision have been perturbed by the whole range of platform innovations that have been implemented in particular jurisdictions. This relates to further issues, including how embryonic public and civil society-owned urban mobility platforms seek to institutionalise and the struggles they face in stabilising and scaling up; how the diffusion of urban platforms across cities involves different patterns of societal embedding or resistance – in terms of public support or opposition, and in terms of regulatory assault or renegotiation; how urban platforms promote or constrain, over time, civil engagement and participation in the planning of urban services; how the 'data exhaust' captured from the use of urban mobility platforms is used variably for either the promotion of a data commons, open government or propriety databases.

It is important to identify sometimes contradictory interests and coalitional politics involved in such efforts, as well as detail the forms of platform organisation and governance that they propose. The chapters in this theme contribute to debates on whether urban platforms represent a new model of urban governance, and if so what the implications are for already established modes.

Sarah Barns develops an approach for understanding processes of platform intermediation which underpin the creation of 'platform ecosystems' as novel modes of urban governance. Using Uber as an emblematic example, Sarah argues that platform intermediation enrols diverse actors in forms of value-sharing which, while not equally distributed, are nevertheless vital to the underlying appeal and scalability of platform-based urban interventions. As such, understanding the influence of Uber as a platform means moving beyond a simple view of platform value extracting towards the view of the platform ecosystem as engineered sociality and governance, enlisting many different parties in value sharing activities.

Through a case study of Airbnb, Niels van Doorn examines its shift from regulatory evasion towards a new role as 'regulatory entrepreneur', where changing laws becomes a central component of its business model. This shift, it is argued, views Airbnb as a new urban institution transforming governance relations between state, market, and civil society. The chapter discusses Airbnb's strategies as a regulatory entrepreneur, especially discursive strategies and the way that it mobilises and instrumentalises its user base of entrepreneurial property owners.

Luke Yates focuses on similar themes but argues for an approach that goes beyond understanding the specificities of particular platform organisations in the context of their associated fields (e.g. Airbnb in relation to urban housing, short-term rental, and gentrification). As such, he casts a wider net to look across platform-based conflicts involving profit and non-profit organisations across a variety of sectors, looking for commonalities and differences. Platform organisations are found to declare themselves as vectors of progress, albeit in very different ways according to the social interests involved. Similarly, platform-based grassroots lobbying is also found to be a common strategy, via creating new advocacy groups, mobilising their user base, and building alliances. As such, commonalities in strategy are revealed, but so are differences in purpose and visions which point to intense political struggles around the meaning of urban platforms and the future of urban infrastructure and its governance.

Richard Heeks and Satyarupa Shekhar also focus beyond the high-profile corporate platforms, in their case on the 'platformisation of slums', specifically the urban data platforms that store, process, visualise, and disseminate data on low-income groups and settlements. They introduce a general 'platform justice' model before proceeding to an analysis of the right for slum communities to be represented, the legibility of slums to external actors, and what this means for urban equality. Slums are often unmapped by the major mapping platforms, thereby rendering their assets, services, voices, and livelihoods of citizens invisible. But, when platforms do cover slum settlements, there are risks that residents are excluded from platform-related labour and services.

3. What kinds of knowledge of urban futures are generated, legitimised, and valued through platforms?

   Platforms produce an enormous amount and variety of 'real-time' data on urban activities that steer urban futures, making some more likely than others (Kitchin, 2014). Data may be used to coordinate all manner of activities and services. These data are often quite particular and partial, however, and many times depend on the platforms themselves for use and interpretation (e.g. urban 'dashboards') (Mattern, 2015; Kitchin et al., 2016). Implicitly, acting on these data brings forth some urban futures as others fall by the wayside. How can we better understand the politics of this data production and scenario-making, and how it contributes to attempts to control (and produce) urban futures?

   The organisation of urban platforms in stacked configurations (Shapiro, 2017) and their governance shapes the data that are produced and, ultimately, the forms of urban knowledge. The power of digital platforms is that they are 'knowledge operators (introducing new conceptions of truth and knowledge...)' (Andersson, 2017, p. 380). One view is that through focusing on how the urban stack is configured 'we might better understand how urban data can be made a valuable commodity' and how 'value production for urban data infrastructures hinges on their producers' ability to enrol heterogeneous elements into their stacked configuration, and then use this configuration to control the flow of data and information' (Shapiro, 2017, p. 76). This refers to how the data are produced and how they are translated into urban knowledge and communicated to citizens and other groups. This means that how the platform is configured, and the thinking behind this configuration, is critical to the kind of data that is produced; whether that be a business-led configuration of digital economic circulation (Langley and Leyshon, 2017), as participatory innovation platforms (Anttiroiko, 2016) or various other configurations.

   Architectures of monitoring and sensor technologies become important and are used to manage, map, and influence people's activities. This dynamism, disruption, and on-going monitoring feed (and reinforce) a market for urban data 'that goes far beyond urban governance, and using public authorities, services and infrastructure to create new streams of data' (Taylor, 2018)[4]. Given that '[n]o human being can make sense of this data unassisted, there is significant reliance on algorithms to mine, parse, sort and configure the data into useful packages' (Danaher et al., 2017, pp. 1–2). How the data is aggregated, processed, and analysed can be linked to algorithms and software systems. Yet, it also becomes a struggle between big data, often controlled and governed by corporate interests, and configurations of interests driven by civic concerns and efforts to generate thick data (Smith, 2018) where embedded, rich, place-based meaning and understanding of existing urban context informs the analysis of data.

A further issue is how such data is communicated either as data or presented as knowledge. Put baldly: 'The selectivity by which information is communicated to different users is derived from a set of decisions made by technology producers to achieve desired effects from users' interactions. Such decisions are thus an important source of control in urban space and a key objective amongst urban technology producers who utilize urban data' (Shapiro, 2017, p. 65). The packaging up of this into city dashboards 'has attracted the attention of researchers quick to raise the spectre of a resurgent positivism and abstraction of urban knowledge' (Barns, 2018, p. 7).

In short, various urban platforms – constituted in different ways as urban stacks – produce different kinds of data and knowledge of the urban. These ways of knowing are far from neutral representations. Rather, they are active and alive with implications for how collective urban problems are addressed and for how everyday urban life unfolds. The four chapters in this part engage with some of these issues around urban knowledge.

In the first chapter in this part, Jathan Sadowski focuses on data as a form of *capital*. Marking a departure from those studies that treat data as a commodity, he argues that this intellectual manoeuvre renders datafication of an active process of perpetual (data) capital accumulation and circulation, flanked and supported by the production of various forms of diagnostics, expertise, knowledge, and understanding. The supposed universality of data reframes everything as falling under the domain of data capitalism, including a growing set of sites and spaces in city and those over which urban processes prevail. Jathan argues that data capitalism constitutes a geographically variable and uneven transition towards a new kind of capital and new methods of accumulation but one that shares characteristics and features with past political-economic regimes.

Maroš Krivý focuses on what he terms the 'cultural fantasies' associated with the platform metaphor to consider the kinds of power it produces and sustains. Exploring what might be thought as the urban knowledge complex, Maroš critiques some of the still-growing academic-cum-populist writings on cities and the urban. He contends that the attraction of the 'platform' in much of this work stems from it appearing to signal control without being controlling, of engaging the totality of the urban in some sort of non-totalising way.

Padmini Ray Murray and Ayona Datta turn to an experiment to reimagine the relations of some of the most marginalised citizens in Indian cities. Drawing upon fieldwork in a slum resettlement colony in Delhi's urban edge, Padmini and Ayona discuss the emancipatory potential of open knowledge platforms, using Wikipedia as a case study. Working with 12 young women, the aim of the Wikipedia editathon was to produce a page on 'Madanpur Khadar JJ Colony', where all had lived since childhood. Through this case they raise three issues of wider significance. First, that the platform represents the city in a manner that reflects how marginal citizens experienced the city in contrast to the 2-dimensional Google map and the geo-located pins connected to the network in real time. Second, the platform allows a degree of local detail and nuance which is often lost in visualisation of 2-dimensional relations in a map. Third, and finally, the curatorial requirements of Wikipedia exclude the sorts of embodied and oral knowledge possessed by those living in the margins – such as the 12 women. This would appear to question the capacity of open knowledge platforms such as Wikipedia to challenge fully the more corporate and restrictive platforms.

In the final paper in this part, Clancy Wilmott turns to blockchain technologies in considering the political consequences of digital mapping in platform urbanism. As a peer-to-peer or peer-to-object exchange, blockchain mapping platforms claim to offer more adaptability, efficiency, and resolution for near real-time tracking of location and movement in cities in which the density and height of buildings challenge more traditional geo-locative mapping. She uses the examples of two blockchain mapping companies, Foam and Hyperion, to think through some of the consequences of shifting from top-down traditional forms of geo-locative mapping towards blockchain mapping within platform urbanisms. In particular, Clancy highlights that for all the promise of a more horizontal way of thinking about digital mapping, the inclusionary and representational practices of blockchain appear strangely familiar. The flattening requirement for a universal language, coupled with partial transparency in the production of data means that thus far blockchain has delivered more accountability or responsibility in the work it is doing in the production of a particular urban future.

4. How are platforms re-shaping everyday urban experiences?

What do urban platforms mean for everyday urban experiences? The sheer multiplicity and variety of urban platforms do not exist in a vacuum but are configured in relation to very different urban contexts and historical legacies. The incorporation of everyday practices into urban platforms fundamentally affects these practices in a number of ways, from the seemingly subtle (ordering delivery through UrbanSpoon instead of over the phone) to the more visibly obvious (replacement of taxis with Uber or Ola cars).

Digital platforms disrupt existing forms of production and collective urban consumption and contribute to the production of new forms of consumption, with variable geographies. As disruptors of traditional systems of service provision, platforms have been shown to exacerbate existing economic inequalities and information asymmetries as well as add new dimensions of inequality (Mazumdar, this volume). Yet, the implications of digital platforms for everyday urban life often produce ambiguities that require contingent understanding and analysis.

Digital platforms have implications for participation and democratic engagement in urban public and civic life. The challenge is whether urban platforms enhance or degrade such engagement. If

> a key feature of the platforms studied is their purported capacity to link different groups, individuals, and organizations, either with each other, or with goods and services...do all users have the same opportunity to connect or transact...are these platforms really as open and democratic as they seem?
>
> *(Nash et al., 2017, p. 370)*

What follows from this is that although there is a much-vaunted 'participatory' principle of opening up of possibilities for new voices to be heard via digital platforms they are also often opaque structures that mask technical (e.g. algorithms) and cultural arrangements (the constitution, organisation, and 'rules' of platform-enabled social groups) that condition and govern user behaviours in ways that may produce new temporalities (Kitchen, 2014) and types and spaces of urban publicness (see Moore and Rodgers, this volume). In this sense, this raises fundamental questions about the ways in which digital platforms, as communications media, may contribute to the reconfiguration of urban public spheres; and also inform the shape of long-standing debates on the shape of the public sphere that have been vibrant since the publication of Jűrgen Habermas' seminal work on this and ongoing responses to it (Habermas, 1989; Keane, 2000).

Platforms also reshape the landscape of urban labour, creating new forms of 'micro-work', free labour, insecurity, and precarity as well as entrepreneurialism (Terranova, 2000; Irani, 2015). These effects are uneven, affecting residents differently along lines of race, class, gender, nationality, ability, and so on, and differently shaping what kind of everyday experiences urban platforms enable and constrain. What is evident is that digital platforms are reconfiguring labour relations. Through processes of Schumpeterian creative destruction the tension is whether this is seen as the disruption of existing services, forms of consumption, and modes of employment, with the threats to livelihoods that this entails, or whether it constitutes an entrepreneurial opportunity to produce new ways of delivering services, forms of consumption, and opportunities for employment. Frequently, this has seen corporate platforms, such as Uber in relation to ride-hailing and Deliveroo in respect to food delivery, mobilising platforms in a top-down way to restructure labour relations. The rhetoric of freedom and opportunity is used to legitimise the flexibility of labour practices. At an aggregate level this has contributed to the emergence of the so-called gig economy. Yet, this focus on structural shifts in the organisation of production, consumption, and labour suggests processes of internalisation of change that says little about the agency and contingent negotiations of platform labour. This suggests a need for analyses that move beyond binaries of exploitation and entrepreneurship to examine the contradictions and dissonances within the everyday practices of platforms (Mazumdar, this volume). This begins to scope out emerging issues in respect of the implications of digital platforms for everyday urban experience. Four chapters, selectively, deal with some of these issues.

In their chapter addressing platform phenomenologies and urban public life, Susan Moore and Scott Rodgers argue that we should address social media as experiential infrastructures of everyday urban communication. Drawing on the example of a controversial cycling programme in Walthamstow, East London, they focus on the role of social media in mediating public exchanges around the programme and the ways in which the technical features of such platforms and everyday practice intersect to produce new forms of urban public life, the consequences of which are ambiguous.

Lizzie Richardson's chapter focuses on the role of meal delivery, in the English city of Newcastle-upon-Tyne, via the platforms of companies such as Deliveroo and UberEats. Such platforms allow a customer to order, pay for, and receive a meal to be delivered to a location of their choice. With consequences for the urban geography of consumption, Lizzie argues that such platforms are contributing to the dispersal of consumption of prepared food from the restaurant to distant sites of consumption and also to the concentration of new physical sites for the collective preparation of meals, in what she terms dark kitchens. She argues that such platforms are being mobilised in the creation of new markets for urban takeaway food delivery. Lizzie shifts focus from the 'on-screen representation' of meal delivery platforms to the contingent material practices that address the role of platforms in market-making.

In his chapter on the everyday politics of ride-hailing taxis, Anurag Mazumdar examines platform drivers, driver networks, and emergent platform governance in the context of India. Whilst there has been much discussion about ride-hailing this has addressed issues such as whether drivers are either micro-entrepreneurs or alternatively exploited in an evolving informal sector; it has also focused on regulatory battles between ride-hailing companies and urban public authorities. Anurag shifts focus to everyday practices of ride-hailing platforms, the dynamics of this, and what this tells us about platform urbanism in India. Using academic research, media narratives, government documents, and ride-hailing blogs, Anurag argues that platforms selectively endorse and dissolve acts of community and that the platform architecture of new communities is not predicated on unidirectional algorithms.

In their chapter, Andrés Luque-Ayala and colleagues address platforms in the making and the role of civic hackers and forms of data activism in such processes. With aims of intervening in a breadth of urban issues, Andrés and colleagues argue that digital interventions in the city prefigure urban platforms both materially and in terms of their political orientation and that there are a range of platforms in the making in the city. Focusing on cities of the global South, particularly the Brazilian cities of Rio de Janeiro, São Paulo, Recife, and Porto Alegre, their argument is made at the intersection of digital tools, data activism, and political asymmetries in both re-imagining and re-making the city's environment. Their analysis looks at the tensions and differences between 'data-led' forms of activism and 'situated' interventions and explores the implications of this in an empirical context.

Notes

1 https://www.mediapolisjournal.com/2018/10/platform-urbanism-an-introduction/ [accessed 16/09/2019].
2 https://www.theguardian.com/technology/2017/oct/21/google-urban-cities-planning-data [accessed 16/09/2019].
3 https://amp.ft.com/content/6d2fe2a8-722c-11e7-93ff-99f383b09ff9 [accessed 16/09/2019].
4 https://www.ijurr.org/spotlight-on/disruptive-urban-technologies/what-can-disruptive-urban-technologies-tell-us-about-power-visibility-and-the-right-to-the-city/ [accessed 16/09/2019].

References

Andersson Schwarz, J., (2017) 'Platform logic: An interdisciplinary approach to the platform-based economy,' *Policy & Internet*, 9, no. 4, 374–394.
Anttiroiko, A., (2016) 'City-as-a-platform: The rise of participatory innovation platforms in Finnish cities,' *Sustainability*, 8, no. 9, 922.
Anttiroiko, A., Valkama, P., and Bailey, S., (2014) 'Smart cities in the new service economy: Building platforms for smart services,' *AI and Society*, 29, 323–334.
Aronoff, K., Battistoni, A., Aldana-Cohen, D., and Riofrancos, T., (2019) *A Planet to Win: Why We Need a Green New Deal*, Verso: London.
Barns, S., (2016) 'Mine your data: Open data, digital strategies and entrepreneurial governance by code,' *Urban Geography*, 37, no. 4, 554–571. Routledge.
Barns, S., (2018) 'Smart cities and urban data platforms: Designing interfaces for smart governance,' *City, Culture and Society*, 12, 5–12.
Barns, S., (2020) *Platform Urbanism: Negotiating Platform Ecosystems in Connected Cities*, Palgrave-Macmillan: London.
Barns, S., Cosgrave, E., Acuto, M., McNeill, D., (2017) 'Digital infrastructures and urban governance,' *Urban Policy and Research*, 35, no. 1, 20–31.
Bell, D., (1976) *The Coming of Postindustrial Society*, Basic Books: New York.
Bratton, B., (2015) *The Stack: On Software and Sovereignty*. MIT Press: Cambridge, MA.
Brenner, R., (2006) *The Economics of Global Turbulence: The Advanced Capitalist Economies from Long Boom to Long Downturn, 1945–2005*, Verso: New York.
Bulkeley, H., and Castan Broto, V., (2013) 'Government by experiment? Global cities and the governing of climate change,' *Transactions of the Institute of British Geographers*, 38, no. 3, 361–375.
Danaher, J., Hogan, M.J., Noone, C., Kennedy, R., Behan, A., De Paor, A., Felzmann, H., Haklay, M., Khoo, S-M, Morison, J., Murphy, M.H., O'Brolchain, N., Schafer, B., and Shankar, K., (2017) 'Algorithmic governance: Developing a research agenda through the power of collective intelligence,' *Big Data and Society*, July–December, 1–21.
Easterling, K., (2014) *Extrastatecraft: The Power of Infrastructure Space*, Verso: London.
Evans, J., (2016) 'Trials and tribulations: Problematizing the city through/as urban experimentation,' *Geography Compass*, 10/10, 429–443.

Florida, R., Adler, P., and Mellander, C., (2017) 'The city as innovation machine,' *Regional Studies*, 51, no. 1, 86–96.
Gawer, A., (2014) 'Bridging differing perspectives on technological platforms: Toward an integrative framework,' *Research Policy*, 43, 1239–1249.
Geissdoerfer, M., Savaget, P., and Bocken, N., (2017) 'The circular economy – a new sustainability paradigm,' *Journal of Cleaner Production*, 143, 757–768.
Geissinger, A., Laurell, C., and Sandström, C., (2018, June 23) 'Digital disruption beyond Uber and Airbnb—tracking the long tail of the sharing economy,' *Technological Forecasting and Social Change*, 155, 119323.
Gibson, J., (1979) *The Ecological Approach to Visual Perception*, Houghton Mifflin: New York.
Habermas, J., (1989) *The Structural Transformation of the Public Sphere: An Inquiry in to a Category of Bourgeois Society*, MIT Press: Cambridge, MA.
Harvey, D., (2007) *The Limits to Capital*. 2nd ed., Verso: London.
Harvey, D., (2008) 'The right to the city,' *New Left Review*, 53, September–October.
Henderson, R.M., and Clark, K.B., (1990) 'Architectural innovation: The reconfiguration of existing product technologies and the failure of established firms,' *Administrative Science Quarterly*, 35, 9–30.
Hill, K., (2019) 'I tried to block Amazon from my life. It was impossible,' *Gizmodo*, January 22. https://gizmodo.com/i-tried-to-block-amazon-from-my-life-it-was impossible-1830565336.
Hodson, M., Geels, F., and McMeekin, A., (2017) 'Reconfiguring urban transitions, embracing multiplicity,' *Sustainability*, 9. doi:10.3390/su9020299.
Hodson, M., and Marvin, S., (2010) 'Urbanism in the anthropocene: Ecological urbanism or premium ecological enclaves?' *City*, 14, no. 3, 298–313.
Irani, L., (2015) 'The cultural work of micro-work,' *New Media and Society*, 17, no. 5, 720–739.
Jonas, A.E.G., Gibbs, D., and While, A., (2011) 'The new urban politics as a politics of carbon control,' *Urban Studies*, 48, no. 12, 2537–2554. doi:10.1177/0042098011411951.
Keane, J., (2000) 'Structural transformations of the public sphere,' in Scammell, M., and Semetko, H., (eds) *The Media, Journalism and Democracy*, Routledge: London.
Kibum, K., Baek, C., and Lee, J.D., (2018, April 1) 'Creative destruction of the sharing economy in action: The case of Uber,' *Transportation Research Part A: Policy and Practice*, 110, 118–27.
Kitchin, R., (2014) 'The real-time city? Big data and smart urbanism,' *GeoJournal*, 79, no. 1, 1–14.
Kitchin, R., Maalsen, S., and McArdle, G., (2016) 'The praxis and politics of building urban dashboards,' *Geoforum*, 77, 93–101.
Krippner, G., (2011) *Capitalizing on Crisis: The Political Origins of the Rise of Finance*, Harvard University Press: Cambridge, MA.
Langley P., and Leyshon A., (2017) 'Platform capitalism: The intermediation and capitalization of digital economic circulation,' *Finance and Society*, 3, no. 1, 11–31.
Lash, S., and Urry, J., (1987) *The End of Organised Capitalism*, Polity: Cambridge.
Lash, S., and Urry, J., (1994), *Economies of Signs and Space*, Sage: London.
Lefebvre, H., (1968) *Le droit à la ville*, Anthopos: Paris.
Lefebvre, H., (2003[1970]) *The Urban Revolution*, University of Minnesota Press: Mineapolis.
Leszczynski, A., (2016) 'Speculative futures: Cities, data and governance beyond smart urbanism,' *Environment and Planning A*, 48, no. 9, 1691–1708.
Marvin, S., and Luque-Ayala, A., (2017) 'Urban operating systems: Diagramming the city,' *International Journal of Urban and Regional Research*, 41, no. 1, 84–103. doi:10.1111/1468-2427.12479.

Mattern, S., (2015) 'Mission control: A history of the urban dashboard,' *Places Journal*, 1, no. 10.

McMeekin, A., Geels, F., and Hodson, M., (2019) 'Mapping the winds of whole system reconfiguration: Analysing low-carbon transformations across production, distribution and consumption in the UK electricity system (1990–2016),' *Research Policy*, 48, 1216–1231.

Meisenzahl, M., 'Mark Zuckerberg dominated people's phones over the decade. Here are the 10 most downloaded apps, nearly half of which Facebook owns,' *Business Insider*, December 20, 2019. https://www.businessinsider.com/most-downloaded-apps-of-decade-facebook-instagram-whatsapp-tiktok-snapchat-2019-12.

Merrifield, A., (2014) *The New Urban Question*, Pluto: London.

Moazed, A., and Johnson, N., (2016) *Modern Monopolies: What Does It Take to Dominate the 21st Century Economy*, St Martin's Press: London.

Morozov, E., (2017) 'Google's plan to revolutionise cities is a takeover in all but name,' *The Guardian*, 22nd October.

Mosco, V., (2017) *Becoming Digital: Towards a Post-Internet Society*, Emerald: Bingley.

Nash, V., Bright, J., Margetts, H., and Lehdonvirta, V., (2017) 'Public policy in the platform society,' *Policy & Internet*, 9, no. 4, 368–373.

Rodgers, S., and Moore, S., (2018a) 'The horizons of platformed urban politics,' *Mediapolis*, 3, no. 4, Roundtables, online version.

Rodgers, S., and Moore, S., (2018b) 'Platform urbanism: An introduction,' *Mediapolis*, 3, no. 4, Roundtables, online version.

Sadowski, J., (2019) 'When data is capital: Datafication, accumulation, and extraction,' *Big Data & Society*, 6, no. 1, 1–12. doi:10.1177/2053951718820549.

Sadowski, J., (2020) 'The internet of landlords: Digital platforms and new mechanisms of rentier capitalism,' *Antipode*, 52, no. 2, 562–580. doi:10.1111/anti.12595.

Sassen, S., (1991) *The Global City: New York, London, Tokyo*, 2nd ed., Princeton University Press: Princeton, NJ.

Scholz, T., and Schneider, N., (2016) *Ours to Hack and to Own: The Rise of Platform Cooperativism, a New Vision for the Future of Work and a Fairer Internet*, OR Books: New York.

Scott, A.J., (2012) *A World in Emergence: Cities and Regions in the 21st Century*, Edward Elgar Publishing: Cheltenham, UK.

Shapiro, A., (2017) 'The urban stack. A topology for urban data infrastructures,' *Tecnoscienza*, 8, no. 2, 61–80.

Shelton, T., Zook, M., and Wiig, A., (2014) 'The actually existing smart city,' *Cambridge Journal of Regions, Economy and Society*, 8, no. 1, 13–25.

Smith, A., (2018) 'Cities need thick data, not big data,' *The Guardian*, 18th April.

Srnicek, N., (2016) *Platform Capitalism*, Polity: Cambridge.

Star, S., (1999) 'The ethnography of infrastructure,' *American Behavioural Scientist*, 43, no. 3, 377–391.

Star, S.L., and Bowker, G.C., (2006) 'How to infrastructure,' in *Handbook of New Media: Social Shaping and Social Consequences of ICTs*, Sage: London, 230–245.

Stehlin, J., Hodson, M., and McMeekin, A., (2020) 'Platform mobilities and the production of urban space: Toward a typology of platformization trajectories,' *Environment and Planning A*, 0308518X19896801.

Stehlin, J.G., (2018) 'Urban platforms, rent, and the digital built environment,' *Mediapolis: A Journal of Cities and Culture*, 4, no. 3. https://www.mediapolisjournal.com/2018/10/urban-platforms-rent-and-the-digital-built-environment/.

Sundarajan, A., (2016) *The Sharing Economy: The End of Employment and the Rise of Crowd Based Capitalism*, MIT Press: Cambridge, MA and London.

Taylor, L., (2018) 'What can "Disruptive Urban Technologies" tell us about power, visibility and the right to the city?' *International Journal of Urban and Regional Research*, Essays on Disruptive Technologies https://www.ijurr.org/spotlight-on/disruptive-urban-technologies/introduction-2/.

Taylor, P.J., Walker, D.R.F., and Beaverstock, J.V., (2000) 'Introducing GaWC: Researching world city network formation,' in *Telematics and Global Cities*, Oxford: Blackwell.

Terranova, T., (2000) 'Free labor: Producing culture for the digital economy,' *Social Text*, 18, no. 2, 33–58.

Thrift, N., (2005) *Knowing Capitalism*, Sage: London.

Tieman, R., (2017) 'Barcelona: Smart city revolution in progress,' *Financial Times*, 26th October.

SECTION 1

What kind of urban infrastructure are platforms?

2

THE URBAN STACK

A topology for urban data infrastructures

Aaron Shapiro

Topologizing urban data infrastructures

To the extent that there's any consensus around the future of our cities, one thing is for sure: the city of tomorrow will be populated by all sorts of digital screens and urban interfaces. Electronic kiosks will give real-time updates on transit delays. Touchscreen maps will occupy sidewalks in central shopping districts. LED displays will rotate advertisements along major thoroughfares, replacing motionless billboards with radiant moving images. Indeed, in cities across the world, an explosion of urban interfaces is already underway. In Paris bus stops are now "smart," with signage informing riders of traffic conditions. In and around Boston's convention centers are thousands of digital screens. And visitors to Chicago's Millennium Park will enjoy an entire wall of LED video overlooking the reflection pool along Michigan Avenue.

What are we to make of this explosion of digital signage and its significance for the politics of urban futures? A superficial analysis might suggest that such interfaces are perfect examples of how new information and communication technologies should be leveraged to improve existing city services and public spaces. On this view, urban interfaces render dynamic what was once static, translucent what was once opaque. They show us what's going on, where we should go, how we might get there; they reduce the complexity and chaos of urban life, making the everyday more manageable. But on a more nuanced read, we'd begin to see that the urban interface shrouds as much as it reveals. As Shannon Mattern (2014) writes, the information made available on urban interfaces comes pre-processed, masking the conditions of its own making. Colorful info-visualizations may boast of efficiency in city services, but they offer "little understanding of how and where the mediation of urban systems takes place within the city itself" (Mattern, 2014). So, while the interface promises to improve urban systems by expanding information access, everyday citizens are prevented from engaging with those systems directly or understanding how they are transformed into consumable data.

How is the data captured, processed, and manipulated for our consumption? What data infrastructures lie beneath the interface? What operating systems and material supports lie hidden, behind the screen? Getting at these mediations requires attention to the mundane—the out-of-view formations of hardware and software whose material and protocological configuration creates the digital assemblages of platform urbanism (Galloway, 2004; McFarlane, 2011). This is what Mattern means when she writes of the *urban stack*. Its hardware includes: switches, wires, and cables; pipes, telephone poles, and gas lines; the transmitters and receivers of mass communication broadcasts as well as wi-fi internet connections and 5G cell networks; the dirt, concrete, plastics, rubber, metal, and flesh that are the city's core materials. Its "software" certainly includes the key elements of urban interfaces—"all those zoomable maps and apps that translate urban data into something useful" (Mattern, 2014)—but it also involves other interfaces, which need be neither public nor digital: the paperwork of the police officer, the ticket punch of the train conductor, the analog clock on top of city hall, the route of the postal worker, and street addresses (Rose-Redwood, 2006; Valverde, 2011). Taken together, all these assemblages of humans and their social practices, objects and their materials, and "infrastructured" technologies and their interfaces are what make the city an *urban* space, "not simply a context for the support or appropriation of specific lives," but "the provisionally stitched together, jigged up intersections of bodies and materials upon which things are both moved and caught" (Simone, 2011, p. 356).

This chapter builds upon the urban stack as a framework for understanding how and why the composition and configuration of urban data infrastructures matter. It analyzes two case studies and their stacked assemblages: LinkNYC, a public wi-fi infrastructure in New York City, and on-demand service platforms such as Uber and Caviar. I conclude by arguing that value production in the urban stack, as a *platform*, hinges on technology producers' capacity to enroll heterogeneous elements into a hierarchical flow of information and, through this enrollment, effect novel forms of control and value extraction.

The urban stack

The concept of the stack is borrowed from software production, where it refers to a specific, hierarchical assemblage of hardware, network protocols, programming, and software design (Solomon, 2013). Theorists of software and power have used the term to map how digital media relate to and affect the material, cultural, legal, and political worlds in which they are embedded (Solomon, 2013; Bratton, 2016; Straube, 2016). The stack itself, however, is a somewhat ambiguous analytic object. As Solomon (2013) writes, the stack topology conflates the "operative structure that exists materially within the program code of software systems" with the "class of diagrams used to explain both these operative structures and software systems more generally." Without being able to fully disentangle these two dimensions of the stack, their slippage productively captures, first, the ways that practitioners conceptualize the integration of software and hardware, and second, the topological relationships within their integration. So, while the stack is a specific type of assemblage, its specificity is revealing for data infrastructures that bridge material-digital divides—exactly what is at stake with the infrastructures enabling the explosion of urban interfaces.

Here I follow Mattern's (2014) more liberal and heuristic use of the term to conceptualize the relationship between data and materiality in urban platforms. As a heuristic, the *urban* stack allows us to see how disparate material infrastructures are juxtaposed in meaningful ways as platformic assemblages—how the objects, materials, human practices, technologies, and systems that make up the city can be enrolled in and assembled as looping systems of data flow (Kitchin & Lauriault, 2014). The stack's emphasis on topology presupposes a distinct spatial and relational way of thinking, in which "spatial problems depend not on the exact shapes of the objects involved but on the ways that they are put together, on their continuities, and cuts" (Secor, 2013, p. 431). My contention is that urban data infrastructures work similarly. They gather together digital and non-digital infrastructural components that, through their topological ordering, accumulate value by mediating and expropriating public activity and materials (Thatcher, O'Sullivan & Mahmoudi, 2016; Sadowski, this volume).

Figure 2.1 shows how practitioners adapt a stack topology from software development for smart city development. At the bottom of the diagram are a number of devices for collecting data about urban populations, spaces, and processes, including RFID tags, infrared sensors, and so on. And while techno-optimists and chauvinists (Broussard, 2018) may claim that making a city "smart" requires the acquisition of these devices, in practice, much data collection involves both digital and analog data as well as combinations of automated and manual collection practices. For example, the New York City Department of Transportation uses traffic cameras equipped with algorithmic detection software to capture bicycle ridership data, but at sites where data collection is deemed necessary but the technology is unavailable they continue to hire people to manually count cyclists (NYC Department of Transportation, 2016).

We can imagine this base layer, then, as a *distributed infrastructure,* tethered together into a coherent network through its configuration within the stack. Crucially, the vast majority of urban data infrastructures rely on already-existing infrastructural networks. Data collection and aggregation is highly opportunistic in this sense (Thatcher, 2014). Infrastructures erected for one purpose are coopted for another. The same goes for networks and protocols, digital and otherwise. A firm doesn't have to reinvent GPS, traffic systems, census tracts, or government bureaucracies to implement a data infrastructure; they need only yoke the data collection assemblage to whatever proprietary components that they happen to bring to the table (cf. Graham & Marvin, 2001).

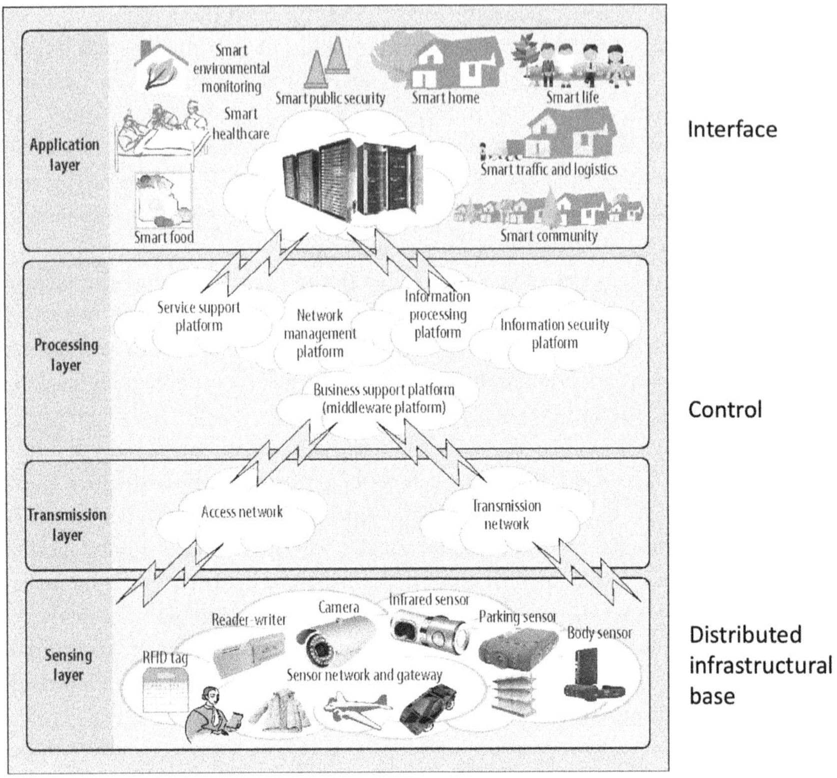

FIGURE 2.1 The urban stack.
Source: Adapted from Liu and Peng (2014).

These proprietary components operate at what we might call a *control* level, which in Figure 2.1 traverses the processing and transmission layers. It is here that analog and native-digital data are aggregated, processed, standardized, and analyzed. No matter the provenance, it is at the control level that data becomes privatized through its processing and manipulation—for instance, with algorithms trained to execute commands based on machine-detected patterns or correlations. It should also be noted that control operations, and the code that executes them, are highly complex and opaque by design. The cloud symbols thus stand in for mutable suites of analytic techniques that ultimately legitimate system operators' claims to exclusivity and ownership (Thatcher, O'Sullivan & Mahmoudi, 2016; Sadowski, 2019).

Finally, at the top of the schematic, processed data are presented to end-users as information across social registers ("smart healthcare," "smart community," and so on). It is here, at the *interface* level, that urban systems are imagined to become "smart." The interface provides a "doubly communicative" access point, at once informational and informatic—a node in the network that both gathers and displays information through each interaction or event (Halpern et al., 2013). It should be noted, however, that this double-communicativity is highly uneven. While it might be obvious that system administrators enjoy greater access to a network than end users, the information selected for display on the users' end—and the ways in which it is displayed—may also vary according to a hierarchical typology of users. For example, within an outdoor signage network, wealthier or more populous neighborhoods may receive targeted, personalized advertising, while less populous or poorer areas may only receive standardized ad rotations. These distinctions may seem subtle, but they nonetheless perform the social distances that they mediate. As "autonomous zones of activity" (Galloway, 2012, p. vii), interfaces do not merely reflect social relationships but contribute to their enactment through informational arbitrage. In this respect, this smart city infrastructure becomes a platform for differentiated value extraction.

The city of the future: two case studies

The urban stack's topology helps us to understand how control is exercised, and value generated, in and through urban platform infrastructures. In this section, I explore how the stack heuristic might illuminate two case studies. In each case, technology producers assemble variegated networks of materials and existing infrastructure as a distributed infrastructural base; protocological and analytic control are wielded to establish ownership over the information system; and highly uneven displays are used to manage and influence end-users' activity in real-time (Kanngieser, 2013; Levy, 2015; Rossiter, 2015). As I will show, it is through the stacking of these elements that technology producers are able to extract value from the activity of urban life.

"The future of public spaces"

In 2014, New York City Mayor Bill de Blasio announced that "the future of public spaces" would be digital (NYC Office of the Mayor, 2014). After years of hype, the City had finally completed an agreement with a consortium of companies to build and operate LinkNYC, a city-wide network of wi-fi hotspots to replace the city's aging payphone infrastructure. Intersection, the lead firm in the consortium, was now licensed to implement and operate this massive project, purported to be one of the fastest municipal wi-fi systems in the country and the largest digital out-of-home advertising network in the world (*Screen Media Daily*, 2016).

In place of the old payphones, LinkNYC now provides free access to gigabit Wi-Fi, domestic phone calls, and, through its touchscreen interface, access to city services and local information, including transit schedules and maps. In exchange for providing these free services, Intersection gained near-monopoly access to digital advertising real estate across New York City's five boroughs—the largest market for out-of-home advertising in the U.S. by nearly double its closest rival's revenue share. Intersection therefore came to be seen as an ideal investment. Tech giants, including Google's urban tech spin-off, Sidewalk Labs, took notice, launching a campaign to finance LinkNYC's roll-out.

But LinkNYC would not be built from scratch. Intersection and its partners took advantage of a number of existing infrastructures, including communications conduits buried beneath the streets of New York City, as well as the old payphones' hardline connections. Construction began in late December 2015 and the first Links went live in February 2016. At the time of writing, nearly 2,000 Links are active across the five boroughs, with up to 7,500 planned by 2026—each equipped with two 55-inch digital displays dedicated exclusively to advertising along with a suite of sensing devices, including microphones, cameras, and Bluetooth beacons.

Although the advertising revenue has been somewhat disappointing to date (Voytko, 2019), proponents remain excited about the data and its potential applications for urban planning (NYC Office of the Mayor, 2014; Hotz, 2015; Fung, 2016). Intersection's executives insist that LinkNYC will produce data on everything from traffic and noise to air quality. These public benefits have become key talking points for promoting the system (cf. Gustin, 2016). As Intersection's Chief Strategist Dave Etherington put it during an interview,

> when you think about LinkNYC and the 7,500 or so fairly evenly distributed nodes across the five boroughs, it represents a really interesting opportunity to learn about the city, the behaviors of the city, that could lead directly to health benefits, more efficient use of traffic—being able to sense, are trucks idling near these things illegally? Is there congestion? Is there a traffic jam? Is there noise pollution, air pollution? All of these things, by micro-location, could empower some really interesting insights about the city that will make it a kind of more enjoyable place to live.
>
> *(eMarketer, 2016)*

All of these possibilities sound generous and productive enough, but they also raised privacy concerns among activists and civil rights advocates, including the New York Civil Liberties Union (NYCLU, 2016). In response, Intersection updated its terms of service to ramp up user protections (NYC Department of Information Technology & Telecommunications, 2016). According to the new terms, data shared over the networked would be encrypted and automatically anonymized; Intersection and its partners also promised not to track users' web browsing.

But LinkNYC's data is valuable not because it knows what you look at online but because the system tracks *where you are, when, and for how long*. As the terms of service put it, while the Links "do not collect information about your precise location," the firms "know where we provide Wi-fi services, so when you use the Services we can determine your general location" (NYC Department of Information Technology & Telecommunications, 2016). And this may be enough. By simply logging onto the Wi-Fi, users are counted, in real-time, wherever they are within the network, which is coextensive with the city's territory. All of this is valuable information, especially given that LinkNYC's profit hinges on the ad sales. Like the data-driven ad auctions pioneered by Google for the web, LinkNYC's data will allow Intersection to charge more for ads that can demonstrably reach more eyeballs. Even the anonymized and encrypted data can be used to price the ad space and capture a greater share of profit.

Perhaps more troubling, however, LinkNYC is also able to count and track people *even when they're not connected to the wi-fi*. This is evident in documentation of LinkNYC's technical capabilities (cf. Kofman, 2018) and corroborated by considering Intersection's behavior prior to LinkNYC. Intersection formed as a merger of Control Group, a boutique design firm, and Titan Outdoor, an advertising company that purchased Verizon's payphone stock in 2009. In 2014, without notifying citizens or the city, Titan installed dozens of Bluetooth low energy (BLE) beacons, which can count any Bluetooth-enabled devices within range, on its payphones. Although the Department of Information Technology and Telecommunications (DoITT) forced Titan to remove the beacons once this was revealed (Bernstein, Singer-Vine, & Ryley, 2014), the same technology is built into each Link in the network (Intersection, n.d.), meaning that Intersection can log any device that comes within range of a Link kiosk.

While people counting is neither new nor limited to LinkNYC, LinkNYC's scale and granularity make it unique. With smartphone penetration at 80% in New York City as of November 2015 (NYC Department of Consumer Affairs, 2015)—and likely even higher now—LinkNYC has the potential to capture data on millions of individuals' movements, every day. So, where the urban planning uses legitimate data collection, advertising revenue—which is shared with the city—legitimates the public-private partnership. In Etherington's words, "advertising concessions" are really just "vehicles for innovation," making it possible for Intersection to "not just increase advertising revenue for cities" but also "bring in new technologies and new innovation" (eMarketer, 2016). What will this innovation look like? Intersection's recent spinoff, Place Exchange, gives us a sense. Place Exchange is devoted to delivering "true real-time [ad] bidding" and "granular device-level attribution." Its patent-pending technology will help clients "buy and measure [out-of-home] media in the same way as they do web, mobile, and other digital media," so media buyers can "enjoy the benefits of massive reach, high-impact, always-viewable messaging that reaches consumers in the physical world" (Intersection, 2019). Taken together then, what LinkNYC's "innovation" means, and what LinkNYC has managed to accomplish, is to recreate the political economy and experience of online advertising platforms in our cities' physical public spaces.

"The future of work"

If LinkNYC heralds how public space will be transformed in the "smart city" of the future, then on-demand service platforms like Uber, Instacart, and Postmates represent "the future of work" (Hanrahan, 2015)—a future in which technology revolutionizes urban service economies. These services are "on-demand" on both sides of the transaction: customers need only pull up an app to order a ride or a sandwich delivery, while workers are promised the flexibility to work whenever they want—labor ondemand. In practice, on-demand platforms simply reproduce age-old practices of worker control and exploitation—and with dramatic effects on our cities. In New York City, for example, Uber's entrance to the transportation market had a dramatic impact on the taxi industry, which was already in the midst of a massive financial bubble. The value of taxi medallions—long required to operate a cab—plummeted from a peak of $1.3 million in 2013, when Uber first started operating, to as low as $160,000 in 2018. And because taxi workers' financial wellbeing and retirement plans had become intertwined with the medallions' value, drivers were hit with an immense debt crisis associated with the suicide deaths of at least eight drivers within a single 12-month span (Fitzsimmons, 2018).

Platforms like Uber, which describes itself as a "cross between lifestyle and logistics" (Tsotsis, 2012), are a distinct breed within the broader platform economy. They claim to be mere facilitators of transactions while in practice intervening in them, using algorithmic calculations to set prices for customers and pay rates for workers, and collecting a fee from each transaction. If described at all, these calculations are explained with vague terminology (distance, demand, etc.) and neither the consumer nor the worker has full access to the relevant information (Kirchner & Mattu, 2015; Rosenblat & Stark, 2016; Shapiro, 2018). Despite such severe information asymmetries, platforms insist on their facilitator function, which serves more than ideological ends. By claiming to be technology firms rather than service providers, the platforms justify their classification of workers as independent contractors rather than employees, getting the firm off the hook for costly expenses, from equipment investment to unemployment and health insurance.

These platforms use a number of disciplining techniques, and a pricing algorithms are essential tools in the managerial toolkit. Price drops, of course, entice customers during slower periods, but discounted prices come out of workers' wages. Consequently, platform managers use wage manipulations to cajole workers—to get them to work when managers want, say, or in a particular area of the city. The food delivery service Caviar, for example, doesn't *require* that its workers commit to a schedule; however, it does give "priority" to couriers who choose to commit ahead of time, meaning that workers who log on without scheduling aren't likely to receive many jobs—and because the pay is piece-rate, these workers won't make much money.

This kind of opacity creates uncertainties on the workers' end. In addition to wage manipulations, platform managers are often unclear about their basic policies. For several years, Uber "deactivated" (read "fired") workers for reasons that fundamentally contradicted workers' rights as independent contractors—from inactivity to low job-acceptance rates—despite contractors' rights to log off at will or reject undesirable jobs (Dough, 2019; see also Rosenblat, 2018). These and other disciplinary techniques are automated and incorporated into the fabric of workers' day-to-day experiences. In another example, Caviar introduced a "scheduling reliability" index: a personalized chart showing workers a ratio of the time they spent logged on during their scheduled shifts. Though workers are not obligated to commit to working at any specific time, the index reminds them of the platform's control over their ability to earn an income. What it doesn't show is how often workers agree to help out the platform by logging on during an unexpected spike in demand (say, during a rainstorm), nor does it reflect how long workers sit idly logged on without being paid. Crucially, it's also unclear to workers whether their "reliability" affects their ranking in the dispatching queue. As one Caviar courier explained during an interview,

> When I first started working . . . I was told that we weren't obligated to accept orders, that it's completely at our discretion when we want to work and what orders we want to accept. That was a big selling point . . . Now, they're doing this reliability system . . . It just feels like they're trying to guilt trip us for not showing up for our shifts, which are not obligatory, and whether or not we're being penalized for showing up for our shifts is kind of unclear. But whether or not they're penalizing us, it seems like they're asking us to penalize ourselves.
>
> *(Interview conducted March 7, 2016)*

Such uncertainty is hardly accidental. Platform managers cultivate it. When Caviar first launched, job orders included key information, such as the restaurant location for pick-up and the customer address for the drop-off, but this was eventually removed—to many workers' dismay. After an update to the app, the drop-off location was removed from workers' screens until they'd indicated that the order was ready and they were on their way to the customer. This strategic removal of key information undercut a common practice of rejecting orders to high-rise apartment or office buildings, which couriers tend to dislike because the pay calculations are based on ground distance, and don't account for the time spent getting up the building to the customer. Without access to the full address (such as Suite 2501, indicating the 25th floor), couriers are less likely to reject the order. Such asymmetries give workers just enough information to complete a task, but obscure enough detail that the company can affect how workers make their decisions.

Infrastructural stacking

These two case studies obviously differ in some significant ways. LinkNYC is a large infrastructural overhaul, managed by private firms and marketed as a public good; on-demand platforms are much more explicitly focused on extracting value from workers. But the cases also share a good deal in common. By using the urban stack as a heuristic, the cases can be "bent" and "stretched" to facilitate comparison (Secor, 2013, p. 431), allowing us to better comprehend the contours of the "actually existing" smart city (Shelton, Zook & Wiig, 2015).

Distributed infrastructures

Both case studies rely on a distributed infrastructural base, upon which other elements are stacked to create small monopolies on data collection, storage, and analysis. This distributed base externalizes costs and mitigates risk by taking advantage of extant infrastructural conditions. With LinkNYC, there are two ways that existing infrastructure is enrolled in the network. First, LinkNYC exploits the sunk costs of telecommunications infrastructure already in place, with hardwired connections laid in New York City's underground conduits. Fiber optic connections can be strung through these conduits quite easily, since they can be accessed by a manhole cover. Second, LinkNYC's data infrastructure relies on user devices to serve as proxies for their owners, in essence creating an entire city's worth of sensors for data generation. Without Bluetooth- and Wi-Fi-enabled smartphones already in citizens' pockets, the system would not be able to count and track audiences in public spaces and use that information to price its ads.

On-demand platforms likewise appropriate workers' devices, not only to send and receive information; they also count on workers, as independent contractors, to pay for the data as part of supplying their own equipment. This shifts massive financial responsibility from employers onto workers, who similarly receive no employment benefits, social security contributions, unemployment insurance, or workers' compensation for any on-the-job injury. It is the outsourcing of the entire social infrastructure of welfare that makes the on-demand economy profitable (Srnicek, 2016). As workers are symbolically and materially left to their own devices (Ticona, 2016), the freedom and autonomy that the flexibility narrative promised them quickly translates to precarity, in which consumer devices are assembled as a networked infrastructure for the platform's profits.

Control

In both cases, too, control is exercised through black-boxed calculative equipment. In the case of LinkNYC, two functions at the control layer are essential. The first is the hidden protocological activity that randomizes or anonymizes user identification as a supposed privacy-enhancing measure. These protocols translate aggregated user data into an informationally meaningful format while simultaneously offering a technical fix to privacy concerns—thereby providing some legitimacy to the data collection. The second function involves the dynamic ad sales based on this information (Intersection, 2019). With the ability to track users through their devices, advertisers can "deliver highly targeted content" similar to the "ad-targeting algorithms [that] users encounter while surfing the Web" (Campbell, 2016). Prices for ads at the most desirable locations—those with the largest audiences—will automatically update to reflect audience size and perhaps even spending capacity, capturing a greater share of value for Intersection and its media-buying clients.

With on-demand platforms, operations at the control layer are strategically shrouded from workers as a technique for achieving desirable managerial outcomes, such as convincing workers to log on when they're needed or encouraging them to move to underserved locations (Rosenblat & Stark, 2016; Shapiro, 2018). Crucially, while platforms justify this opacity by invoking trade secrecy, algorithmic management also justifies their claims to be tech companies, not service providers, which justifies the contractor classification. As one Uber engineer explained, "Uber is not a taxi company"; it "deals with building data centers, running real-time software services, facilitating payment and conducting research into the economics of real-time transportation automation, among solving all sorts of other interesting technological problems" (Tal, 2015). In this narrative, tech companies produce the opaque cloud-like operations of the control layer, not the actual service.

Interface

Finally, the interface's selective and uneven display of information is key to both LinkNYC and on-demand platforms. On the one hand, system managers strategically omit information to achieve desirable effects, particularly with on-demand platforms, where information is released on a "need-to-know" basis—workers get just enough information to complete the task, but not enough to "game" the system by rejecting undesirable jobs. But the same could be said of LinkNYC and its arbitrage of data transmission (Wi-Fi) and capture (sensors). From one angle, system usage generates data; but from another, the data is for industrial use (in ad sales), to which end users never get access.

On the other hand, while certain information is strategically omitted, other information is strategically included. On demand platforms, for example, pro cess and communicate information to discipline or sway workers—for example, the scheduling reliability index discussed above, which workers describe as "mind games" or "guilt trips." Such techniques are common in sites where data-driven surveillance mechanisms have been integrated into the workflow as "soft" forms of control (Kanngieser, 2013; Levy, 2015; Rossiter, 2015; Rosenblat & Stark, 2016). Such "contexts of control" (Levy, 2015) rely less on the exclusion of information than its selective inclusion. LinkNYC is likewise engaged in influencing end users—not their labor but their consumption. Data-driven ad targeting, adapted from the interactivity of web activity into the physical world of our public spaces, works to steer potential customers into consumption situations. "You can expect the kiosks to start telling you there's a table for two open at the French bistro down the street, for instance" (Fung, 2016); in 2018, Intersection partnered with Marvel Studios to broadcast showtimes for *Avengers:*

Infinity War at the closest theater (Kushner, 2018). This kind of hyperlocal notification takes the raw data of people-counts and turns that information into valuable, "actionable" information: market value.

Conclusions

As a heuristic, the urban stack helps us to extend analysis "beneath" the interface to examine how heterogenous elements are strung together as a platform infrastructure of urban datafication. To be sure, the framework has its limits; it certainly doesn't go far enough toward ensuring that urban data infrastructures serve citizens rather than institutional or corporate imperatives (Mattern, 2014). The utility derives from the stack's descriptive value, its potential to highlight the labor involved in assembling all the disparate elements together and to emphasize the public provenance of many of the infrastructures appropriated for private gain. These are critical exercises if we citizens wish to engage more deeply in the construction of urban futures. If we cede the power to define the "city of tomorrow" to profit-seeking firms, and then we risk our public spaces becoming milieus of passive info-consumerism, our livelihoods becoming increasingly precarious. When we use the urban stack to see that it is not only clever algorithms and shiny digital screens that make up the platform city, but also the "soft" infrastructures of legal designations, franchise agreements, privacy policies, and info-graphics, then we can begin to contest these manufactured futures, to reclaim the publicness of the value that urban data infrastructures capture.

References

Bernstein, J., Singer-Vine, J. & Ryley, S. (2014) 'Exclusive: Hundreds of devices hidden inside New York City phone booths'. *Buzzfeed*, 6 October. Available at: http://www.buzzfeed.com/josephbernstein/exclusive-hundreds-of-devices-hidden-insidenewyork-cityph (Accessed 9 December 2019).

Bratton, B. H. (2016) *The Stack: On Software and Sovereignty*. Cambridge, MA: The MIT Press.

Broussard, M. (2018) *Artificial Unintelligence: How Computers Misunderstand the World*. Cambridge, MA: The MIT Press.

Campbell, C. (2016) 'LinkNYC kiosks provide free internet, but for a price'. *Government Technology*, 9 February. Available at: http://www.govtech.com/dc/articles/LinkNYC-KiosksProvide-Free-InternetBut-for-a-Price.html (Accessed 9 December 2019).

Digital Signage Connection, 15 May. Available at: https://www.digitalsignageconnection.com/linknyc-kioskshelp-boost-box-office-numbers-for-avengers-infinity-war (Accessed 9 December 2019).

Dough (2019) 'Fired from Uber: Why drivers get deactivated, and how to get reactivated'. *Ride Sharing Driver*, 4 October. Available at: http://www.ridesharingdriver.com/fired-uber-drivers-get-deactivatedand-reactivated/ (Accessed 9 December 2019).

eMarketer (2016) 'Listen in: Blending tech with media to make cities smarter'. *Behind the Numbers Podcast*, 8 August. Available at: https://www.emarketer.com/Article/

Listen-In-Blending-Tech-withMedia-Make-CitiesSmarter/1014312 (Accessed 9 December 2019).

Fitzsimmons, E. G. (2018) 'Why are taxi drivers in New York killing themselves?' *The New York Times*, 2 December. Available at: https://www.nytimes.com/2018/12/02/nyregion/taxi-drivers-suicide-nyc.html (Accessed 9 December 2019).

Fung, B. (2016) 'The tremendous ambitions behind New York City's free WiFi'. *Washington Post*, April 8. Available at: https://www.washingtonpost.com/news/the-switch/wp/2016/04/08/the-tremendousambitions-behindnew-york-citys-free-wifi/ (Accessed 9 December 2019).

Galloway, A. R. (2004) *Protocol: How Control Exists After Decentralization*. Cambridge, MA: The MIT Press.

Galloway, A. R. (2012) *The Interface Effect*. Malden, MA: Polity Press.

Graham, S. & Marvin, S. (2001) *Splintering Urbanism: Networked Infrastructures, Technological Mobilities and the Urban Condition*. New York: Routledge.

Gustin, S. (2016) 'LinkNYC's new free network is blazing fast. But at what cost to privacy?' *Motherboard*, 19 Feburary. Available at: http://motherboard.vice.com/read/linknycs-new-free-networkis-blazing-fast-but-at-what-cost-to-privacy (Accessed 9 December 2019).

Halpern, O., LeCavalier, J., Calvillo, N. & Pietsch, W. (2013) 'Test-bed urbanism'. *Public Culture*, 25(2), pp. 272–306.

Hanrahan, O. (2015) 'We must protect the on-demand economy to protect the future of work'. *Wired*, 9 November. Available at: http://www.wired.com/2015/11/we-must-protect-the-on-demand-economy-toprotect-the-future-of-work/ (Accessed 9 December 2019).

Hotz, R. L. (2015) 'As world crowds in, cities become digital laboratories'. *Wall Street Journal*, 11 December. Available at: http://www.wsj.com/articles/as-world-crowds-in-cities-become-digitallaboratories-1449850244 (Accessed 9 December 2019).

Intersection (2019) 'Place exchange: Real programmatic out of home'. Available at: http://placeexchange.com (Accessed 9 December 2019).

Intersection (n.d.) 'LinkNYC fact sheet'. Available at: https://www.link.nyc/assets/downloads/LinkNYCFact-Sheet.pdf (Accessed 9 December 2019).

Kirchner, L. & Mattu, S. (2015) 'Uber's surge pricing may not lead to a surge in drivers'. *ProPublica*, 29 October. Available at: https://www.propublica.org/article/uber-surge-pricing-may-not-lead-to-a-surge-indrivers (Accessed 9 December 2019).

Kitchin, R. & Lauriault, T. P. (2014) 'Towards critical data studies: Charting and unpacking data assemblages and their work'. *The programmable city: Working paper 2*. National University of Ireland, Maynooth. Available at: http://mural.maynoothuniversity.ie/5683/1/KitchinLauriault_CriticalDataStudies_ProgrammableCity_WorkingPaper2_SSRN-id2474112.pdf (Accessed 9 December 2019).

Kofman, A. (2018) 'Are New York's free LinkNYC internet kiosks tracking your movements?' *The Intercept*, 8 September. Available at: https://theintercept.com/2018/09/08/linknyc-free-wifikiosks/ (Accessed 9 December 2019).

Kushner, J. (2018) 'LinkNYC kiosks help boost box-office numbers for *Avengers: Infinity War*'. *Digital Signage Connection*, 15 May. Available at: https://www.digitalsignageconnection.com/linknyc-kiosks-help-boost-box-office-numbers-for-avengers-infinity-war (Accessed 29 August 2020).

Levy, K. E. C. (2015) 'The contexts of control: Information, power, and truck-driving work'. *The Information Society*, 31, pp. 160–174.

Liu, P. & Peng, Z. (2014) 'China's smart city pilots: A progress report'. *Computer*, 47, pp. 72–81.

Mattern, S. (2014) 'Interfacing urban intelligence. *Places Journal*, April. Available at: https://placesjournal.org/article/interfacing-urban-intelligence/ (Accessed 9 December 2019).

McFarlane, C. (2011) 'The city as assemblage: Dwelling and urban space'. *Environment & Planning D: Society & Space*, 29(4), pp. 649–671.

NYC Department of Consumer Affairs (2015) *New York City Mobile Services Study*. Available at: https://www1.nyc.gov/assets/dca/MobileServicesStudy/Research-Brief.pdf (Accessed 9 December 2019).

NYC Department of Transportation (2016) 'Bicycle counts'. Available at: http://www.nyc.gov/html/dot/html/bicyclists/bike-counts.shtml (Accessed 9 December 2019).

NYC Department of Information Technology & Telecommunications (2016) 'Public communications structure franchise agreement: Exhibit 2—CityBridge privacy policy'. *Official Website of the City of New York*. Available at: http://www1.nyc.gov/assets/doitt/downloads/pdf/Proposed-PCS-Franchise-Exhibit-2CityBridgePrivacy-Policy.pdf (Accessed 9 December 2019).

NYC Office of the Mayor (2014) 'Support pours in for LinkNYC'. *Official Website of the City of New York*. Available at: www1.nyc.gov/office-of-the-mayor/news/944-14/support-pours-for-linknyc (Accessed 9 December 2019).

NYCLU (2016) 'City's public wi-fi raises privacy concerns'. *New York Civil Liberties Union*. Available at: http://www.nyclu.org/news/citys-public-wi-fi-raises-privacy-concerns (Accessed 9 December 2019).

Rose-Redwood, R. S. (2006) 'Governmentality, geography, and the geo-coded world'. *Progress in Human Geography*, 30(4), pp. 469–486.

Rosenblat, A. (2018) *Uberland: How Algorithms Are Rewriting the Rules of Work*. Oakland, CA: University of California Press.

Rosenblat, A. & Stark, L. (2016) 'Algorithmic labor and information asymmetries: A case study of Uber's drivers'. *International Journal of Communication*, 10, pp. 3758–3784.

Rossiter, N. (2015) 'Coded vanilla: Logistical media and the determination of action'. *South Atlantic Quarterly*, 114, pp. 135–152.

Screen Media Daily (2014) 'CityBridge to launch LinkNYC, largest urban digital ad network'. Available at: http://screenmediadaily.com/citybridge-to-launch-linknyc-largest-urban-digital-ad-network/ (Accessed 9 December 2019).

Secor, A. (2013) 'Topological city: 2012 *Urban Geography* plenary lecture'. *Urban Geography*, 34(4), pp. 430–444.

Shapiro, A. (2018) 'Between autonomy and control: Strategies of arbitrage in the "on-demand" economy'. *New Media & Society*, 20(8), pp. 2954–2971.

Shelton, T., Zook, M. & Wiig, A. (2015) 'The "actually existing" smart city'. *Cambridge Journal of Regions, Economy & Society*, 8, pp. 13–25.

Simone, A. (2011) 'The surfacing of urban life'. *City*, 15(3–4), pp. 355–364.

Solomon, R. (2013) 'Last in, first out: Network archaeology of/as the stack'. *Amodern: Network Archaeologies 2*. Available at: http://amodern.net/article/last-in-first-out/ (Accessed 9 December 2019).

Srnicek, N. (2016) *Platform Capitalism*. Malden, MA: Polity Press.

Straube, T. (2016) 'Stacked spaces: Mapping digital infrastructures'. *Big Data & Society,* July–December. doi: 10.1177/2053951716642456

Tal, R. (2015) 'Is Uber really not a taxi service?' *Quora*. Available at: https://www.quora.com/Is-Uberreally-not-a-taxi-service (Accessed 9 December 2019).

Thatcher, J. (2014) 'Living on fumes: Digital footprints, data fumes, and the limitations of spatial Big data'. *International Journal of Communication*, 8, pp. 1765–1783.

Thatcher, J., O'Sullivan, D. & Mahmoudi, D. (2016) 'Data colonialism through accumulation by dispossession: New metaphors for daily data'. *Environment & Planning D*, 34(6), pp. 990–1006.

Ticona, J. (2016) *Left to Our Own Devices: Navigating the Risks of Work and Love with Personal Technologies*. Dissertation. University of Virginia, 27 April. doi: 10.18130/v3vg4d.

Tsotsis, A. (2012) 'Uber opens up platform to non-limo vehicles'. *TechCrunch*, July 2. Available at: http://social.techcrunch.com/2012/07/01/uber-opens-up-platform-to-non-limo-vehicles-with-uber-xservice-will-be-35-less-expensive/ (Accessed 9 December 2019).

Valverde, M. (2011) 'Seeing like a city: The dialectic of modern and premodern ways of seeing in urban governance'. *Law & Society Review*, 45(2), pp. 277–312.

Voytko, L. (2019) 'Payphone-replacing LinkNYC kiosks not generating projected revenue'. Available at: https://www.gothamgazette.com/city/8502-city-s-much-heralded-link-kiosks-not-generating-projectedrevenue (Accessed 9 December 2019).

3

POLITICAL ECOLOGIES OF PLATFORM URBANISM

Digital labor and data infrastructures

Dillon Mahmoudi, Anthony M. Levenda and John G. Stehlin

Toward a political ecology of platform urbanism

> All that is solid melts into tweets.
>
> *(Wyly, 2013, p. 391)*

In the contemporary networked city, an integrated machinic complex of information and communication technologies (ICTs) represents a new moment in capitalist urbanization, a phenomenon exemplified by the proliferation of urban platforms (Graham and Marvin, 2001, Amin and Thrift, 2002). As urbanism and digital platforms become a way of life, the city and the platform become increasingly conjoined as the joint medium of capital accumulation and sociality (Zip et al., 2013). The co-evolution was not necessarily unforeseen. At the turn of the 21st century, broad changes in technology, social life, and urbanization led many scholars to theorize a shift toward a new phase of capitalism based on immaterial labor. Both Autonomist Marxists and economic geographers have argued that "cognitive" or "cognitive-cultural" capitalism is marked by an accumulation process centered on immaterial inputs, immaterial and digital labor processes, and the production of immaterial goods, such as services, cultural products, knowledge, or communication (Hardt and Negri, 2004, Scott, 2009, 2014, Peters and Bulut, 2011). More recently, platform urbanism theorists have made similar arguments about the non-material digital processes that tap into existing circuits of urbanization (Artioli, 2018, Rodgers and Moore, 2018, Wyly et al., 2018). Yet these analyses often take for granted the material networks and physical infrastructure required as inputs into this reconfiguration of space and daily life.

This widespread focus on the immaterial aspects of contemporary digital capitalism, particularly the framing of platforms from search engines to ride-hailing apps as "services" (cf. Walker, 1985, pp. 50–51), obfuscates the materiality and socio-environmental foundations of capital accumulation and circulation that are increasingly mediated through digital platforms. In this chapter, we argue that the labor associated with the production of digital platforms, the labor associated with their use as "machinery," and the data on whose circulation this work depends, are all quite material. Our goal is to highlight how the high-value work of the "tech" economy and the precarious work of the gig economy are digitally interlinked, not just through an app but also an entire apparatus of energy-intensive data transmission and storage stretching far beyond the "city."

Our approach builds from digital political ecology (DPE) to understand the physical infrastructures and digital components of platform urbanism. While there has been significant scholarship focused on the political ecologies of urban biophysical processes (water, vegetation, waste, etc.; cf. Meehan, 2014), communication and information infrastructures have seen less attention, even though they likewise facilitate material flows and capital circulation. DPE scholarship materializes the immense hidden digital and energy infrastructures necessary for advanced computing, such as cryptocurrency mining (Lally et al., 2019) and e-waste processing (Pickren, 2014). This chapter combines these insights to examine the infrastructures that undergird platform urbanism, with a focus on data centers in the Pacific Northwest of the United States, to understand how a new division of labor (re)inscribes social disparities in the uneven geographies of the city and landscapes beyond.

Platform urbanism and the restructuring of capitalism

Platform urbanism is an essential part of a broader shift in capitalist urbanization toward what Scott calls "cognitive-cultural capitalism," which has three defining features. First, calculation, communication, information storage, and process design are performed using digital methods, reducing communication times and transportation/storage costs and enabling new forms of production, business organization, and collective consumption (Castells, 1983). Second, a new division of labor between two distinct class fragments—highly qualified "symbolic analysts," and a low-wage service underclass or precariat (Sassen, 1988, Scott, 2011)—has been spatially co-embedded by processes of urbanization. The former performs non-routine functions using knowledge, cognition, and symbols, while the latter performs service functions as either deskilled manual labor or menial service labor. Lastly, these productive changes are also reflected in consumption patterns that have shifted toward "experiential" goods and services clustered in urban areas (Markusen and Schrock, 2009, Currid-Halkett and Scott, 2013). These trends have been marked by changes in the "urbanization of capital" (Harvey, 1981, 1989).

Today, the temporal, spatial, and technological complexity of digital and industrial production necessitates technologically advanced cognitive-cultural labor to produce digital platforms that function as machinery for advanced production and logistics, complex targeted and individualized sales and advertising, and advanced consumer tracking and surveillance. Platforms require deskilled operators involved in menial tasks working in deskilled distribution centers, deskilled transportation, nearly automated advertising, and so forth. The data produced through the operation of this digital machinery by deskilled labor, and the surveillance of the "consumer," is, in turn, used to generate a "behavioral surplus" (Zuboff, 2019), in which data on users' activities is used to create new digital data commodities and/or apply to logistics processes that further deskill menial labor. In short, the enormous superstructure shaping and shaped by digital capitalism continues to become more complex and more urban as the benefits of data production agglomerate in cities, creating a positive feedback cycle that encourages further digital urbanization. Platform urbanism represents the co-evolution of the productive apparatuses of both technology and urbanization.

Contemporary urban development logics create pressure to expand digital, cultural, and/or informational economies—the "cynosures of the so-called 'new' economy" (Scott, 2011, p. 290)—and position cities as key nodes in the global "network society" (Castells, 2000). As a result, the concentration of people and businesses create an agglomerative site of data production spanning social networks, informal labor platforms, ride-hailing, check-ins, geolocation-based advertising, and so on. These social, economic, cultural, and informational changes afforded by digital ICTs correspond to rearrangements in the primary, secondary, and tertiary circuits of capital: commodity production, fixed capital (built environment for production, e.g. roads, rail, and other infrastructures) or a consumption fund (built environment for consumption), and long-term expenditures like health care or state-sponsored research and development that enhance labor's productivity, respectively. Platform urbanism speaks to a blurring of these circuits, as computational research in the tertiary circuit and secondary circuit elements like housing and transport infrastructure become, through platforms like Airbnb and Uber, drivers of *data* production that fuels the realization of value in the primary circuit.

Digital labor and platform value production

Theorists of digital capitalism like Wyly see the co-evolution of technological innovation and *urbanization* as the underpinning of a system of collection, surveillance, and value production based on

> [...] billions of smartphones, RFID (radio frequency identification) chips and QR (Quick Response) codes, and trillions of social-media data trails on preferences and purchases of physical commodities, services and media content. Data flood in, and the pattern-recognition algorithms optimize and monetize attention, creativity and communication amidst the neoliberal wind that capitalizes, commodifies, classes, and marketizes everything.

> Social reality is ransacked, but not for theory: click-throughs, page views, eyeballs, and ad revenue are what matter.
>
> *(2013, p. 392)*

Similarly, Terranova asserts that the technological innovation of the internet is "animated by cultural and technical labor through a continuous production of value that is completely immanent to the flows of the network society at large" (2000, pp. 33–34). This cultural and technical labor requires spatial structures in which "the physical conditions of exchange" (Marx, 1993, 444–448, 472)—or the urbanization process—become ever more important as the infrastructure of production. Thus, technological developments are increasingly intertwined with "advancements" in urbanization, reproducing urban space as part of the affordances of the production system.

The shifts in processes of urbanization and capital accumulation corresponding to platform urbanism can be demonstrated using the example of Uber. The Uber app is developed by cognitive-cultural programmers to track the locations of cars and users, and the Uber server back-end is programmed to make transportation calculations and store this data. When a user requests a ride, an Uber server makes the necessary calculation and communication to hail an available nearby driver operating their vehicle. The app computes the fastest route to the rider and the fastest route to the rider's destination and calculates a fare in advance. The driver performs the menial labor of driving following turn-by-turn in-app directions—a deskilled version of a taxi driver that required the craft of finding potential riders, knowing when and where people in the city might need a ride, and knowing what routes are fastest at particular times of the day. Uber riders are encouraged in the app to splurge and upgrade, as an experiential service, to luxury Uber Black or Uber Black SUV services. These interactions, along with in-app ads, formal tie-ins to other apps, or other informal forms of digital footprint sleuthing, provide rich accounts of user behavior (Thatcher, 2014, Thatcher et al., 2016, Couldry and Mejias, 2019). These processes began as consumer-oriented services but are also increasingly part of corporate operations; as of 2018, Uber and Lyft accounted for 71% of the market share in ground transportation for business travelers, and Uber alone has expanded to over 75 countries (Kerr, 2018). What looks like a service from the consumer's perspective is thus also a process of producing data as capital—digital machinery used in the production and realization of value (Sadowski, this volume).

Uber drivers are very aware of the value that their work generates in the form of data on both the user and the city, including location, times, traffic flows, and any corresponding significant events such as sporting events, concerts, or rallies that may affect demand for travel, and that this data may be used to further deskill driving or even replace drivers entirely with autonomous vehicles (Attoh et al., 2019). Further, the data collected is used to create a behavioral surplus, a form of value unique to digital platforms, and a necessary input into a new circuit of producing surplus value (Zuboff, 2019). In this case, the behavioral surplus stems from either using the data collected as an input into machine-learning algorithms that direct drivers to certain places at certain times (for either pickup or routes) or by connecting Uber accounts to social media accounts, which refines advertising profiles through data complementarity. For example, by connecting profiles across devices or browsers, Uber is in principle able to create profiles of all Uber users that have ever taken a Uber to the Moda Center in Portland, Oregon, which has a dedicated Uber Zone for dropoff and pickups (Uber.com, 2020), within an hour of the start of a Portland Trail Blazers basketball game and have a Facebook account, from which age, relationship status, and recent restaurant check-ins might be used to identify single 25- to 34-year-old men who recently ate at Burger King and went to the game.

In this example, the Uber platform forms a hinge between the urban built environment and the physical infrastructure of data circulation on the one hand and between dead labor embedded in algorithm production and the living but deskilled labor of driving on the other. The output of this function is not just a mobility service but also increasingly valuable data "fumes" (Thatcher, 2014). Scholars, therefore, must question how the data is being transmitted, where it is stored and copied, who has access to it, and how it is used to create or add to an advertising profile. Equally, they must ask about the division of labor involved in producing the platform itself: who uses this data to provide a service under what conditions of deskilling, automation, or punitive "reskilling" and who programmed the platform architecture that structures this labor process. Finally, scholars must ask how the infrastructure of the built environment *affords* the collection of data through situated platform services, its circulation through physical ICT infrastructure, and the materials and energy on which this process depends.

Thus, where platforms are typically framed as immaterial or as simple services, we see them as material parts of the process of producing value. Platform urbanism, as an exemplar of cognitive-cultural capitalism and the co-evolution of technological change and urbanization, reveals how the cognitive work of digital laborers and the manual labor of deskilled laborers are interlinked through the digital machinery of the platform. But this is made possible only by their necessary connection to massive data storage and processing centers, and the greater the data collected, transmitted, and processed, the greater the storage, transmission, and computing requirements. Fixed capital investment in data-related infrastructure is thus used to support these modes of production both in the reshaping of the urban environment and in the so-called hinterlands through data center expansion. This raises several questions: What are the socio-spatial characteristics and impacts of these digital infrastructures? Where are data centers located, and why in those specific locations? What are the socio-material impacts and benefits of data centers, and how are they distributed? To answer these questions requires theorizing the infrastructures of digital ICTs both beyond the screen and beyond the city.

The geography of data centers

A DPE analysis of platform urbanism, much like that of UPE, requires the examination of the material infrastructure and flows that make possible the expansion of surplus value through digitally mediated circulation as a moment of production. The on-going processes of urbanization, and their vital connection to the circulation of capital, reaches beyond the bounds of the city, aiding the seemingly immaterial forms of labor associated with cognitive-cultural production and the mundanely material labors of the gig economy alike. Thus, a focus on "the screen"—a phone, tablet, computer, or other digital ICT devices—experience of platforms misses their socio-environmental impacts, including the life cycle of the "smart" device from production to disposal, the fixed capital infrastructure that enables the networked connectivity vital to user-screen interactions, and the material flows that mediate these two moments. As Marx and Engels explain in *The German Ideology*, "The greatest division of material and mental labor is the separation of town and country" (Marx and Engels, 1978, p. 176). The materiality of "mental labor"—or cognitive and cultural labor—reaches beyond the city, invades the lifeworlds of a planet of urban residents, and excretes concrete, silicon, bits, servers, and energy waste, producing an "urban landscape" or "second nature" beyond the city.

From this perspective, one critical infrastructure of platform urbanism is the data center. Some firms own data centers, while others outsource storage and computing power to "cloud services" providers like Amazon, Google, and Microsoft. For example, Facebook owns its servers, while Uber and Twitter rent from Amazon. Some firms, like Amazon, are both data infrastructure providers, through Amazon's cloud services, and platforms themselves, with increasingly urban-oriented services like Amazon Fresh (food delivery), Amazon Ring (home security), and Amazon Prime Now (on-demand product delivery). In the era of "big data," where data is leveraged to solve all manner of social and environmental problems, data center capacity and growth are necessary requirements (boyd and Crawford, 2012, Ash et al., 2016). And as the data accumulated by urban platforms grows, driven by location detection and the capacity to generate dynamically interlinked consumer data profiles, this storage and processing capacity is increasingly essential to the continued functioning of the platform-based city itself.

Data centers are far from cloud-like auras. They are massive structures housing thousands of servers for storing data, advanced mechanical cooling and ventilation equipment, batteries and diesel generators for backup power and redundancy, and (depending on the location and owner) a highly securitized shell of fencing and walls with limited access areas and surveillance systems.

By design, data centers are also energy intensive. In 2012, a widely shared *New York Times* story drew attention to the energy requirements of these facilities, pointing to problems of overheating, space limitations, and memory limitations Facebook encountered with its 10 million users at the time (Babcock,

2012, Glanz, 2012). As the story reported, these "cloud factories" used about 30 billion watts of electricity worldwide, roughly the same as generated by 30 medium-sized nuclear or coal-fired power plants. Some data centers required "more power than a medium-size town" (Glanz, 2012), and for this reason, "data centers are among [electric] utilities' most prized customers." (Compared to Facebook's scale today, and the immensity of the data produced by one billion worldwide users requiring storage on its data servers, these quaint beginnings seem almost comically small.) While paling in comparison to "dirtier" industries like paper production, the polluting impacts of the immense, steady demand on predominately coal-fired power facilities, using two percent of all energy in the United States, exposed big data's "dark side" (Oremus, 2012), and even worse, the *New York Times* investigation showed that up to 90% of the energy consumed was wasted.

The data center industry responded first by addressing minor numerical errors in the *New York Times* analysis (Wilhelm, 2012), and second, by improving energy efficiency and investing in renewable energy sources, effectively, or at least discursively, "greening" their data center operations (cf. Amazon.com, 2014, Google, 2015a). These "modern" data centers have much-improved power usage effectiveness (PUE, or energy used overall divided by energy used for computing) from approximately 2.0 to near 1.07 (Babcock, 2012). The high percentage of hydropower capacity in the Pacific Northwest aids in the purported environmental sustainability of data center industries. The technical characteristics of data centers, including their energy and land requirements, have shaped locational choices by data center owners like Facebook and Amazon: free air-cooling, low electricity rates, inexpensive land, and enterprise zones that limit taxation are key decision points. This poses further questions about the politics of investment in places struggling to attract capital for economic development.

Non-governmental organizations have also stepped in to advocate for advancements in reducing polluting impacts and intensive energy consumption of data centers (McMillan, 2014). Greenpeace, in particular, promoted "clicking clean" as an environmental strategy to influence companies like Amazon Web Services to use cleaner sources of energy. Despite attempts to increase the efficiency of data centers, however, the overall growth in data storage needs represents something of a Jevon's paradox: increased computing efficiency affords, and possibly spurs, additional computing needs, potentially fueling more consumption and production of data and energy.

The geography of data centers in the Pacific Northwest displays their locational logic. The state of Oregon hosts large data centers for Facebook, Google, and Amazon, mainly in rural areas. Facebook has a large data center, exemplary of modern, high-efficiency facilities, in Prineville, a town of roughly 10,000 in central Oregon. Apple does not disclose all of its locations but also has a data center next to Facebook's Prineville facility. Google has developed a data center just east of Portland in The Dalles, adjacent to hydro-power facilities in the Columbia River Gorge dividing Oregon and Washington. It is one of only a handful of data centers valued at over $1 billion USD (Miller, 2013) and regularly featured by the company because of its aesthetically pleasing interior design (Google, 2015b). Amazon does not disclose the specific locations of its data centers, but at least one is located in Boardman, Oregon (Rogoway, 2011), and the company does confirm that it owns caching centers—small collections of servers that store data in locations more proximate to its users— outside major metropolitan areas throughout the West Coast of the US (Amazon.com, 2015). Amazon and Apple continue to expand in rural Oregon (Rogoway, 2015a). Finally, Quincy, Washington is home to one of the world's largest data centers, owned by Microsoft, as well as other large data facilities owned by Dell and Yahoo.

Cheap electricity is a major draw, with the Columbia River Basin providing over 40% of all US hydro-power electricity (Lillis, 2014). In addition to access to inexpensive rural land and electricity, the Columbia River Basin has access to high-bandwidth fiber optic cables (Miller, 2012a). The area provides links to numerous intra- and international long-haul cable connecting the region to other cable connections, providing high-bandwidth access to points across the globe. Regulatory changes have pushed these changes along as well. Rising interest in building data centers in Oregon led the state government to reduce or remove property taxes on "intangible" and "hard to quantify" assets like company branding and computer equipment, a clear nod to the tech industry later emulated by Washington (Miller, 2012b). During state legislative hearings, Google and Amazon representatives testified that the previous tax regime had prevented the companies from expanding their technical infrastructure. Google claimed that without the tax break, it could not develop its Google Fiber internet infrastructure in the city of Portland (Rogoway, 2015b). Shortly after the change in tax code, Amazon announced plans to build eleven more data centers in the region (Rogoway, 2015a).

The tax breaks also made it possible to build data centers closer to cities. Hillsboro, within the Portland metropolitan region, is the future site of a reasonably sized 18,500 square meter data center (Rogoway, 2015c). Hillsboro is also the terminus of three major long-haul cable submarine lines (Tyco Global Network Pacific, Southern Cross, and Trans-Pacific Express) connected to sites in Northern California, Japan, and other places in Southeast Asia. Each cable line is over 20,000 km long (Submarine Cable Networks, 2015). Within the Portland region, there are numerous land-based high-capacity long-haul cable connections to: Seattle and Tacoma in Washington; Boise, Idaho; Palo Alto, San Jose and Santa Clara in California; Cheyenne, Wyoming; and Kansas City, Missouri. A loop system connects the Oregon coast and central Oregon's data center's runs through a connection in Medford, Oregon (TR, 2014).

This digital machinery of platform urbanism does not necessarily benefit local communities. These massive data centers do not provide superior service to the populations of the small municipalities in which they are located, nor are they designed to serve *consumers* in the nearest large metropolitan areas, such as the "second-tier" tech hub of Portland, Oregon (Mayer, 2012). Instead, the regional digital labor and global reach of digital products produced necessitate data centers near, but not at, the site of digital *workers*. Urban high-tech and information technology firms, as well as consumers, require tight links to these data centers located in the "hinterland" that offer little or no direct benefit to rural municipalities or their residents (Glanz, 2013). On the basis of this infrastructural capacity, Portland is home to the annual Open Source Software Conference, the inventors of Linux and the "wiki," Intel's largest manufacturing site and patents, and a growing software and technology scene (Rogoway, 2014). Despite discursive appeals to local development made by large tech firms, the reality is that the benefits are not seen locally nor is their location driven by local demand.

Conclusion

In this chapter, we have highlighted how platform urbanism brings together the cognitive-cultural economy and the precarious service economy through underlying data and energy infrastructure that stretches far into the urban hinterland. We argue that an examination of platform urbanism necessitates the materialization of the digital infrastructure in the form of the urban built environment and its linkages to primarily rural data centers. We show that the clustering of the data centers of Amazon, Facebook, and Google in rural Oregon and the broader Pacific Northwest—powering other platform urbanism firms, like Uber, through their data centers—contrasts sharply with the image of these firms, both popular and academic, as constitutively "urban," just as their mobilization of precarious labor in the gig economy contrasts with the notion that they herald an age of "immaterial" work. Of interest for questions of platform urbanism is not just the way in which ICT infrastructures replace labor, but instead how these infrastructures *deploy* labor on an ever-expanding scale and an increasingly precarious basis.

This deployment is the key to the connection between urban and rural. To return to the example of Uber, assessing the environmental impact of an Uber ride in Portland requires understanding the impacts of Uber's back-end computation and storage on an Amazon server in rural Oregon. Activities facilitated by platforms such as Uber implicate any number of other rural Oregon data centers or subcontracted digital platform companies. For example, verifying an Uber account using Facebook allows for data sharing between Uber and Facebook, connecting trips with social media profiles. Paying for Uber, or Uber Eats, with Google Pay or Apple Pay connects trips or restaurant orders to respective user accounts at Google and Apple. A seemingly isolated platform action might involve an entire ecosystem of digital platforms and numerous separate data centers.

Further, we show how urban platforms rely on a growing class divide. On the one hand, agglomerations of cognitive-cultural workers concentrate in urban areas. Amazon, Facebook, and Google are, again, emblematic of cognitive-cultural capitalist production and broadly underpin the proliferation of digital platforms. At the same time, the deskilled laborers who rely on this new digital machinery must also concentrate in these areas because of the density of demand and the availability of gig labor. In other words, cognitive-cultural workers are employed to create the digital machinery that increases the rate that capital is realized, while deskilled platform labor in the gig economy actually realizes physical tasks that generate essential data about consumer desires (Attoh et al., 2019, Payne and O'Sullivan, 2020).

Platform urbanism combines sophisticated manipulations of nature and intensification of urbanization processes that link together both cognitive-cultural labor with deskilled platform labor, and the data production of the city with the computation and storage of rural data centers. Borrowing from DPE, we suggest that platform urbanism, as an appendage of the growing complexity of third-phase digital capitalist industry and urbanization, masks these types of labor and the necessary material infrastructure that enables them. This massive infrastructure both makes digitally mediated labor possible and positions rural localities as the bearers of new energy-intensive industries with little in the way of local benefits like employment growth, environmental improvement, or digital inclusion. Platform urbanism embodies the dialectic and material representation of both dead labor and the general intellect—shaping new, and uneven, socio-material natures and futures.

Acknowledgement

This chapter is adapted from "Beyond the Screen: Uneven Geographies, Digital Labor, and the City of Cognitive-Cultural Capitalism." *TripleC: Communication, Capitalism* 14 (1): 99–120 doi:10.31269/triplec.v14i1.699 (Mahmoudi and Levenda, 2016). The authors would like to thank the attendees and organizers of the "Urban platforms and the future city" workshop at the University of Manchester in February of 2019 for their support and helpful feedback on an earlier adaptation of this paper. Authorship order reflects the winner of rock, paper, scissors at a West Didsbury & Chorlton Football Club match.

References

Amazon.com (2014). AWS and sustainable energy. *Amazon Web Services, Inc.* Available at: http://aws.amazon.com/about-aws/sustainable-energy/ (Accessed 22 Jun 2015).

Amazon.com (2015). Global infrastructure. *Amazon Web Services, Inc.* Available at: http://aws.amazon.com/about-aws/global-infrastructure/ (Accessed 1 Mar 2015).

Amin, A. and Thrift, N., (2002). *Cities: Reimagining the Urban.* 1st edition. Cambridge: Polity.

Artioli, F., (2018). Digital platforms and cities: a literature review for urban research. Available at: https://core.ac.uk/display/160063276.

Ash, J., Kitchin, R., and Leszczynski, A., (2016). Digital turn, digital geographies? *Progress in Human Geography*, 42(1), pp. 25–43.

Attoh, K., Wells, K., and Cullen, D., (2019). "We're building their data": Labor, alienation, and idiocy in the smart city. *Environment and Planning D: Society and Space*, 37(6), pp. 1007–1024.

Babcock, C., (2012). N.Y. Times data center indictment misses big picture. *InformationWeek*. Available at: http://www.informationweek.com/cloud-computing/infrastructure/ny-times-data-center-indictment-misses-b/240007880 (Accessed 22 Jun 2015).

boyd, danah and Crawford, K., (2012). Critical questions for big data: Provocations for a cultural, technological, and scholarly phenomenon. *Information, Communication & Society*, 15 (5), pp. 662–679.

Castells, M., (2000). *The Rise of the Network Society Volume 1*. 2nd edition. Oxford: Blackwell.

Couldry, N. and Mejias, U.A., (2019). Data colonialism: Rethinking big data's relation to the contemporary subject. *Television & New Media*, 20(4), pp. 336–349.

Currid-Halkett, E. and Scott, A.J., (2013). The geography of celebrity and glamour: Reflections on economy, culture, and desire in the city. *City, Culture and Society*, 4(1), pp. 2–11.

Glanz, J., (2012). Data centers waste vast amounts of energy, belying industry image. *The New York Times*, 22 Sep. Available at: https://www.nytimes.com/2012/09/23/technology/data-centers-waste-vast-amounts-of-energy-belying-industry-image.html (Accessed 17 Aug 2020).

Glanz, J., (2013). Is big data an economic big dud? *The New York Times*, 17 Aug. Available at: https://www.nytimes.com/2013/08/18/sunday-review/is-big-data-an-economic-big-dud.html (Accessed 17 Aug 2020).

Google, (2015a). The big picture, Google Green. *Google.com*. Available at: http://www.google.com/green/bigpicture/ (Accessed 22 Jun 2015).

Google, (2015b). Data centers. *Google.com*. Available at: http://www.google.com/about/datacenters/ (Accessed 22 Jun 2015).

Graham, S. and Marvin, S., (2001). *Splintering Urbanism: Networked Infrastructures, Technological Mobilities and the Urban Condition*. London: Routledge.

Hardt, M. and Negri, A., (2004). *Multitude: War and Democracy in the Age of Empire*. New York: Penguin Books.

Harvey, D., (1981). The spatial fix–Hegel, von Thunen, and Marx. *Antipode*, 13(3), pp. 1–12.

Harvey, D., (1989). *The Urban Experience*. Baltimore, MD: The Johns Hopkins University Press.

Kerr, D., (2018). Lyft edges in on Uber when it comes to business travelers, report says. *CNET*. Available at: https://www.cnet.com/news/lyft-edges-in-on-uber-when-it-comes-to-business-travelers/ (Accessed 27 Feb 2020).

Lally, N., Kay, K., and Thatcher, J., (2019). Computational parasites and hydropower: A political ecology of Bitcoin mining on the Columbia River. *Environment and Planning E: Nature and Space*, 251484861986760.

Lillis, K., (2014). The Columbia River Basin provides more than 40% of total U.S. hydroelectric generation. *Today in Energy - U.S. Energy Information Administration*, 27 June. Available at: https://www.eia.gov/todayinenergy/detail.php?id=16891 (Accessed 24 Feb 2020).

Mahmoudi, D. and Levenda, A., (2016). Beyond the screen: Uneven geographies, digital labour, and the city of cognitive-cultural capitalism. *tripleC: Communication, Capitalism & Critique. Open Access Journal for a Global Sustainable Information Society*, 14(1), pp. 99–120.

Markusen, A. and Schrock, G., (2009). Consumption-driven urban development. *Urban Geography*, 30(4), pp. 344–367.

Marx, K. and Engels, F., (1978). The German Ideology. In R. Tucker (Ed.) *The Marx-Engels Reader*. 2nd edition (pp. 146–200). New York: W. W. Norton & Company.

Marx, K., (1993). *Capital: Volume 2: A Critique of Political Economy*. London: Penguin Classics.

Mayer, H., (2012). *Entrepreneurship and Innovation in Second Tier Regions*. Northampton, MA: Edward Elgar Pub.

McMillan, R., (2014). Amazon and Twitter's data centers flunk Greenpeace report. *WIRED*. Available at: http://www.wired.com/2014/04/greenpeace/ (Accessed 22 Jun 2015).

Meehan, K.M., (2014). Tool-power: Water infrastructure as wellsprings of state power. *Geoforum*, 57, pp. 215–224.

Miller, R., (2012a). The new data center geography. *Data Center Knowledge*, 28 February. Available at: https://www.datacenterknowledge.com/archives/2012/02/28/the-new-data-center-geography (Accessed 17 Aug 2020).

Miller, R., (2012b). Washington state passes data center tax breaks. *Data Center Knowledge*, 16 April. Available at: https://www.datacenterknowledge.com/archives/2012/04/16/washington-state-restores-data-center-tax-breaks/ (Accessed 17 Aug 2020).

Miller, R., (2013). The billion dollar data centers. *Data Center Knowledge*, 29 April. Available at: https://www.datacenterknowledge.com/archives/2013/04/29/the-billion-dollar-data-centers (Accessed 17 Aug 2020).

Oremus, W., (2012). Big data's dark side: A massive, polluting drain on the nation's power supply. *Slate*, 24 September. Available at: https://slate.com/technology/2012/09/big-data-pollution-cloud-servers-waste-electricity-on-massive-scale-new-york-times-finds.html (Accessed 17 Aug 2020).

Payne, W.B. and O'Sullivan, D., (2020). Exploding the phone book: Spatial data arbitrage in the 1990s internet boom. *Annals of the American Association of Geographers*, 110(2), pp. 391–398.

Peters, M.A. and Bulut, E., eds., (2011). *Cognitive Capitalism, Education and Digital Labor*. 1st edition. New York: Peter Lang Publishing Inc.

Pickren, G., (2014). Geographies of e-waste: Towards a political ecology approach to e-waste and digital technologies. *Geography Compass*, 8(2), pp. 111–124.

Rodgers, S. and Moore, S., (2018). Platform urbanism: An introduction. *Mediapolis*, 23 (4). Available at: https://www.mediapolisjournal.com/2018/10/platform-urbanism-an-introduction/ (Accessed 17 Aug 2020).

Rogoway, M., (2011). Amazon confirms its data center near Boardman has begun operating. *OregonLive.com*, 9 Nov. Available at: https://www.oregonlive.com/silicon-forest/2011/11/amazon_confirms_its_data_cente.html (Accessed 17 Aug 2020).

Rogoway, M., (2014). Oregon tech employment hits 12-year high as software plays a growing role. *Oregon Live*, 16 Sep. Available at: https://www.oregonlive.com/silicon-forest/2014/09/oregon_tech_employment_hits_12.html (Accessed 17 Aug 2020).

Rogoway, M., (2015a). Amazon plans up to 11 more Oregon data centers if tax situation addressed. *OregonLive.com*, 8 Feb. Available at: https://www.oregonlive.com/silicon-forest/2015/02/amazon_plans_up_to_11_more_ore.html (Accessed 17 Aug 2020).

Rogoway, M., (2015b). Tax cuts for Google Fiber, Comcast and data centers move forward in Oregon Senate, but cities object. *OregonLive.com*, 27 Feb. Available at: https://www.oregonlive.com/silicon-forest/2015/02/tax_changes_for_google_fiber_c.html (Accessed 17 Aug 2020).

Rogoway, M., (2015c). ViaWest building new, 200,000-square-foot data center in Hillsboro. *OregonLive.com*, 9 Feb. Available at: https://www.oregonlive.com/silicon-forest/2015/02/viawest_building_new_200000-sq.html (Accessed 17 Aug 2020).

Sassen, S., (1988). *The Mobility of Capital and Labor*. New York: Oxford University Press.

Scott, A.J., (2009). *Social Economy of the Metropolis: Cognitive-Cultural Capitalism and the Global Resurgence of Cities*. Oxford; New York: Oxford University Press.
Scott, A.J., (2011). Emerging cities of the third wave. *City*, 15(3–4), pp. 289–321.
Scott, A.J., (2014). Beyond the creative city: Cognitive–cultural capitalism and the new urbanism. *Regional Studies*, 48(4), pp. 565–578.
Submarine Cable Networks, (2015). TPE Hillsboro cable landing state. | *The World of Submarine Cable Systems and Networks*. Available at: http://www.submarinenetworks.com/stations/north-america/usa-west/hillsboro-tpe (Accessed 19 Jun 2015).
Terranova, T., (2000). Free labor: Producing culture for the digital economy. *Social Text*, 18(22), pp. 33–58.
Thatcher, J., (2014). Big data, big questions | Living on fumes: Digital footprints, data fumes, and the limitations of spatial big data. *International Journal of Communication*, 8, pp. 1765–1783.
Thatcher, J., O'Sullivan, D., and Mahmoudi, D., (2016). Data colonialism through accumulation by dispossession: New metaphors for daily data. *Environment and Planning D: Society and Space*, 34(6), pp. 990–1006.
TR, (2014). Network maps: USA Longhaul. *Telecom Ramblings*. Available at: https://www.telecomramblings.com/network-maps/usa-fiber-backbone-map-resources/ (Accessed 17 Aug 2020).
Uber.com, (2020). Driver Instructions for Pickups at the Moda Center. Available at: https://www.uber.com/drive/portland/venues/portland-moda-center/ (Accessed 10 Feb 2020).
Walker, R., (1985). Is there a service economy? The changing capitalist division of labor. *Science & Society*, 49(1), pp. 42–83.
Wilhelm, A., (2012). Microsoft responds to the NYTimes. *The Next Web*. Available at: http://thenextweb.com/microsoft/2012/09/25/microsoft-responds-nytimes-data-center-article-gently-pointing-its-bunk/ (Accessed 22 Jun 2015).
Wyly, E., (2013). The city of cognitive–cultural capitalism. *City*, 17(3), pp. 387–394.
Wyly, E., Daniels, J., Dhanani, T., and Yeung, C., (2018). Hayek in the cloud: Conservative cognition and the evolution of the smart city. *City*, 22(5–6), pp. 820–842.
Zip, L., Parker, R., and Wyly, E., (2013). Facebook as a Way of life: Louis Wirth in the social network. *Geographical Bulletin*, 54(2), pp. 77–98.
Zuboff, P.S., (2019). *The Age of Surveillance Capitalism: The Fight for a Human Future at the New Frontier of Power*. London: Profile Books.

4

UNICORNS, PLATFORMS, AND GLOBAL CITIES

The economic geography of ride-hailing

Shauna Brail[1]

Introduction

Ride-hailing is dynamic, evolving, and contentious. In less than a decade, ride-hailing has gone from a mere idea to a global phenomenon, attracting venture capital funding, embedding itself in urban innovation and startup ecosystems, and reshaping urban transportation—possibly the most dramatic change that cities have witnessed since the widespread introduction and adoption of the automobile.

Ride-hailing—the practice of digitally matching nonprofessional drivers of private vehicles with paying customers—began in San Francisco with Uber, founded in 2009. From about 2012, the proliferation of ride-hailing firms accelerated globally alongside the widespread adoption of smartphones, GPS, and wireless internet services. By 2018, there were 11 ride-hailing unicorns (privately held technology companies worth $1 billion US or more) around the world.

A growing body of scholarly work, both critical and laudatory, analyzes the urban impacts of ride-hailing as a disruptive form of mobility (see Clewlow and Mishra, 2017; Stocker and Shaheen, 2017; Sumantran, Fine and Gonsalvez, 2017; Sperling, 2018). However, there is a lack of scholarship on the location patterns of ride-hailing firms, particularly with respect to their headquarters and their secondary office locations used for engineering and research and development (R&D). Examining the economic geography of ride-hailing is important because it helps us understand not only the spatial organization of new firms and industries in the digital platform economy but also the impacts of ride-hailing production on cities. Furthermore, ride-hailing as a production process, and the concentration of production in certain global cities, have implications for the way we assess how urban platforms are proceeding to leverage and redirect transformations in cities and in society.

The remainder of this chapter examines the economic geography of ride-hailing, underscoring the challenges and opportunities that the growth of platform urbanism presents for cities and urban infrastructure. Ride-hailing depends on public, physical infrastructure such as streets in order to operate. At the same time, the presence of ride-hailing's key office locations latches onto intangible infrastructures that derive from their embeddedness in dense urban agglomeration economies. This case study highlights how ride-hailing's economic geography contextualizes the role of a select number of global cities as both hosts to ride-hailing firms' headquarters, engineering, and R&D locations and as sites of government advocacy and lobbying. The next section considers the contributions of scholarship to understanding the global organization of ride-hailing firms and the production and reproduction of concentrated activity in a select subset of world cities. The literature review reveals connections between two perspectives: first, traditional urban theory on the role of cities as centres of command and control in the global economy, and second, contemporary factors associated with technological change and innovation that influence patterns of urban investment and growth. These factors include the rise of an intangibles economy, the subsequent emergence of the digital platform economy, and the concept of platform urbanism. Following this review, the chapter analyzes a novel database of ride-hailing's 11 global unicorns. By identifying the location patterns associated with each firm's headquarters and secondary offices (for engineering and R&D activity), and by considering the cities in which these firms offer ride-hailing services, the analysis presents insights on platform urbanism from the perspective of a firm's location decisions. Ultimately, this chapter suggests that the urban economic geography of ride-hailing shapes our understanding of platform urbanism at both intraurban and interurban scales.

World cities, intangibles, and the platform economy in an era of platform urbanism

The world cities literature illuminates the role of a small set of cities, predominantly in the global North, as key global centres of economic and political power. These cities, it is argued, play a disproportionately significant role as centres of concentration for multinational headquarters (Hall, 1966) and corporate decision-making (Cohen, 1981; Friedmann, 1986) and as sites where producer services firms agglomerate (Sassen, 1991). Beaverstock, Smith, and Taylor's (1999) study moved beyond the notion of a single tier of world cities towards a world cities hierarchy. By examining world cities—as defined by the concentration of global finance, insurance, and real estate firms; legal firms; and advertising and accounting firms—Beaverstock et al. challenge us to examine a wider economic geography of world cities, albeit one that as recently as 20 years ago remained concentrated in the global North.

Until the turn of the 21st century, world cities scholarship focused on a small, select number of cities, typically New York, London, Paris, Hong Kong, and

Tokyo. Swaths of the world with the largest absolute populations, including cities in China and India, were completely off the map in these earliest iterations. By analyzing world cities through a tiered roster (Beaverstock et al., 1999; Kratke and Taylor, 2004; Benton-Short, Price, and Friedman, 2005), the scholarship offers a lens through which we can interpret the spatial patterns and urban economic geography of a new, technology-driven industry with a penchant for urban concentration. This approach fundamentally opens up our thinking about not only the roles of world cities but also the potential redistribution of economic command and control functions to more cities globally.

Economic geographers and urban economists have established that innovative firms and industries benefit from being in close proximity to one another, thanks to agglomeration economies (Glaeser, 2010; Scott and Storper, 2015). Particular urban locations provide a number of supports, especially through a strong network of actors and associations, frequently referred to as an ecosystem. This ecosystem typically includes shared knowledge and specialized labour pools, research activity, and training programs stemming from universities, supportive institutions, and a governance structure that facilitates continued economic development. At the top of the world cities roster, firms can choose highly skilled talent from a global labour market and justify paying these workers high wages (Beaverstock, Smith, and Taylor, 2000). Firms actively seek out new locations outside their home country base in order to access talent pools across a range of wage and skill levels (Branstetter, Glennon, and Jensen, 2018).

If we connect our knowledge of cities in which firms concentrate decision-making activities with our understanding of the changes in firm organization precipitated by the digital turn (Ash, Kitchin, and Leszczynski, 2018), we can better understand the rise of urban platforms. First, we can recognize the increasing relevance of intangible assets as a precursor to understanding the emergence of new sectors. The term "intangible economy" is a relatively recent economic concept that refers to a shift in the way economic value is measured. Traditionally, companies were valued based on their tangible assets, including property and physical goods. Shortly after the start of the 21st century, however, economists began to consider the significance of intangible and knowledge-based assets, such as ideas, brands, software, and networks, in the overall value of a firm or organization (Haskel and Westlake, 2017). According to Haskel and Westlake (2017, 7), "the steady move to intangible investment helps us understand some of the key issues facing us today: innovation and growth, inequality, the role of management, and financial and policy reform." While accounting systems do not always recognize the true contribution of these knowledge-based assets to overall firm value, the growing intangible nature of value has significant implications for competitiveness and policy.

Four characteristics of the intangibles economy lead to uncertainty and challenges: the nature of sunk costs, spillovers, scalability, and synergies (Haskel and Westlake, 2017). These characteristics are also particularly relevant to geographers. Notably, the way in which intangibles emphasize the importance of high-skilled talent, regional spillovers, and agglomeration economies contributes to increased concentration in cities where relevant networks are located. Network effects that arise with the growth of intangibles may privilege early entrants and regions, especially in places where entrepreneurship and innovation concentrate and flourish. Finally, intangible assets are distinct from tangible assets because they can be used over and over again, concurrently across geographies, without depletion. This characteristic contributes to a winner-take-all-model associated with the rise of intangible capital and further aggravates the negative consequences that result when economic benefits are distributed unequally (Asselin and Speer, 2019).

Thus, the growth of intangibles and the rise of platform economy firms go hand in hand. The term platform economy refers to economic activity enabled through digitization, in which the production of a good or service depends not on ownership of the means of production but on a digital network that provides a matching service (Kenney and Zysman, 2016). On the surface, platform economy firms profit through the monetization of underused resources, such as property and vehicles. These firms can scale rapidly, make minimal local investments, and cause significant local disruption. Made possible through digitization, the widespread adoption of smartphones and the ubiquity of the internet, the platform economy offers a significant economy of scale. Value is derived not just by a product or service offerings but by rapid scaling, network effects, and the monetization of secondary attributes, such as data.

The platform economy, its proliferation, and its associated urban geographies present opportunities and challenges for cities, places, infrastructures, and people. Over a short period of time, the platform economy has demonstrated that it can impact quality of life and economic opportunity in cities and communities—both those chosen as central places of investment and those left behind. Indeed, as this impact suggests, the "sharing economy is a particularly urban phenomenon" (Davidson and Infranca, 2018, 208). It is not just the case that cities engender platform economy firms (Davidson and Infranca, 2016), but, importantly, that ongoing regulatory debates will likely continue to affect these places the most (Davidson and Infranca, 2018).

Although the connection is often overlooked, the growth of intangibles as a source of value, alongside the associated rise of the platform economy, encourages platform urbanism. Understood as the way digital platform economy firms restructure urban spaces, lives, and livelihoods, platform urbanism seems to be an outgrowth, or mutation even, of the smart cities concept. According to Barns (2019), as private sector firms aspire to profit from leveraging digital technologies to solve urban challenges, and as budget-strapped governments seek less expensive ways of improving life for inhabitants, digitization expands to all aspects of city life, for better or for worse. This extension necessitates new forms of partnerships, negotiations, collaborations, and exchanges between private and public sectors and further widens the influence of private firms on urban spaces and assets (Kitchin, 2014; Taylor Buck and While, 2017; van der Graaf and Ballon,

2019; Caprotti and Liu, 2020). With respect to the confluence of public and private interests, van der Graaf and Ballon (2019, 371) suggest:

> The idea of a private commercial public sphere is not new, an increase in inclusivity of the public sphere is not new, and ideological isolation is not new either. What may perhaps be new is the pervasiveness of the gatekeeping that complex multi-stakeholder platform-based ecosystems make possible, the invisibility of that filtering and the lack of transparency.

Amongst other characteristics, platform urbanism contributes to increasing difficulty in identifying and separating private from public.

In light of these considerations, the connection between smart or digital cities and platform urbanism becomes evident. According to Barns (2019, 2), "it is not always clear how a pivot towards platforms, embodied by concepts like platform urbanism, might be distinguished from existing domains of digital geography, including critical interest in the sharing economy, smart cities, and algorithmic geographies." We can see platform urbanism as an extension, and predictable by-product, of the transformative role that private sector firms play in city-building in an era of hyper-digitization. Urban platforms leverage a form of shadow rent enabled via the combination of physical and digital infrastructure (Stehlin, 2018). As such, they draw upon, and indeed reshape, the physical and social infrastructures of cities. Furthermore, the combination of platform urbanism's digital nature and urban embeddedness results in at least four perplexing outcomes: firms that benefit from local urban infrastructure absent local investment (Barns et al., 2017; Taylor Buck and While, 2017; van der Graaf and Ballon, 2019); the reorganization and recalibration of relationships between public and private sectors (Kitchin, 2014; Barns et al., 2017; van der Graaf and Ballon, 2019; Caprotti and Liu, 2020); the exacerbation of digital, socioeconomic, and spatial divides (Ash, Kitchin, and Leszczynski, 2018); and uncertain, yet certainly challenging prospects for organized labour, wage disparity and precarious employment (Bates et al, 2019).

Ride-hailing's economic geography

Since Uber's founding in 2009, 11 ride-hailing firms have emerged with a market valuation of $1 billion or more. Together, these firms produce ride-hailing services in 85 countries around the world, spanning more than 2,600 municipalities. Within less than a decade, ride-hailing has reshaped a traditionally local service offered by locally embedded firms into a global service supported by nationally and/or internationally recognized brands.

Examining the global economic geography of these firms underscores three connected and significant findings regarding how the digital platform economy affects cities. First, and despite the global reach of ride-hailing apps, the production of ride-hailing is highly concentrated in a select number of cities. Second, the location patterns of ride-hailing unicorn headquarters and secondary offices demonstrate that the geography of world cities is expanding to previously disregarded cities and regions, which are rising as new centres. Third, the findings suggest a continued need for policy interventions at various scales to manage the advancement of new industries, the development of entrepreneurship, and the expansion of the world cities roster.

Worldwide reach/concentrated production

The concentration of decision-making and wealth creation in a small number of world cities has global implications for labour, investment, partnerships with government, local ground transportation, and more. With power residing in ride-hailing unicorns, this case study establishes that the digital turn continues to privilege urban agglomeration economies. While consumption of ride-hailing is global and distributed, control over production is steadfastly concentrated.

Table 4.1[2] presents an overview of the 11 ride-hailing unicorns, highlighting their market valuation, the year in which they reached unicorn status, their headquarters location, and the number of cities and countries in which each one operates. Collectively, the market value of these 11 firms is $171.1 billion USD (at the time of data collection). Uber's and Didi Chuxing's market valuations of $72 and $56 billion, respectively, represent nearly 75% of the total market valuation for

TABLE 4.1 Global footprint of ride-hailing unicorns

Firm	*Market valuation ($billion USD)*	*Year $1 billion valuation reached*	*Headquarters location*	*Number of cities*	*Number of countries*	*Geography*
Total	**$174.8**			**2,600+**	**85**	
Uber	$72	2013	San Francisco	685	66	Global
Didi Chuxing	$56	2014	Beijing	710	3	China
Lyft	$15.1	2014	San Francisco	648	2	US/Canada
Grab	$10	2014	Singapore	160	8	Southeast Asia
Ola	$7	2017	Bangalore	109	2	India
Gojek	$5	2016	Jakarta	50	1	Indonesia
Cabify	$1.4	2018	Madrid	39	11	Spain, South America
Gett	$1.4	2018	Tel Aviv	103	4	Israel, Europe, North America
Careem	$1.2	2016	Dubai	77	14	Middle East (excluding Israel)
99	$1	2018	Sao Paolo	572	1	Brazil
Bolt	$1	2018	Tallinn	43	27	Europe, Africa

Source: Author compilation of public data. Data accurate as of August 2018.

the 11 firms. These two firms dwarf the market valuations of the remaining companies, which range from $15.1 billion (Lyft) to $1 billion (99 and Bolt[3]).

The global reach of ride-hailing is indisputable. These 11 firms provide ride-hailing services in 85 countries and in more than 2,600 municipalities[4] around the world (see Figure 4.1). Ride-hailing firms operate in many cities. However, these activities alone do not represent a significant economic opportunity in terms of investment or job creation. Instead, the locations of high-level activities in ride-hailing—including headquarters, secondary offices (see Figure 4.2), and to a lesser extent regional and national operations—are highly concentrated. Table 4.2 identifies the cities that fall under each category.

While digital platforms enable firms to scale and expand rapidly across borders, these same processes negatively impact local employment. If a ride-hailing unicorn merely opens a local office that recruits drivers, conducts marketing campaigns, and employs a small administrative staff, then we cannot claim that the city receives real direct economic benefits in the form of job growth, talent development, and local investment. Local offices are typically small, connected locally but not necessarily globally. Uber, the largest firm, employed 16,000 people directly at the end of 2017 (Bhuiyan, 2018). That same year, reports indicated there were over 2 million Uber drivers (Camp, 2017), none of whom were considered or treated as employees. While this case study does not examine labour issues in ride-hailing, it is crucial to acknowledge that ride-hailing's impacts include difficult and sometimes destructive consequences for labour.

Headquarters cities

Whereas ride-hailing operations are expansive and distributed, headquarters activities are limited to ten cities. Locations of ride-hailing headquarters fall into one or more of the following categories: established or emerging world cities for technology startups (Adler et al., 2019) (San Francisco, Beijing, Singapore); cities with national or regional policies that attract innovation-oriented firm formation (Crescenzi et al., 2019) (Bangalore, Singapore, Tel Aviv, Tallinn); and gateway cities (Taylor et al., 2002), places that benefit from cultural, linguistic, and/or region-specific qualities, including access to large populations (Bangalore, Beijing, Dubai, Jakarta, Madrid, São Paolo, Singapore).

If we examine the cities hosting a firm's secondary offices (R&D activities, engineering), the list not only replicates the sites of headquarters cities but also adds another layer of locations. Table 4.2 highlights the 29 cities where ride-hailing firms strategically situate their headquarters and secondary offices, revealing that each firm co-locates engineering and R&D activities in its headquarters location. Table 4.2 suggests that firms benefit from agglomeration economies as well as knowledge and talent spillovers based on intrafirm and interfirm location patterns.

Significantly, of the 29 cities hosting secondary offices, five act as headquarter locations to other ride-hailing firms: Bangalore, Beijing, Jakarta, San Francisco,

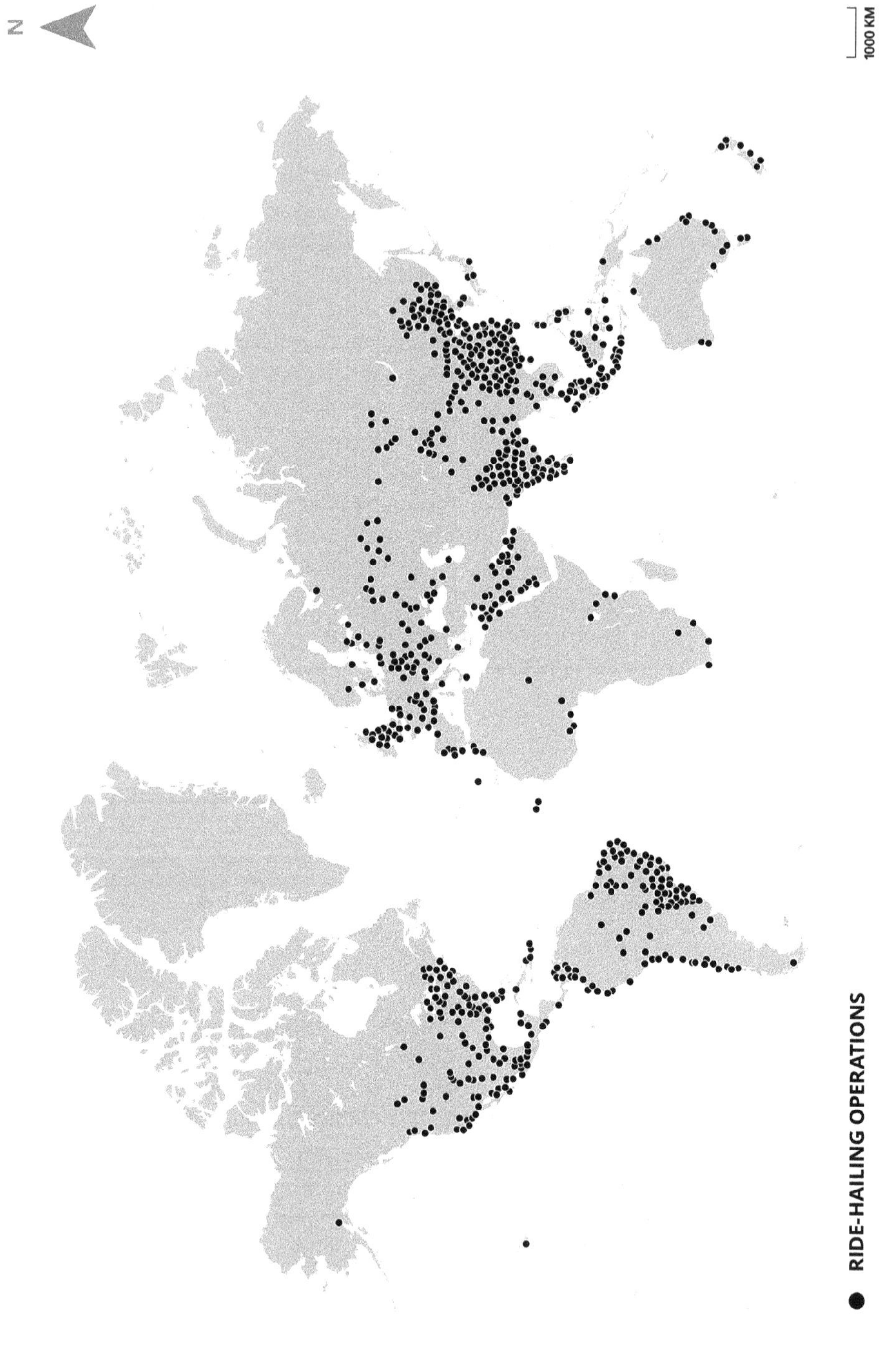

FIGURE 4.1 Ride-hailing's global reach.

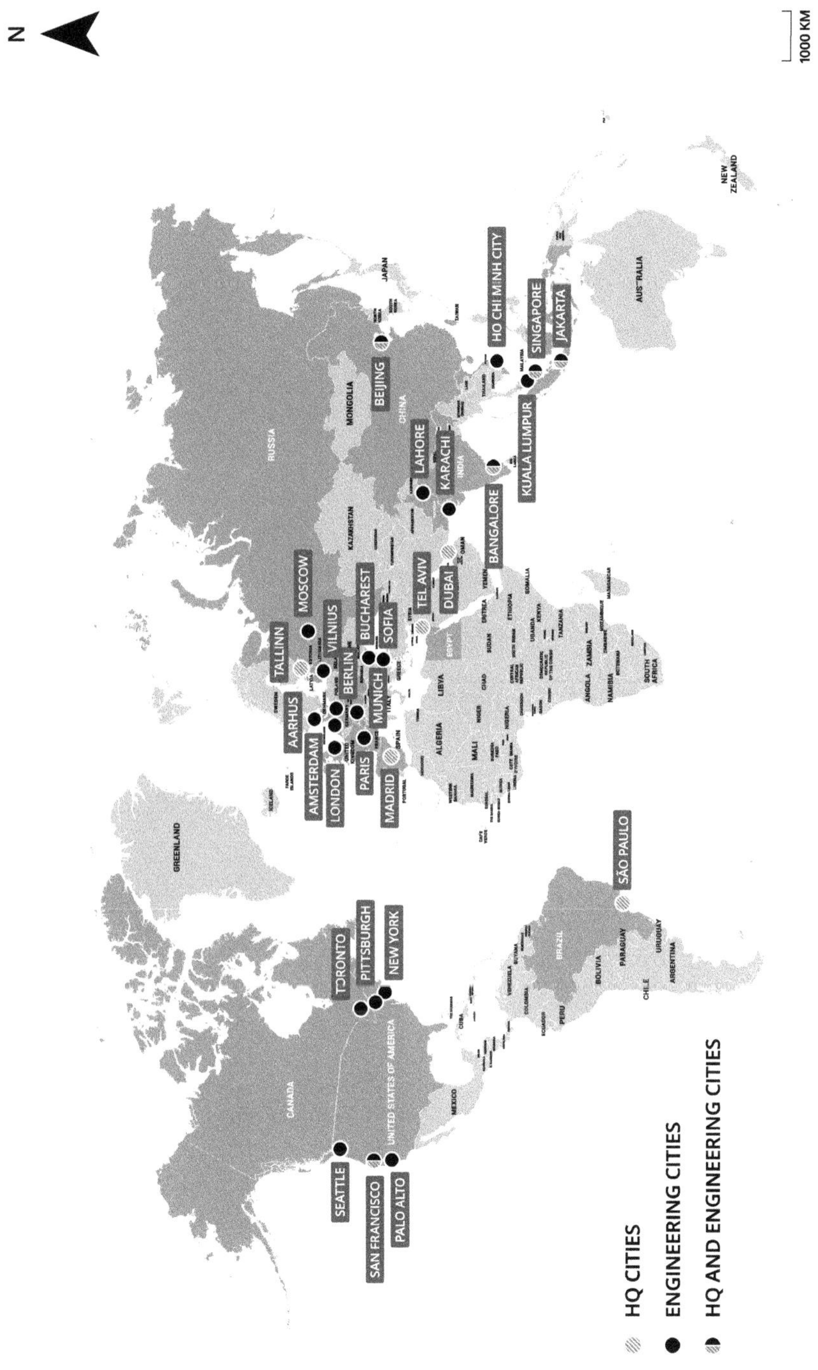

FIGURE 4.2 World cities of ride-hailing.

TABLE 4.2 World cities of ride-hailing by headquarters and secondary office (R&D/engineering) locations

City	*Headquarters*	*Secondary offices*
Aarhus		
Amsterdam		
Bangalore		
Beijing		
Berlin		
Bucharest		
Dubai		
Ho Chi Minh City		
Jakarta		
Karachi		
Kuala Lumpur		
Lahore		
London		
Madrid		
Moscow		
Munich		
New York		
Palo Alto		
Paris		
Pittsburgh		
San Francisco Bay Area		
São Paulo		
Seattle		
Singapore		
Sofia		
Tallinn		
Tel Aviv		
Toronto		
Vilnius		

Source: Author compilation of public data.

and Singapore. Interviewees acknowledge that situating secondary locations in the headquarters cities of competitor firms is strategic. Firms locate in the same cities as competitors to access a pool of experienced, specialized tech talent and to benefit from the presence of startup ecosystems. For example, Jakarta-based Gojek opened research offices in Singapore and Bangalore before expanding operations beyond Indonesia. Both cities provide Gojek with engineering talent—at different wage points—and access to each city's tech ecosystem, recent university graduates, and talent spillovers. This international expansion of ride-hailing production signals two trends: the globalization of highly skilled workers and the determination of firms to tap into the spaces where they pool.

Ride-hailing unicorns tend to locate offices in cities where they also operate ride-hailing services. Eleven cities, however, are sites of secondary offices in

TABLE 4.3 Locations of ride-hailing firm offices with no operations

City	*Firm(s)*
Aarhus, Den.	Uber
Bangalore	Gojek, Grab
Beijing	Grab
Berlin	Careem
London	Lyft
Munich	Lyft
San Francisco Bay Area	Didi Chuxing
Seattle	Grab
Singapore	Gojek, Uber
Sofia, Bul.	Uber
Toronto	Didi Chuxing

Source: Author compilation of public data.

which firms have an administrative presence but no ride-hailing operations (see Table 4.3). Four of these cities (Bangalore, Beijing, San Francisco/Bay Area, and Singapore) are headquarters locations of ride-hailing competitors. The remaining cities (Aarhaus, Berlin, London, Munich, Seattle, Sofia, and Toronto) are attractive for unique reasons, mainly related to specialized talent. For instance, Didi Chuxing opened a research lab in Toronto, Canada, in 2018—the firm's second international location. Explaining this choice, Didi Chuxing highlights the "city's inclusive environment for innovation and entrepreneurship" as well as the location's proximity to the University of Toronto with its strengths in artificial intelligence and smart transportation ("DiDi Launches Labs in Toronto, Expanding Global Research Network to Canada," 2018). Similarly, Lyft acquired London-based Blue Vision Labs in 2018, thus expanding its autonomous vehicle efforts (Rudgard, 2018) to that city. With these office locations, firms gain access to the talent they need to produce ride-hailing services globally.

The dynamism of world cities

The spread of ride-hailing's global headquarters reveals that the geography of world cities is dynamic. Ride-hailing emphasizes the shifting concentration of global cities over the past two decades. If we compare ride-hailing headquarters cities to the alpha world cities identified by Beaverstock et al. (1999), Singapore is the only city that qualifies. Three world cities of ride-hailing were classified as secondary world cities 20 years ago: Madrid, San Francisco, and São Paolo (Beaverstock et al., 1999). A further two cities were identified as third-tier world cities (Jakarta and Beijing) while three others—Tel Aviv, Dubai, and Bangalore—demonstrated evidence of world city formation but could not be classified as world cities (Beaverstock et al., 1999).

An updated list of world cities developed in 2016 by the GaWC (2017) emphasizes the scale, shift, and expansion of cities between 1999 and 2016. The location of ride-hailing firms in newly emergent top-tier cities parallels the expansion of world cities illuminated in recent scholarship. Based on the GaWC 2016 world cities classification, headquarters for nine of 11 ride-hailing unicorns are located in top-tier world cities. Two locations—Bangalore and Tallinn, home to Ola and Bolt, respectively—are second- and third-tier world cities according to this classification. The presence and growth of ride-hailing unicorns in Bangalore and Tallinn may signal the increased significance of both cities as global centres.

Policy-making and policy interventions

Local, regional, and national policies can encourage the establishment, retention, and/or mobility of urban platforms. For instance, Grab's founders moved the company's headquarters from Kuala Lumpur, Malaysia, to Singapore to benefit from various advantages, including government subsidies and tax breaks, an internationally oriented ecosystem of startups, and a strong reputation for economic stability ("Grab leaving Malaysia is Kuala Lumpur's loss to Singapore", 2017). San Francisco and the Bay Area are dominant players and the site of headquarters for technology-based firms in part because, as Saxenian (1983) and Mazzucato (2013) remind us, large waves of government investment in R&D established the region as a centre of innovative firms and industries. This earlier public investment continues to influence the regional concentration of innovation and entrepreneurship in particular regions.

More recently, government initiatives and supports seem connected to the growth of ride-hailing unicorns in relatively small countries, such as Israel and Estonia, where policies focused on innovation have attracted firms to the largest cities, Tel Aviv and Tallinn. In Tel Aviv, state-led R&D and innovation policies have been key to the city's strength in spawning startups (Breznitz, 2007) while significant military spending and training have inspired post-military service entrepreneurship (Avnimelech and Teubal, 2004). In Estonia, cities have benefited from the government's plan to become one of the world's most digitally advanced countries. Over the past two decades, following the country's independence from the former Soviet Union, the Estonian government has developed regulations, e-government services, and innovation-focused strategies to attract investment (Hammersley, 2017). Based in Estonia's largest city, Bolt became the country's fourth unicorn in 2018.

When ride-hailing firms establish an office focused on government liaison, they often choose a nation's capital as the location. This choice gives a company easier access to politicians and policymakers, especially important for lobbying activities. In the United States, Uber's lobbying efforts are well-documented (Flores and Rayle, 2017) and credited with the implementation of municipal and state regulations that permit ride-hailing. Ride-hailing firms are dependent on government permission and regulation in order to offer their services, and this can result in the sector's uncertainty and instability. For instance, in November 2015, Indonesia's minister of transportation banned ride-hailing outright.

However, the president quickly reversed this ruling (Johnson, 2016). Regulatory unpredictability presents both a problem and an opportunity for ride-hailing firms.

Ride-hailing firms operate in a regulatory and policy environment with distinctive municipal, national, and international characteristics. Because legislators work in local and therefore fragmented ways, ride-hailing firms know more about national and international regulatory practices than any individual government as a result of the interurban infrastructure created and leveraged through the development of each firm's own network of headquarters and engineering/R&D locations. The uneven knowledge environment in which governments operate, relative to the multinational operations of ride-hailing firms, produces an asymmetry of knowledge. Governing ride-hailing is critical in ensuring that policy directions and goals are carefully directed. This is particularly important as ride-hailing grows and practices continue to evolve. Concurrently, policies need to manage and redirect challenging impacts associated with an increasingly digital economy.

The urban economic geography of ride-hailing unicorns reinforces the notion of world cities as places marked by the globalizing flows of capital and people. The locations of these firms' headquarters demonstrate a global spread of world cities beyond North America and Western Europe. Yet the prominence of these new world cities—as sites of concentrated investment, as magnets drawing in highly skilled talent, and as centres developing innovative solutions to urban mobility challenges—is not accidental. Rather, these locations exemplify successful and deliberate government-led strategies focused on encouraging entrepreneurship, inspiring innovation, and attracting venture capital in order to amplify urban agglomeration in the digital platform economy.

Conclusion

While cities adapt to an evolving digital economy, ride-hailing illustrates the shifting dynamics of world cities in the digital era. Ride hailing has rapidly emerged as a global urban force with disruptive impacts. This chapter sets out the parameters for studying a new industry characterized by urban, innovation-oriented platform economies. Though global in terms of operations, ride-hailing unicorns concentrate strategic decision-making and engineering activities in a small, select group of cities. In this respect, the case of ride-hailing mirrors observed shifts taking place across cities and industries in the 21st century. These changes include the increasing global spread of world cities (GaWC, 2017); the inclusion of cities located in less developed regions with large and growing populations, notably in Asia (Derudder and Taylor, 2005); the concentration of economic activity in a small number of superstar urban centres (Muro and Whiton, 2018); and the continued leadership of cities in the digital economy, especially with respect to digital platform economy activities (Davidson and Infranca, 2016). The location of ride-hailing unicorns provides further evidence that the concentration of this industry in certain world cities reflects factors of agglomeration, including government policy, knowledge spillover, and access to talent. This underscores the pull of local social infrastructure, and the significance of tacit knowledge, that draws ride-hailing firms to their respective locations.

From a policy perspective, there is both good news and bad news. The good news for cities is that there is no single route to capturing the knowledge-based jobs and investment activities that go along with ride-hailing. The patterns of location suggest that ride-hailing companies prefer to locate their offices in established centres with high concentrations of tech firms and talent, thus leveraging a range of tangible and intangible infrastructures associated with place, public investment and private ventures. Yet these patterns also reveal another story: emerging tech centres, aspiring cities with innovation-oriented policies, and centres with unique cultural features are also potential leaders. On the flip side, the ride-hailing industry involves a limited number of participants, in terms of both cities and high-paid, highly skilled workers. It involves only a few cities and creates only a small number of high-wage, highly skilled jobs. From this perspective, the future of work and prospects for greater and shared prosperity connected to the growth of ride-hailing look bleak.

The concentration of ride-hailing activities contributes to a growing concern that these 29 world cities are absorbing an increasing proportion of investment, employment, and wealth creation at the expense of other places. For cities and people not included in this bounty, the implications include rising inequality on interurban and intraurban scales. Calls for policy interventions that help to spread economic opportunity are one attempt to redress imbalances, extending well beyond the scope of ride-hailing (Lee, 2018). Insofar as ride-hailing represents platform urbanism more broadly, these findings may imply an even more significant challenge to cities as the growth of platform economy firms continues.

This research suggests that the digital platform economy, and ride-hailing in particular, have expanded opportunities for participation in the global economy for a select group of cities and a select cohort of talent. Accordingly, the urban economic geography of ride-hailing is emblematic of the 21st century: highly concentrated, highly uneven, driven by talent and innovation, dynamic, and challenging.

Notes

1 Many thanks to Monica Brondholt Sorensen for her excellent research assistance. This research was conducted as part of the Creating Digital Opportunity (CDO) research project with financial support provided by the Social Sciences and Humanities Research Council. Any errors remain the responsibility of the author.

2 To enable analysis of ride-hailing firms' location patterns, a novel database was built to enable this research. The database was created through compilation of data from numerous sources, including ride-hailing company websites, news media, research reports and company profiles produced by business intelligence firms. Data collection took place from February to December 2018. The database provides firm, sector and city-specific insights that build new knowledge about the geography of ride-hailing.

Furthermore, the database creates an understanding of city-level location decisions and the role of agglomeration and knowledge spillovers in particular cities. In addition, interviews with four of the six largest ride-hailing unicorns carried out by the author between 2016 and 2018 provide additional context and qualitative insight on location patterns, firm strategy and decision-making.

3 Formerly Taxify, the firm changed their name to Bolt in early 2019.

4 There are limitations with respect to calculating the number of municipalities in which firms operate. First, the list undergoes near-constant change, as firms continue to expand and, in some cases, contract. Second, for some firms it is challenging to accurately collect the names of all cities of operation, particularly firms operating in non-English speaking countries. Third, most firms list the names of municipalities in which they operate, although sometimes the city name represents a metropolitan area and not a municipality.

References

Adler, P., Florida, R., King, K., & Mellander, C. (2019) The city and high tech startups: The spatial organization of Schumpetarian entrpreneurship. *Cities*, 87, pp. 121–130. doi:10.1016/j.cities.2018.12.013.

Ash, J., Kitchin, R., & Leszczynski, A. (2018). Digital turn, digital geographies? *Progress in Human Geography*, *42*(1), pp. 25–43. doi:10.1177/0309132516664800.

Asselin, R. and Speer, S. (2019, April). *A New North Star: Canadian Competitiveness in an Intangibles Economy*. Ottawa, ON: Public Policy Forum.

Avnimelech, G., & Teubal, M. (2004). Venture capital start-up co-evolution and the emergence & development of Israel's new high tech cluster. *Economics of Innovation and New Technology*, *13*(1), pp. 33–60.

Barns, S., 2019. Negotiating the platform pivot: From participatory digital ecosystems to infrastructures of everyday life. *Geography Compass*, *13*(9), pp. e12464. doi:10.1111/gec3.12464.

Barns, S., Cosgrave, E., Acuto, M., & Mcneill, D. (2017). Digital infrastructures and urban governance. *Urban Policy and Research*, *35*(1), pp. 20–31. doi:10.1080/08111146.2016.1235032.

Bates, L., Zwick, A., Spicer, Z., Kerzhner, T., Joo Kim, A., Baber, A., Green, J. W, & Moulden, D.T., (2019). Gigs, side hustles, freelance: What work means in the platform economy city/ Blight or remedy: Understanding ridehailing's role in the precarious "gig economy"/ Labour, gender and making rent with Airbnb/ The gentrification of 'sharing': From bandit cab to ride share tech/ The 'sharing economy'? Precarious labor in neoliberal cities/ Where is economic development in the platform city?/ Shared economy: WeWork or We Work together, *Planning Theory & Practice*, *20*(3), pp. 423–446. doi:10.1080/14649357.2019.1629197.

Beaverstock, J. V., Smith, R. G., & Taylor, P. J. (2000). World-city network: A new metageography? *Annals of the Association of American Geographers*, *90*(1), pp. 123–134.

Beaverstock, J., Taylor, P. & Smith, R. (1999). A roster of world cities. *Cities*, *16*(6), pp. 445–458.

Benton-Short, L., Price, D., & Friedman, S. (2005). Globalization from below: The ranking of global immigrant cities. *International Journal of Urban and Regional Research*, *29*(4), pp. 945–959.

Bhuiyan, J. (2018, April 25). Uber's first diversity report under CEO Dara Khosrowshahi shows Uber is still mostly white and male. *Recode*. Available at: https://www.vox.com/2018/4/25/17275010/uber-executive-diversity-black-latinx (Accessed 17 Aug 2020).

Branstetter, L. G., Glennon, B., & Jensen, J. B. (2018). The IT revolution and the globalization of R&D. *Innovation Policy and the Economy, 19*, pp. 1–37.
Breznitz, D. (2007). Industrial R&D as a national policy: Horizontal technology policies and industry-state co-evolution in the growth of the Israeli software industry. *Research Policy, 36*(9), pp. 1465–1482.
Camp, G. (2017, June 20). Uber's path forward. *Medium*. Available at: https://medium.com/@gc/ubers-path-forward-b59ec9bd4ef6 (Accessed 17 Aug 2020).
Caprotti, F., & Liu, D. (2020). Emerging platform urbanism in China: Reconfigurations of data, citizenship and materialities. *Technological Forecasting and Social Change, 151*(119690). doi:10.1016/j.techfore.2019.06.016.
Clewlow, R. R., & Mishra, G. S. (2017). *Disruptive Transportation: The Adoption, Utilization, and Impacts of Ride-Hailing in the United States* (Research Report 17-07). Institute of Transportation Studies, University of California, Davis.
Cohen, R. B. (1981). The new international division of labor, multi-national corporations and urban hierarchy. In M. Dear & A. J. Scott (Eds.), *Urbanization and Urban Planning in Capitalist Society* (pp. 287–316). New York: Methuen, Inc.
Crescenzi, R., Iammarino, S., Ioramashvili, C., Rodriguez-Pose, A., & Storper, M. (2019, November). The geography of innovation: Local hotspots and global innovation networks. *World Intellectual Property Organization*. Available at: https://www.wipo.int/wipo_magazine/en/2019/06/article_0001.html (Accessed 1 Sep 2020).
Davidson, N., & Infranca, J. (2016). The sharing economy as an urban phenomenon. *Yale Law Policy Review, 34*(2), pp. 215–280.
Davidson, N., & Infranca, J. (2018). The place of the sharing economy. In N. Davidson, M. Finck, & J. Infranca (Eds.), *The Cambridge Handbook of the Law of the Sharing Economy* (pp. 205–219). Cambridge: Cambridge University Press.
Derudder, B., & Taylor, P. (2005). The cliquishness of world cities. *Global Networks, 5*(1), pp. 71 91.
DiDi Launches Labs in Toronto, Expanding Global Research Network to Canada. (2018, November 19). *Financial Post*.
Flores, O., & Rayle, L. (2017). How cities use regulation for innovation: the case of Uber, Lyft and Sidecar in San Francisco. *Transportation Research Procedia, 25*, pp. 3756–3768.
Friedmann, J. (1986). The World City Hypothesis. *Development and Change, 17*(1), pp. 69–83.
GaWC (2017). The World According to GaWC 2016. Available at: https://www.lboro.ac.uk/gawc/world2016.html (Accessed 17 Aug 2020).
Glaeser, E. (2010). Introduction. In Edward L. Glaeser (Ed.), *Agglomeration Economics* (pp. 1–14). Chicago, IL: The University of Chicago Press.
Grab leaving Malaysia is Kuala Lumpur's loss to Singapore (2017, June 6). *The Independent (SG)*. Available at: http://theindependent.sg/grab-leaving-malaysia-is-kuala-lumpurs-loss-to-singapore/ (Accessed 17 Aug 2020).
Hall, P. (1966). *The World Cities*. London: Weidenfeld and Nicolson.
Hammersley, B. (2017, March 27). Concerned about Brexit? Why not become an e-resident of Estonia. *Wired UK*. Available at: https://www.wired.co.uk/article/estonia-e-resident (Accessed 17 Aug 2020).
Haskel, J., & Westlake, S. (2017). *Capitalism without Capital: The Rise of the Intangible Economy*. Princeton, NJ: Princeton University Press.
Johnson, C. (2016, March 22). Indonesia: Dispute over ride share companies to be resolved. *Global Legal Monitor*. Available at: https://www.loc.gov/law/foreign-news/article/indonesia-dispute-over-ride-share-companies-to-be-resolved/ (Accessed 17 Aug 2020).
Kitchin, R. (2014). The real-time city? Big data and smart urbanism. *GeoJournal, 79*(1), pp. 1–14.

Kratke, S., & Taylor, P. (2004). A world geography of global media cities. *European Planning Studies, 12*(4), pp. 459–477.

Lee, N. (2018). Inclusive growth in cities: a sympathetic critique. *Regional Studies, 53*(3), pp. 424–434.

Lee, N., & Rodríguez-Pose, A. (2013). Innovation and spatial inequality in Europe and USA. *Journal of Economic Geography, 13*(1), pp. 1–22.

May, J., Wills, J., Datta, K., Evans, Y., Herbert, J. and McIlwaine, C.J. (2007). Keeping London working: Global cities, the British state and London's migrant division of labour. *Transactions of the Institute of British Geographers, 32*, pp. 151–167.

Mazzucato, M. (2013). *The Entrepreneurial State: Debunking the Public vs Private Myth in Risk and Innovation*. London: Anthem Press.

Muro, M. and Whiton, J. (2018, December 17). Tech is (still) concentrating in the Bay Area: An update on America's winner-take-most economic phenomenon. *Brookings*. Available at: https://www.brookings.edu/blog/the-avenue/2018/12/17/tech-is-still-concentrating-in-the-bay-area-an-update-on-americas-winner-take-most-economic-phenomenon/ (Accessed 17 Aug 2020).

Rudgard, O. (2018, October 23). Uber rival Lyft buys London start-up to boost driverless carplans. *The Telegraph*. Available at: https://www.telegraph.co.uk/technology/2018/10/23/uber-rival-lyft-buys-london-start-up-boost-driverless-car-plans/ (Accessed 17 Aug 2020).

Sassen, S. (2001). *The Global City: New York, London, Tokyo*. Princeton: Princeton University Press.

Saxenian, A. (1983). The genesis of silicon valley. *Built Environment, 9*(1), pp. 7–17.

Scott, A. J., & Storper, M. (2015). The nature of cities: The scope and limits of urban theory. *International Journal of Urban and Regional Research, 39*(1), pp. 1–15. doi:10.1111/1468-2427.12134.

Sperling, D. (2018). *Three Revolutions: Steering Automated, Shared, and Electric Vehicles to a Better Future*. Washington, DC: Island Press.

Stehlin, J. (2018, October 26). Urban platforms, rent, and the digital built environment. *Mediapolis*, 4: Roundtables, Vol. 3. Available at: https://www.mediapolisjournal.com/2018/10/urban-platforms-rent-and-the-digital-built-environment/.

Stocker, A., & Shaheen S. (2017). *Shared Automated Vehicles: Review of Business Models* (Discussion Paper No.2017–09). Berkeley: University of California, Berkeley, Transportation Sustainability Research Center.

Sumantran, V., Fine, C. and Gonsalvez, D. (2017). *Faster, Smarter, Greener: The Future of the Car and Urban Mobility*. Cambridge, MA: MIT Press.

Taylor, P. J., Walker, D.R.F., Catalano, G. and Hoyler, M. (2002). Diversity and power in the world city network. *Cities, 19*(4), pp. 231–41. doi:10.1016/S0264-2751(02)00020-3.

Taylor Buck, N., & While, A. (2017). Competitive urbanism and the limits to smart city innovation: The UK future cities initiative. *Urban Studies, 54*(2), pp. 501–519. doi:10.1177/0042098015597162.

Van Audenhove, F.-J., Korniichuk, O., Dauby, L., & Pourbaix, J. (2014). *The Future of Urban Mobility 2.0: Imperatives to Shape Extended Mobility Ecosystems of Tomorrow*. Washington, DC: The National Academies of Sciences, Engineering, and Medicine.

Van der Graaf, S., & Ballon, P. (2019). Navigating platform urbanism. *Technological Forecasting and Social Change, 142*, pp. 364–372. doi:10.1016/j.techfore.2018.07.027.

Wills, J., May, J., Datta, K., Evans, Y., Herbert, J., & McIlwaine, C. (2009). London's migrant division of labour. *European Urban and Regional Studies, 16*(3), pp. 257–271.

5

DIGITAL INFRASTRUCTURES, SERVICES, AND SPACES

The geography of platform urbanism

Alan Wiig and Michele Masucci

Introduction

Platform urbanism may look toward the high-tech city of the future, but it is also already here, embedded in—and reliant on—the ubiquitous digital systems permeating the landscape. Platformed services have become commonplace over the last few years, offering the promise of satiating needs and wants with the swipe of a finger across a smartphone screen. These services are supported by a rapidly expanding infrastructure for information connectivity. Recent forecasts of the widespread adoption of 5G networking technologies highlight that this new class of digital infrastructure will increase the speed of data transfer and enable the anticipated 42–75 billion Internet of Things (IoT) connected devices to work much more seamlessly and interoperably. In turn, 5G and IoT will gather, share, and analyze information; provide decision support; and enable a new class of technology autonomy by 2025 (Brown, 2016; Help Net Security, 2019; IoT Playbook, 2019; Petrov, 2019). This proliferation of devices is predicted to create a world in which the average person is affected by nearly 5,000 IoT transactions per day by 2025 (Reinsel et al., 2018). The connectivity will in turn provide information access to autonomous vehicle movement on smart transportation grids; real-time monitoring systems for their homes, health, and environments; and distance transactions (like health care diagnostics and educational access). This city of the future is quickly arriving.

And yet, the rapidly emerging technologies and conveniences of platform urbanism typically fail to address the long-standing, material decline of post-industrial cities, sidestepping significant investment in public services (Wiig, 2016) in favor of partnering with industry to instead expand particular information platforms that, often, do not serve a wide cross-section of a city's diverse population (Masucci et al., 2019). Expanding this line of argument, we theorize platform urbanism through an empirically motivated case study of Lower North Philadelphia, an industrial-era, inner-city district emblematic of this platformed layering of digital and material, of infrastructure, services, and spaces. By situating this chapter on six square miles of the city (City of Philadelphia, 2014), we place this emergent geography within the rise and fall of the 20th-century city's ambition for universalizing, networked urbanization (Graham and Marvin, 2001; Berger et al., 2011; Zylstra, 2013) and the 21st-century challenges of socio-economic disenfranchisement (Fairbanks, 2011) amid the transition to a globalized, high-tech economy (Hodos, 2002). Lower North Philadelphia is emblematic of the post-industrial decline as well as long-standing poverty that still divides the contemporary city, even as the city's economy has experienced a distinct revival and the services of platform capitalism (Langley and Leyshon, 2017; Srnicek, 2017) permeate the landscape. As Philadelphia re-industrializes around new information - and innovation-focused enterprise (Wiig, 2019), we argue that, ultimately, platform urbanization has enacted a further fragmentation of the city, recreating an uneven landscape of economic change that reconstitutes past inequalities rather than stimulating a wholesale or even piecemeal restructuring toward a more socially and geographically inclusive platformed city.

With this chapter we theorize platform urbanism broadly, as the form, function, and governance of cities through digital augmentation. In employing this definition, we specifically include earlier efforts at digitally driven urban revitalization that precede the wireless, mobile Internet, the smartphone, and its app-based services. To craft this geography of platform urbanism, we synthesize our own scholarship focused on issues of youth and digital inclusion efforts, cellular and telecommunications infrastructure, economic restructuring, and smart city policies in Philadelphia. Of particular utility to this chapter, our research has examined how digital issues affect young people, focusing on how they frame the impacts of digital technologies and infrastructure on the city, connect that to acquiring skills and knowledge through both formal and informal educational experiences, serve as conduits of information technology know-how for their communities and families, and poise themselves to pursue advanced education and jobs in the midst of the rapid development and adoption of digital infrastructure and smart technologies (Masucci et al., 2016; Gilbert and Masucci, 2018; Masucci et al., 2019).

Our focus on youth allows us to view digital and urban inequalities through the lens of their experiences but also to track the confounding ways in which digital policy, educational praxis, and social inequalities reinforce the geographic and economic divides of the city even as the technologies promise to mitigate those divergences. Nowhere is this more evident than in the Lower North Philadelphia district, which is home to Temple University, respectively our current (Masucci) and former (Wiig) institutional home, and the location of much of our jointly conducted research noted above. Ultimately, our focus on youth has shown that while they powerfully envision the platform computing environment that could evolve in a techno-utopian version of the city, they observe that the geographic, structural inequalities are unmoved thus far by the smartening, platforming city's ubiquitous cloud computing software and smartphone apps, which have done little to address longstanding urban issues in Philadelphia. The youth we work with care most about lessening crime and homelessness, reducing drug abuse, navigating a flawed education system, and the lack of job opportunities. They use platform services like Uber Eats, and work entry-level jobs like fast food that have a degree of automation, but at the same time they do not see these platform technologies benefiting their neighborhoods directly (Masucci et al., 2019).

Understanding platform urbanization as a process necessitates considering the local, contextual, and historical evolution of particular neighborhoods in specific cities. To this end, the chapter proceeds as follows. We first offer an overview of Lower North Philadelphia's evolution and present condition to contextualize platform spaces. Then we discuss the digital infrastructure that underlies platform services and platform spaces' need for connectivity and information processing, before turning to several key municipal and private platform services directly. We conclude by drawing attention to the unevenness of platform urbanization, arguing that this "era" of the city is marked by a patchwork evolution of ubiquitous connectivity and haphazard service accessibility, compounded by the emergence of exclusive spaces of platform capitalism amid the ongoing disinvestment in neighborhood-based civic infrastructure, services, and spaces.

The evolution of Lower North Philadelphia into platform space

With this emphasis on Lower North Philadelphia,[1] we center our analysis on a contiguous group of neighborhoods that encompass the trajectory of North American urbanism from the 19th-century rise of industrial manufacturing (Weigley et al., 1982) through the decline of this economy in the latter half of the 20th century (Robbins, 1981), to the highly uneven revitalization of the 21st century (Wiig, 2019). Today, Lower North Philadelphia is home to 95,000 residents, down from 272,000 at its peak in 1950. While the population was stable at the 2010 U.S. Census, the district's majority African American population has been declining since 1980, while the white population increased slightly (City of Philadelphia 2014, pp. 8–9). The district sits directly adjacent to Philadelphia's central business district, and hence spatially proximate to a prominent node of globally oriented enterprise and its high-skill workforce.[2] That said, the district suffers from high unemployment, heavy reliance on public transportation for personal mobility—60% of residents do not have access to an automobile—and difficulty accessing grocery stores, as well as the presence of high levels of obesity in both children and adults (City of Philadelphia 2014, pp. 11–12 and p. 86). These factors all speak to the continuing challenges faced by the residents of the neighborhood: job attainment, individual and community health, and mobility and accessibility within and beyond the immediate area.

Lower North Philadelphia was once considered the "Workshop of the World," home to numerous global manufacturing headquarters, with a strong

FIGURE 5.1 The intersection of Germantown Ave. and Cecil B. Moore Ave. in Lower North Philadelphia. On the right, the site of the Stetson Hat Factory, which employed 5,000 people in the 1920s before closing in 1971 (http://www.philaplace.org/story/326/). The location is now Honor Foods' refrigerated food logistics and distribution center (https://www.honorfoods.com). Note in the left background the cellular antenna sticking above the roof of the Factory Lofts apartment building, providing platform connectivity to the neighborhood. Photo by Michele Masucci, June 2019.

representation of textile industries. Among the products that were manufactured in the area were automobiles, clothing and hats, sheet metal, beer, and processed foods.[3] The expansion of the city during the industrial revolution was designed around the exploitation of new transportation and energy technologies in the form of streetcars and canals to connect new settings with the infrastructure for production and the housing to provide the labor market needed for the plants. Today's Lower North Philadelphia is typified by the neighborhoods of row houses in various states of repair, the remnants of manufacturing facilities in the form of old warehouses and production facilities, and vacant lots where either or both used to be located.

Those remnants of the industrial past have become the nexus of housing, technology, and economic policy that forms the backdrop for the economic resurgence in the city during the past two decades (Wiig, 2019). With the downturn in manufacturing, Philadelphia lost over 500,000 residents from 1950 to 2000 (Rappaport, 2003). In the process, the city became among the most polarized in the U.S. with respect to wealth, race, and class disparities (Ding et al., 2016). Like many formerly industrial U.S. cities, the revitalization thus far has favored a redistribution of people and wealth toward the center of the city where policies directed much of the reinvestments, leaving many other areas to lag behind (Beauregard, 2013; Weaver, 2016). Today, the story is changing for many of those later-to-develop locations, with a new influx of housing, retail, shared co-working spaces, and other signs of new economic investments (City of Philadelphia, 2014). A number of large-scale redevelopment projects meant to anchor additional growth and transformation have been implemented in and near Lower North Philadelphia, such as the American Street corridor focused on new commercial and industrial development. This includes the Honor Foods distribution hub shown in Figure 5.1, the identification of the North Philadelphia rail station as a hub for transit-oriented redevelopment within the next ten years, the launch of biotech research and development activities; and many mixed-use housing projects meant to attract new, middle-class residents who have been priced out of the Center City revitalization efforts. Driving these changes have been a trio of policies focused on repurposing aging physical infrastructure through technology advances, leveraging the remaining anchor institutions for workforce development including education and health care sectors, and drawing on the creative sector to enable the place-making necessary to congeal the growth trajectory (Delaware Valley Regional Planning Commission, 2019). We argue that, taken together, Philadelphia's foundational economic development policies have fueled the city's current renaissance, but have not made significant inroads into addressing the underlying structural inequalities of the city that were codified in the industrial period. It is into this legacy of decline and uneven return that platform urbanism is emerging.

Lower North Philadelphia contains many of the places where these policies have not yet provoked the kinds of reinvestment patterns being found across the city as a whole, in particular within the Center City resurgence. We suggest that our focus on Lower North Philadelphia provides a place-based example that is representative of the unequal redevelopment found in many other places in the city region (Gilbert and Masucci, 2011; Wiig, 2018, 2019). We observe that Philadelphia's geographically uneven growth, and its failure to mitigate intractable poverty and place-based inequalities, stems from the more ubiquitous trends of neoliberal reform and the shape-shifting effects of global capital, and that the policies that aimed to mitigate those trends, in particular digital infrastructure initiatives, have reinforced the spatial inequalities already in place (Gilbert and Masucci, 2011; Gilbert and Masucci, 2018). Early policies aimed at digital inclusion, discussed in greater detail below, have evolved into broader economic development policies to support the construction of the city as a digital platform. We suggest that without rethinking both in parallel, such an approach will continue to further develop along the geographic and social inequalities already built into the local socio-economic structures. Ultimately, we argue that the social and economic transformations that platform urbanism has wrought must be considered in the context of the material geographies of post-industrial disinvestment and decline, and the unevenness of recent economic recovery and re-industrialization.

Digital infrastructure: the foundation of platform urbanism

The infrastructure underlying platform urbanism is digital, fulfilling the need for pervasive, wireless information and communication technology networks to connect users and devices into and beyond the city-region. In Lower North Philadelphia as elsewhere, unless an individual is within a home or other space with a Wi-Fi hotspot, they connect their smartphone (or other device) to the Internet through a cellular, mobile communication network (Wiig, 2013), that is likely to soon be upgraded to the faster 5G data standards. Within a city, this service is typically provided by private, global telecommunication corporations like AT&T, Verizon, or Philadelphia-based Comcast, whose two skyscrapers incidentally define the tallest points on the downtown skyline. Telecommunication firms build and maintain cellular antenna sites that route into city, regional, and trans-continental telecommunications trunk lines, and then connect to the myriad data centers where the information processing of cloud, and now "edge", computing occurs (Ai et al., 2018). Smartphones and all other connected devices function through this ethereal, pervasive connectivity. It enables cloud computing, artificial intelligence, the Internet of Things, and so on: the "platformed" digital and computational, information-processing services. The ethereality is important here: wireless connectivity is composed of electromagnetic waves bouncing through the air, back and forth from personal devices to cellular antenna that then transmits the information to and from a data center. Without a localized signal connecting a smartphone to the globally linked information and communication technology network, platform services cannot function.

The prevalent use of smartphones as a strategy for connectivity is characteristic of lower-income users such as those of Lower North Philadelphia, where contingent, intermittent, and proxy Internet use is the norm (Gilbert and Masucci, 2011; Gilbert and Masucci, 2018). Further, while cellular technology is a slower and less complete means of connecting to the Internet than a desktop or a laptop computer, it does provide the advantage of mobile access, which overcomes some of the structural problems faced in the same low-income communities, like lack of dedicated spaces for computing, older, slower home computers, little to no wired network access, and surveilled access at work or school. The new computing systems, including 5G and "edge computing" solutions to contain data storage within city-regions (Ai et al., 2018), will further accelerate the trend toward decentralized network access and cloud computing, providing more comprehensive coverage but also more localized entry and data transfer points. In turn, these advancements will produce a further proliferation of new devices that are connected to the Internet, driven by an intensified commercialization of the devices themselves, the data they use and access, and the analytics embedded to provide decision and consumer support (Ai et al. 2018), all utilizing the platforms co-developed by public and private interests. These new connected devices include consumer-based, smart home devices such as Internet-enabled thermostats and doorbells with security cameras linked through a smartphone app, to citywide

CCTV surveillance camera networks, to air or water pollution sensor networks, all transmitting and receiving data constantly.

The solution to interoperability at the edge of cellular connectivity will come as the outcome of a vast investment in the capabilities of the devices themselves to transfer information across the array of the IoT device population, the insertion of transmission technologies within the devices themselves, and the proliferation of localized wireless antennas to tether the local use to the larger cloud. This technology, commonly referred to as fog computing (Ai et al., 2018, pp. 81–84), is being engineered to create a Massive Internet of Things (MIoT) data transmission, analytics, and use context that will fuel the next stage of platform urbanism. Such infrastructure will enable autonomous transportation using real-time analytics from sensors (to avoid crashes, anticipate road and weather conditions, and adapt to driver preferences); highly localized business analytics based on real-time data on consumer preferences that will enable drone deliveries, just-in-time manufacturing, and real-time optimization of access to services; highly complex support systems like distance diagnostics and robotic surgery; and a massive uptick in the proliferation of information and analytics to support machine learning and artificial intelligence applications in the workforce, education, energy grid management, and environmental monitoring sectors. Those on the lowest socio-economic end of usership will be relegated to consumer status. As the analytical and optimization end of this implementation will be where workforce development and entrepreneurial opportunities lie, this infrastructural evolution will likely privilege the already-well-connected segments of the economy and technology-use communities. While all connectivity will increase, we can expect a continuation of the gap between the consumer use at one scale of adoption and the innovative use at another, all tethered to the underlying social, economic, and spatial inequalities of today.

Platform urbanism can ostensibly be found anywhere a cellular signal is present. It is consequently present among the crumbling old factories, warehouses, and vacant lots facing onto potholed streets of Lower North Philadelphia, spaces not necessarily included in the imaginary of the city of the future. Due to the density of the built environment in the neighborhood, Lower North Philadelphia's cellular antennas are typically affixed to the rooftops, smokestacks, and water towers of warehouses and other tall buildings left over from the industrial era, or mounted onto utility poles like electricity masts or streetlights. This leads to a juxtaposition between the detritus of the industrial city and its repurposing to support these platform infrastructures. Given Philadelphia's redevelopment strategies, highly networked enclaves of the new, globalized information and innovation economy sit adjacent to disinvested neighborhoods like Lower North Philadelphia that are no longer the sites of innovation and job creation they were a century ago. Instead, such districts host wireless connectivity pulsing through the ether as a mobile, communicative convenience, but not an engine of opportunity. In fact, there are few job opportunities in the new economy in the district, perpetuating a spatial mismatch between jobs—especially for African Americans—and existing, post-industrial neighborhoods (Stoll, 2005).

As an example, platform services are not just the app and the transaction between rideshare driver and passenger but are also the infrastructure distributed across the whole of a city (and beyond). For an individual to request a rideshare, and for that rideshare driver to accept the fare, requires both parties to have a smartphone with the rideshare service's app installed, a cellular data plan, and a connection to the network. The personal mobility is the service being sold through the driver's vehicle, but the transaction is made possible through a constant, wireless connection to the rideshare company's app. The transition to autonomous vehicles only furthers the reliance on this digital infrastructure (Alvarez León, 2018). While the form of the neighborhood, the grid of streets as well as the shape and utility of the built environment, does not actively change, the presence of wireless digital connectivity, translated into information and services through a smartphone and its data plan, ties platform urbanism together.

Beyond the convenience of mobile connection, the layering of digital infrastructure into the built environment facilitates municipal systems reliant on automated monitoring. To take one notable example, dispersed police surveillance camera networks are organized through these infrastructures. In 2012, Philadelphia brought online a Real-Time Crime Center that coordinates 1,800 surveillance cameras as well as automated license plate readers out of one control room (Reyes, 2013). Additionally, much of the Lower North district overlaps with the 22nd Philadelphia Police district that in 2014 piloted police body cameras due to the district being one of the "busier" ones in the city (Whelan, 2014). City officials and police leadership considered the pilot a success. According to the mayor, it increased public confidence in law enforcement, leading to the citywide rollout of body cameras onto all officers (Allyn, 2017). This body camera footage is managed through a private contract, but as with the Real-Time Crime Center, this digital automation of real-time policing and surveillance relies on localized and dispersed mobile telecommunications infrastructure. This surveillance system is specifically mentioned in the Lower North Philadelphia district plan as a means to improve the security of its commercial corridors (City of Philadelphia, 2014, p. 53). Automating policing into a digital platform is seen as a way to reduce crime and improve the perception of a neighborhood as safe, but it also opens up these poor, disenfranchised neighborhoods to the potential for privacy violations and an increase in police ticketing and arrests for minor infractions like riding a bike at night without a headlight, in addition to generally living under constant surveillance (Wiig, 2018). The dispersed but connected nature of these systems means software upgrades to add, for instance, artificial intelligence, facial recognition software to the analysis of video footage can happen remotely and often without substantive, democratic oversight (Stanley, 2019). The politics of platform urbanism are many. These debates may start with infrastructure provision, accessibility, and use, but extend into much broader concerns about collective power and control over these systems and networks that are central to social life (Zuboff, 2019).

Platform services: digitizing municipal and privatized exchange

The provision of platform services falls into two broad categories: public or municipal, and private. As outlined in the previous section, the platforming of services in a city relies on the ubiquitous connectivity provided through digital infrastructure. The digitization of services is an ongoing attempt to overcome existing or perceived deficiencies in a city, or to cultivate new—and ideally profitable—markets. Below, we offer an overview of platform services in Philadelphia, focusing on Philly311, a complaint and information service, and Digital On-Ramps, a workforce education app, and lastly the rise of private, app-based rideshare, automotive mobility services. This is not meant to be an exhaustive discussion of all the platform services available in the district or in Philadelphia as a whole, instead a summary of notable services that in their layering contribute to understanding the geography of platform urbanism.

The digitization of municipal services has been ongoing since the late 1990s. This began with electronic government efforts, commonly termed "e-gov", to bring municipal offices online, making services and information accessible through websites and email communication between officials and residents (Kaylor 2005). At this early stage, the platform services would be accessed through a personal computer at home or in a public space like a library or school (Berger et al., 2011). Through the late 2000s and early 2010s, the adoption of pocketable smartphones grew so much and so fast that most everyone—rich and poor alike—carried the Internet in their pocket (Anderson, 2015). Indeed, by 2019 81% of US residents owned a smartphone (Anderson, 2019).[4]

Even if many individuals were unable to afford a data plan and relied on free Wi-Fi to access the Internet (such as at a library), this moment of computing and connection in the late 2000s signified the first stages of widespread "platforming" of the city. Philadelphia was the first city in the U.S. to roll out free municipal Wi-Fi to all residents and visitors, but the system was difficult to maintain due to its cost and unforeseen complexities of running a citywide network, and was closed down in 2009 (Jassem, 2010, pp. 7–8). Over the same period, the cost and accessibility of monthly data plans dropped alongside the price of smartphones themselves. As a result, the municipal Wi-Fi program dissipated and commercial carriers like Comcast's Internet Essentials program entered to fill the accessibility void, providing lower cost, slower connections for those of the city's poorest residents who had the interest, capacity, and need to be connected.

By the late 2000s into the early 2010s, alongside the growing use of smartphones nationwide, Philadelphia transitioned from promoting e-gov policies that aimed to lessen the digital divide to focusing on smart city, digital inclusion, and engagement efforts (Wiig, 2016). This shift was notable in that it signaled the assumption that access to the Internet's digitized resources was widespread, and that civic efforts should shift to providing services through this platform. We highlight the two prominent efforts of the time. The first was the implementation of Philly311, a phone-based service similar to dialing 911 for emergency services in the U.S. 311 systems act as a contact point for municipal information and complaints, also with a physical office in city hall as well as a smartphone app and web portal (Nam and Pardo, 2012). As a platform, 311 shifts information services to a single phone call or the online portal via the smartphone app or the website, providing a resident or visitor with the information they need without having to search for a particular city office's contact details or sorting out traveling to a particular municipal office in person. Large and small cities across the U.S. and Canada have rolled out similar 311 systems; the service is an established element of municipal management today (Newcombe, 2017).

Emerging in 2011 out of participation in IBM's Smarter Cities Challenge, Philadelphia's second smart city effort was called Digital On-Ramps. This service offered a social media-styled workforce education and job search app. The app intended to train low-skill residents for, then connect them to, entry-level jobs in emerging, high-tech fields through an accessible, "anytime, anywhere learning" platform (Wiig, 2016). The app failed to live up to its promises and was plagued by challenges. These were technical, notably difficulty in resetting users' passwords, but also social and economic. The persistent legacy of underinvestment in public schools meant that during its pilot with 700 high school students, Digital On-Ramps was only accessible in computer labs with outdated, slow desktops, which in 2011 reinforced a digital divide not addressed by the smart-city policymaking. These issues combined with general difficulties from the support and development team in identifying the proper industries to focus the app's lessons around. While the app strove to provide relevant and innovative workforce development skills for seven years, it shut down in 2019 (Digital On-Ramps, 2019).

Both of these services relied on platform connectivity to, at least in policy intent, reduce the digital and material divides—in education and health, job attainment, resource allocation and delivery, and other concerns between the industrial city and the platform city by using smartphone apps to improve everyday urban life, increase government accountability, and offer services through a device most residents carried all of the time: their smartphone. These services aimed to rectify the legacies of underprovision of neighborhood-based utilities and civic assistance as well as the more general disinvestment from inner-city neighborhoods like those within Lower North Philadelphia. Platforms like these offer a means of bridging city offices and resource providers that, due to four-plus decades of neoliberal cost-cutting, on top of deindustrialization and post-war decline (Hackworth, 2007), struggle to reach their constituencies. These municipal services have shifted established, "brick-and-mortar" city offices into digital facsimiles, operating alongside privately managed platform enterprise.

Beyond these public platform services, the app-based conveniences emblematic of platform capitalism—rideshare, food delivery, global logistics, and expedited shipping—are designed and maintained by firms intent on delivering future profits to investors. Below, we focus on rideshare apps because of the popularity of the services and their prominent status in research conducted through group conversations with local youth in 2018.[5] The ability of an app-based enterprise to provide transportation, easily accessed online shopping and entertainment, and other conveniences has quickly allowed users to overcome deficiencies in the urban landscape (see Borowiak, 2019 for discussion of Uber's impact on Philadelphia's transportation economy). Rideshare firms like Uber or Lyft, or Uber's spin-off Uber Eats food delivery, offer expedient access to services that may not be easily available in Lower North Philadelphia's neighborhoods. While they provide convenient services, they don't do much to alter the infrastructural base of the persistent divide between these neighborhoods. This is not to imply that private platform technology corporations like Uber, Lyft, Amazon, or Google's parent company Alphabet should necessarily be expected to manage or reduce the digital divide, let alone longstanding socio-economic divides, but when these same firms aspire to design, build, and manage cities, and their technologies are foundational to everyday life in the platformed city, then conceptualizing platform urbanism needs to take into account which broader functions are seeing investment or disinvestment. This point was reinforced by local Philadelphia youth, who recognized that digital connectivity and platformed services may lead to a job but could not address what they considered the most pressing issues facing the city, namely homelessness, crime, and drug abuse.

While platform services have the ability to transform aspects of the city, they largely sidestep dealing with intractable social and economic problems. Even with all the conveniences of platform services, the digital divide in internet connectivity persists, manifesting today in *slower* connectivity as opposed to the absence of connectivity seen in previous years (Gilbert and Masucci, 2018). And, the geographic divides of connectivity (the lower connectivity of rural areas, poor urban residents, elderly, and non-literate users) produces additional inequalities, including human-digital divides related to skills, access, efficacy use and values; intra- and inter-institutional divides (like urban schools having vastly underperforming networks and capabilities than suburban counterparts in the same urban regions); and information-use divides in all sectors. Consequently, scholarly conceptualizations of the geography of platform urbanism, such as this chapter, take seriously what services are platformed and for whom, within the broader understanding of where employment opportunities are locating in the city, where revitalization efforts are transforming neighborhoods, and what are the outcomes of platforming the city. In our last section we turn to considering the emergence of new technological platforms in the context of their impact on Lower North Philadelphia.

Conclusion: city as platform, platform as city

As platform infrastructures, services, and spaces layer into cities, they still remain piecemeal, subject to competing business plans, the constant evolution of software and hardware, and the need to continually develop new uses and applications. To evaluate the impacts of the technological changes requires examining the underlying assumptions about the value of digital technologies and the perceptions of their utility among users and policymakers alongside the already-prevalent corporate assessments of profit, efficiency, and optimization in both the public and private spheres. Scholars need a much more intentional assessment of a differentiated platform urbanization that includes perspectives from different generations of users, different race, place, gender, and ability backgrounds, and different geographically contextual settings. This emergent critique needs a much deeper examination of the biases that are embedded within the network strategies and the code, analytics, and optimization rubrics themselves. What defines efficiency for a lower-income mother with two children in two different public schools in Lower North Philadelphia is potentially dramatically different than that for a systems architect at a high-tech firm deploying new IoT or 5G technologies. Further, we need to understand much more about how cultural and socio-economic divides on information use efficacy, as well as how gender, race, sexuality, age, and health, all shape the need for and perspectives on current as well as future digital technologies that comprise platform urbanism. The geographic inequalities of today pose questions about the spatialities associated with both current as well as next-generation technologies. These concerns need to be broadened to examine the multiple scales and contexts that are—and will be—impacted by the more-networked, artificial intelligence and machine learning-enabled system that is currently being developed and implemented globally. Within this critique, we call for much more attention to the interconnected concerns of privacy, security, and fairness as the focus of critical studies of platform urbanism to come.

Perhaps more pertinent is the urgent need to consider the places left behind and the pace by which today's platforms are created and discarded. IoT is rapidly making established, already well-connected digital infrastructure completely obsolete, along with the Wi-Fi and cellular telecommunications networks that enabled the first generations of smartphones and mobile computing. As we move into a world in which every human can have a genome map, every community at every scale deploys a real-time GIS-enabled optimization grid for all services, and where artificial intelligence enables mass analytics for non-user-defined functions like autonomous driving and drone delivery systems, we need to constantly consider and reconsider how the geographic landscape is literally becoming digitized.

Acknowledgments

Thanks to Renee Tapp, Kelly Kay, Chris Knudson, Carlos Dobbler, Hamil Pearsall, TJ Seningen and Jonathan Silver for early conversations about this project, as well as the organizers of and participants in the "Urban platforms and the future city: Transformations in infrastructure, governance, knowledge, and everyday life" workshop at the University of Manchester in February 2019.

Notes

1 As part of the "Citywide Vision" Philadelphia 2035 comprehensive, long-term plan adopted in 2011, the city produced 18 "strategic district plans", one of which was for the Lower North district (City of Philadelphia 2014). These district plans are intended to guide land use decision making and zoning changes to address 21st-century social, economic, and ecological demands. Our identification of Lower North Philadelphia as a district, rather than naming the many neighborhoods in the district, is motivated and informed by this document.
2 Versus a deindustrialized, disenfranchised, and blighted area on the periphery of Philadelphia that has not successfully re-urbanized, such as Chester, Pennsylvania or Camden, New Jersey (Mele, 2013; Wiig, 2018).
3 More on what was produced in the district and where can be found at: https://www.workshopoftheworld.com/north_phila/north_phila.html (accessed 17 Aug 2020).
4 Data on device ownership in Lower North Philadelphia is not available but is likely similar to the national percentages.
5 The research process and findings from these conversations is are discussed in detail in Masucci et al., 2019.

Bibliography

Ai, Y., M. Peng, & K. Zhang (2018). Edge computing technologies for internet of things: a primer. *Digital Communications and Networks*, 4(2), pp. 77–86.

Allyn, B. (2017). Philly reaches $12.5M deal with Taser maker for police body cameras. *WHYY Radio*. Source: https://whyy.org/articles/philly-reaches-12-5m-deal-taser-maker-police-body-cameras/ Last accessed 17 June 2019.

Alvarez León, L. (2018). Counter-mapping the spaces of autonomous driving. *Cartographic Perspectives*, 92, pp. 10–23.

Anderson, M. (2015). Technology device ownership: 2015. *Pew Research Center.* Source: http://www.pewinternet.org/2015/10/29/technology-device-ownership-2015/ Last accessed: 13 June 2019.

Anderson, M. (2019). Mobile technology and home broadband (2019). *Pew Research Center.* Source: https://www.pewinternet.org/2019/06/13/mobile-technology-and-home-broadband-2019/ Last accessed 13 June 2019.

Beauregard, R. (2013). *Voices of Decline: The Postwar Fate of US Cities*. New York: Routledge.

Berger, D., Funke, and T. Wolfson (2011). Communications networks, movements and the neoliberal city: the media mobilizing project in Philadelphia. *Transforming Anthropology*, 19(2), pp. 187–201.

Brown, P. (2016). 75.4 Billion devices connected to the internet of things by 2025. *Electronics 360,* April 13. Source: https://electronics360.globalspec.com/article/6551/75-4-billion-devices-connected-to-the-internet-of-things-by-2025 Last accessed 28 June 2019.

Borowiak, C. (2019). Poverty in transit: Uber, taxi coops, and the struggle over Philadelphia's transportation economy. *Antipode*, 51(4), pp. 1079–1100.

City of Philadelphia (2014). *Lower North District Plan*. Philadelphia: City of Philadelphia. Source: https://www.phila2035.org/lower-north Last accessed: 10 June 2019.

Digital On-Ramps (2019). *Innovating Workforce Development*. Source: https://www.digitalonramps.com Last accessed: 17 June 2019.

Ding, L., J. Hwang, and E. Divringi (2016). Gentrification and residential mobility in Philadelphia. *Regional Science and Urban Economics*, 61, pp. 38–51.

Delaware Valley Regional Planning Commission (2019). *Comprehensive Economic Development Strategy*. Source: https://www.dvrpc.org/Economic/CEDS/ Last accessed: 24 June 2019.

Fairbanks, R. (2011). The politics of urban informality in Philadelphia's Recovery House Movement. *Urban Studies*, 48(12), pp. 2555–2570.

Gilbert, M., and M. Masucci (2011). *Information and Communication Technology Geographies: Strategies for Bridging the Digital Divide*. Vancouver: Praxis (e)Press, University of British Columbia.

Gilbert, M. and M. Masucci (2018). Defining the geographic and policy dynamics of the digital divide. In *Handbook of the Changing World Language Map*. Cham, Switzerland: Springer, pp. 1–19.

Graham, S. and S. Marvin (2001) *Splintering Urbanism: Networked Infrastructures, Technological Mobilities and the Urban Condition*. New York: Routledge.

Hackworth, J. (2007). *The Neoliberal City: Governance, Ideology, and Development in American Urbanism*. Ithaca: Cornell University Press.

Help Security Net (2019). Connected IoT devices forecast. June 21, 2019. Source: https://www.helpnetsecurity.com/2019/06/21/connected-iot-devices-forecast/ Last accessed: 28 June 2019.

Hodos, J. (2002). Globalization, regionalism, and urban restructuring: the case of Philadelphia. *Urban Affairs Review*, 37(3), pp. 358–379.

IOT Playbook (2019). The growth in connected IoT devices is expected to generate 79.4ZB of data in 2025, according to a new IDC forecast. *IOT Playbook*. Source: https://www.iotplaybook.com/growth-connected-iot-devices-expected-generate-794zb-data-2025-according-new-idc-forecast Last accessed: 28 June 2019.

Jassem, H. C. (2010). Municipal Wi-Fi: the coda. *Journal of Urban Technology*, 17(2), pp. 3–20.

Kaylor, C. (2005). E-government. The next wave of e-government: the challenges of data architecture. *Bulletin of the American Society for Information Science and Technology*, 31(2), pp. 18–22.

Langley, P., & A. Leyshon (2017). Platform capitalism: the intermediation and capitalisation of digital economic circulation. *Finance and society*, 3(1), pp. 11–31.

Licht, W. (n.d). Workshop of the World. *The Encyclopedia of Greater Philadelphia*. Source: https://philadelphiaencyclopedia.org/test-category-page/ Last accessed 17 August 2020.

Masucci, M., D. Organ, A. Wiig (2016). Libraries at the crossroads of the digital content divide: pathways for information continuity in a youth-led geospatial technology program. *Journal of Map & Geography Libraries*, 12(3), pp. 295–317.

Masucci, M., H. Pearsall, A. Wiig (2019, early release). The smart city conundrum for social justice: youth perspectives on digital technologies and urban transformations. *Annals of the American Association of Geographers*, pp. 1–22. doi:10.1080/24694452.2019.1617101.

Mele, C. (2013) Neoliberalism, race and the redefining of urban redevelopment: Color-blind redevelopment in Pennsylvania. *International Journal of Urban and Regional Research*, 37(2), pp. 598–617.

Nam, T., & T. Pardo (2012). Transforming city government: a case study of Philly311. *Proceedings of the 6th International Conference on Theory and Practice of Electronic Governance*, pp. 310–319.

Newcombe, T. (2017). 311: From a hotline to a platform for citizen engagement. *Government Technology*. Source: https://www.govtech.com/dc/articles/311-From-a-Hotline-to-a-Platform-for-Citizen-Engagement.html Last accessed 25 June 2019.

Petrov, C. (2019). Internet of Things Statistics 2019 [The Rise of IoT]. *TechJury*, March 22, 2019. Source: https://techjury.net/stats-about/internet-of-things-statistics/ Last accessed: 28 June 2019.

Rappaport, J. (2003). US urban decline and growth, 1950 to 2000. *Economic Review-Federal Reserve Bank of Kansas City*, 88(3), pp. 15–44.

Reinsel, D., J. Gantz, & J. Rydning (2018). Data age 2025: the digitization of the world from edge to core. *Seagate Data Age*. Source: https://www.seagate.com/files/www-content/our-story/trends/files/idc-seagate-dataage-whitepaper.pdf Last accessed: 28 June 2019.

Reyes, J. (2013). Real time crime center: 1 year after launch, 24-hour support center will move to Delaware valley intelligence center this spring. *Technical.ly Philly*. Source: https://technical.ly/philly/2013/02/20/real-time-crime-center-1-year-after-launch-24-hour-support-center-will-move-to-delaware-valley-intelligence-center-this-spring/ Last accessed: 17 June 2019.

Robbins, W. (1981). Philadelphia suffers in manufacturing job exodus. *The New York Times*. 15 August. Source: https://www.nytimes.com/1981/08/15/us/philadelphia-suffers-in-manufacturing-job-exodus.html Last accessed 30 May 2019.

Srnicek, N. (2017). *Platform Capitalism*. Malden, MA: Polity Press.

Stanley, J. (2019). *The Dawn of Robot Surveillance*. American Civil Liberties Union. Source: https://www.aclu.org/report/dawn-robot-surveillance Last accessed 17 June 2019.

Stoll, M. (2005). Job sprawl and the spatial mismatch between blacks and jobs. *Brookings Institution Metropolitan Policy Program*. Source: https://www.brookings.edu/research/job-sprawl-and-the-spatial-mismatch-between-blacks-and-jobs/, pp. 1–15.

Weaver, R., S. Bagchi-Sen, J. Knight, & A. Frazier (2016). *Shrinking Cities: Understanding Urban Decline in the United States*. New York: Routledge.

Weigley, R., N. Wainwright, and E. Wolf (1982). *Philadelphia: A 300 Year History*. New York City: W. W. Norton & Company.

Whelan, A. (2014). Police body cameras coming to Philadelphia. *Governing.com*. Source: https://www.governing.com/topics/public-justice-safety/tns-philadelphia-police-body-cameras.html Last accessed: 17 June 2019.

Wiig, A. (2013) Everyday landmarks of networked urbanism: cellular antenna sites and the infrastructure of mobile communication in Philadelphia. *Journal of Urban Technology*, 20(3), pp. 21–37.

Wiig, A. (2016) The empty rhetoric of the smart city: From digital inclusion to economic promotion in Philadelphia. *Urban Geography*, 37(4), pp. 535–553.

Wiig, A. (2018). Secure the city, revitalize the zone: Smart urbanization in Camden, New Jersey. *Environment and Planning C: Politics and Space*, 36(3), pp. 403–422.

Wiig, A. (2019). Incentivized urbanization in Philadelphia: The local politics of globalized zones. *Journal of Urban Technology*. 26(3), pp. 111–129.

Zylstra, G. (2013). Struggle over the streets: industrialization and the fight over the corporate control of street space in Philadelphia, 1830–1860. *Journal of Urban Technology*, 20(3), pp. 3–19.

SECTION 2

Do platforms represent a new model of urban governance?

6

JOINING THE DOTS

Platform intermediation and the recombinatory governance of Uber's ecosystem

Sarah Barns

Introduction

Uber is a big company – and it wants to get bigger. A decade after its launch as "UberCab" in 2009, by 2019 Uber reported operating in 65 countries and over 600 cities. During that decade, the company was able to attract some US $22.2 billion from investors, leading to a valuation, in mid-2018, at over $62 billion, after rising at one point close to $70 billion (Crunchbase 2019). By mid-2019, Uber was reporting an average of 14 million trips being conducted through the platform each day. There were a total of 91 million active platform users, and of those, 3.9 million are drivers (Uber 2019a). From a total of 159 employees on December 31, 2012, it grew to some 22,263 global employees as of December 31, 2018, with around half of these employees based in the United States (Uber 2019b: 39).

Having begun its journey as a black cab service you could order via a well-designed app, Uber quickly became known as a more generic "ride-hailing" service, which gave drivers the ability to use their own cars to pick up passengers wherever they were using the Uber app. By 2019, in just ten years, some 10 billion rides had been facilitated this way (Uber 2019a). In 2018, Uber derived 24% of its ride-sharing bookings from five metropolitan areas – Los Angeles, New York City and the San Francisco Bay Area in the United States; London in the United Kingdom; and São Paulo in Brazil. Fifteen percent of these trips associated with trips that take place to or from an airport (Uber 2019b: 45). Uber can report these numbers because it has the data on every one of the rides that are facilitated via its platforms. Some of its rides have taken place in cities in which Uber is deemed illegal, or operates without a licence. These two dimensions of the company – the data it is able to collect via its service offering, and its combative approach to existing regulations – are critical to the way Uber operates as an urban platform.

Political skirmishes and company scandals have been a well-documented feature of the meteoric rise of this young company, co-founded by tech entrepreneurs Travis Kalanick and Garrett Camp. Posing a direct challenge to the highly regulated taxi industry, which during the 20th century expanded throughout the world under license conditions imposed by local authorities (Cramer and Krueger 2016), Uber has worked hard to evade local regulations by strategically positioning itself as a *new* kind of transport service, one that operates somewhere between what Kalanick once described as "lifestyle and logistics" (Hartmans and Leskin 2019). By 2019, Uber had become a platform providing offerings in the "personal mobility, meal delivery, and logistics industries" (Uber 2019b: 32).

The constant strategic repositioning of the company's service offering relies heavily on Uber being *Uber Technologies Inc.* – that is, primarily a technology platform designed to facilitate mobility services. When it first launched, this was a platform for the "sharing" of rides, but then it became about meals, then logistics as well, and eventually, it hopes, multi-modal transport planning (Hartmans and Leskin 2019). The discursive positioning afforded by being a "platform" provides the basis for launching new services that constantly seek to disrupt existing industries. Being a "new" kind of mobility company, known in some jurisdictions as a "Transportation Network Company" (TNC), has allowed Uber to not only introduce new technology, but to also recalibrate existing service delivery models.

Most controversially, Uber has sought to re-engineer the provision of what are ostensibly taxi services. Positioned as a technology platform, it does not classify Uber Drivers as employees of the company, but rather "Driver-Partners", who operate as self-employed small businesses, responsible for maintaining their own insurances and business affairs. Driver-Partners have no contractual relationship with Uber itself; rather, they enter into contract agreements with an Uber subsidiary known as Raiser LLC (Uber n.d.). Driver-Partners are, of course, critical to the service offering made available by Uber, but they are not classified as being part of the company, they are instead designated "third party" users of the platform. As Uber must continually and controversially assert, the company's primary service offer is that of a technology platform, designed to assist those seeking to either offer, or accept, rides from strangers. As put by Uber in its legal documentation published online to inform private parties in their dealing with Uber: "Uber is a technology company that has developed an app that connects users (riders) with third party transportation providers" (Uber n.d.).

Because Uber is a big company, and receives lots of publicity, good and bad, having achieved global network scale at a very rapid rate, much of the controversy and attention it provokes tends to be framed as highly specific to Uber. Investigative journalists such as the *New York Times'* Mike Isaac have published books on the company that focus attention on the culture of Uber, and the corporate battles that have taken place within the company (Isaac 2019), shaped by the personality of its high-profile CEO Travis Kalanick. The impact of a blog post by former employee Susan Fowler in 2017, reporting on her experience of sexual discrimination and workplace harassment, has helped to reinforce public scrutiny towards the company's problematic corporate culture (Fowler 2017). At the same time, Uber has also been frequently portrayed as the iconic example of a new kind of company, an "asset light" business, one demonstrative of how digital technologies are transforming the world of work, its scale vindication of the potent potentialities of the "sharing economy" as a new way of structuring economic relations (Mohlmann and Geissinger, 2018; Botsman and Rodger 2010; Sundararajan 2016; Dupuis 2018). As with other "asset light" digital companies launching in the late 2000s, including Airbnb, Zipcar and Lyft, Uber has been discursively positioned as an enabler of disruptive innovation which favoured and rewarded the nimble entrepreneur over the legacy corporation (Sundararajan 2016: 237; Pollio 2019).

When the company launched one of the world's most publicised IPOs in 2019, attention pivoted towards the company's current profitability (or lack thereof), prompting increasingly negative press coverage. According to business news sites such as *Forbes*, its business model is "fundamentally broken" (Sherman 2017, 2019). Its vulnerability is seen to lie in the relatively low barriers to entry (e.g., 'multi-homing'), impacting profit margins, and protracted legal battles over the employment status of drivers (Domurath 2019; Pollio 2019).

Much critical attention towards Uber to date has also focused on the practices of driver precarity promoted by the company which are, by virtue of its expansion across global urban markets, also undermining labour and licensing conditions in more established transportation services such as the taxi industry (Isaac 2014, Isaac 2019). Through this lens, Uber is a company that has used the veil of new digital technology to, in the words of Graham and Shaw, "engineer away" the awkward politics of labour conditions (2017).

But Uber is not only an iconic example of a wholly *new* kind of company, good or bad, the genius innovation of its brash, unconventional business leaders, "blitzscaled" to global scale via the ambitions of its high-profile, maverick investors. Uber also represents the manifestation of processes and tactics of digital intermediation that now, by virtue of the unleashing of GPS-enabled smartphones across urban populations, combined with the investment strategies of financial intermediaries, are recalibrating the intimate fabric of everyday urban relationships. Methods of digital intermediation can, particularly when achieved at scale, be used to co-ordinate the possibilities of relational public encounter, via locational attributes of users harvested and co-ordinated in real-time. Decades of technology R&D, including, not least, US Government investment in relatively "open" data made available via Global Positioning Satellite (GPS) infrastructure (Mazzucato 2016) have unleashed the potentials of real-time location tracking and the myriad services this data can be used to enhance, via machine learning techniques. A series of strategic design decisions are also made by platform companies like Uber which, in turn, ensure that data generated via these methods of digital intermediation competitively advance the functional design and performance of their app-based point-to-point customer services.

As I discuss in this chapter, it is useful to reflect on how Uber makes manifest conditions of "platform intermediation", as the positive externalities that result from the translation of non-digital relationalities into data points are used to continuously modulate and expand platform reach. As a technology platform, like other major technology platforms, Uber enfolds many historical innovations achieved through open source technologies within an Uber-intermediated "platform ecosystem" (Tiwana 2013; van Djick et al. 2018), which enables it to deliver an in-demand and on-demand service to urban citizens around the world, and by aggregating the data generated by the transactions it offers, it seeks to leapfrog into the world of automated mobility services. Uber can, through this lens, be considered both one company, *Uber Technologies Inc.*, but also, as implied by its full name, an ecosystem of technologies, platforms, data points, user interfaces, relational infrastructures and application programming interfaces (APIs), service providers, investor-financiers and customers. The scale it seeks to achieve is not only in the numbers of customers using its platform, but also in the reach of its "platform ecosystem".

Forms of relationality operating within Uber's ecosystem makes manifest conditions of data-driven platform governance (Gorwa 2019; van Doorn 2019; Barns 2020) which operate, quite deliberately, beyond the state (Swyngedouw 2005). Opaque conditions of platform governance set in play across an ecosystem like Uber's present conceptual challenges when trying to understand how platforms are recalibrating urban relationships, through the lens of an emergent platform urbanism (van der Graaf and Ballon 2018; Barns 2019; van Doorn 2019). As a growing body of literature attending to the consequences of platform scale has underscored, "the platform" is not only a particular computational structure but also a discursive innovation, which works on behalf of its business owners to recast conditions of value-sharing between companies and their marketplace as ostensibly "open" (Gillespie 2010; Tiwana 2013; Srnicek 2016; Langley and Leyshon 2017; Mackenzie 2018). These ecosystem-relationships make possible conditions of algorithmic control, subtly shaping, nudging and "steering" urban behaviours from infinitesimal to global scales, in ways that point to new asymmetries of urban governance in a world of connected devices, things, people and infrastructures.

While Uber is but one of a number of global digital platforms, its opaque systems of globally integrated data governance, which seek to intervene in the everyday, intimate socialities of urban experience and mobility, point to an emergent urban condition – a platform*ed* urbanism – shaped by conditions of control and communication that operate in co-ordinated and constantly recombinant ways. As such, the morphing and scaling of Uber evidences the challenges of trying to understand the platforming of urban relationalities, in ways that are neither purely and abstractly global, nor relationally intimate, but recombinations of the two, facilitated via constantly modulated and opaque algorithmic integrations.

Making sense of Uber

The year 2019 saw the investing marketplace being asked to appraise the strength of Uber as a company. The company's S-1 filing, in advance of its much-publicised initial public offering (IPO), provided the opportunity for public scrutiny of a company that has spared no efforts to maximise the scale of its operations. Adopting a kind of "reverse innovation process" (Pelzer, Frenken et al. 2019), sometimes described as "permissionless innovation" (Chesborough and Alstyne 2015), Uber never sought the permission of authorities to operate in the cities it launched in. Instead, it entered these markets by directly enlisting drivers and riders to start using its app. Local transport authorities needed in turn to respond to the services being traded through the Uber app, often by introducing new license categories specific to the Uber's categorisation as a TNC (as with other ride-sharing services such as Lyft and Didi). Some jurisdictions have made life hard for Uber by introducing licence fees or other conditions that the company has judged unworkable, prompting it to withdraw the service from the city. Other cities, including Australian cities Melbourne and Sydney, have introduced new levies on Uber rides, to compensate taxi drivers who have struggled to respond to the influx of Uber Drivers willing to deliver what are essentially taxi services delivered via a more sophisticated digital interface and at a lower cost (Bainbridge 2019).

The mission to scale has seen Uber engage in highly aggressive tactics to attract more users to its platform and to detract users from competing platforms such as Lyft in the US and Didi in China. It has also used its software to deliberately "greyball" officials seeking to use the app to fine drivers who may be illegally using the Uber app in cities where it has been banned (Wong 2017). Undermining the validity of existing regulatory frameworks has been one key strategy to ensure Uber scales quickly. The other has been to attracting large volumes of investor finance. Uber's capacity to grow at the rate it has done over the past decade reflects the volume of investment poured into the platform, in order to scale at any cost. In 2017, for example, Uber reportedly lost a total of $6 billion as a consequence of its rapid expansion that year. In the months following the announcement of this loss, venture capital fund Softbank invested more than $8 billion in Uber (Griswald 2018), taking total venture capital investment to $11.5 billion (Sherman 2017).

The S-1 filing meant that Uber's persistent losses became a central focus during 2019. Its S-1 document stated that Uber will continue to "lower its prices [...] and may continue to offer significant Driver incentives, consumer discounts and promotions, which may adversely affect" its financial performance (Uber 2019b: 33). It also faces an uncertain future in many major urban markets. Along with class actions by displaced taxi drivers and Uber drivers who claim poor employment practices, city governments, most notably the City of London, continue to lead legal challenges against the company. In the context of an increasingly crowded marketplace of ride-sharing platforms, and mounting legal challenges, many business reporters and commentators now question the long-term profitability of the company (Domurath 2018). Their concerns seem to be reinforced by Uber, which stated in 2019: "Ridesharing and other categories in which we compete are nascent, and we cannot guarantee that they will stabilise at a competitive equilibrium that will allow us to achieve profitability" (2019b: 34).

Self-professed doubts over the future profitability of ride-sharing served to dampen Uber's post-IPO trading price. According to business news sites such as *Forbes*, its business model is "fundamentally broken" (Sherman 2017). Its vulnerability is seen to lie in the relatively low barriers to entry, impacting profit margins, and growing legal battles over the employment status of drivers (Pollio 2019). But with these concerns in mind, it remains important to address the underlying discursive logics and business tactics that have enabled Uber to scale so rapidly as an urban platform. These point to a more complex set of urban governance conditions enacted by the company that can be expected to prevail, regardless of Uber's fate or market positioning as an individual company. This is because Uber is not *only* a company, it is also the manifestation of a set of processes and discursive tactics constituted by conditions of platform intermediation. Though at the expense of quarterly profits, rapid expansion is designed primarily to accelerate the process of platform intermediation which, to key investors, is the ultimate goal.

For SoftBank, whose $100 billion Vision Fund has been backed by Saudi Arabia's Public Investment Fund ($45 billion), Apple ($1 billion) and Qualcomm ($1 billion), the strategy driving multi-billion dollar investments into platforms such as Uber is to invest in "businesses and foundational platforms that SoftBank believes to revolutionise and innovate the world tomorrow" (Crunchbase, n.d.). Ultimately, the value of Uber as a "foundational platform" lies in the long-term value it can accrue, by delivering an underlying technology platform from which to evolve not only ride-sharing, but future mobility services, in a world that is rapidly increasing its reliance on autonomous, data-driven services. Uber's ultimate goal, as revealed in 2019 by Uber's new CEO Dara Khosrowshahi, is simply to become "the brokerage of all human movement in cities" (Hawkins 2019: para 4). As a platform, serving as a "match maker" between diverse forms of trade and association, Uber Technologies Inc. can activate its platform ecosystem of data-rich transactions to diversify into new service domains, just as other platforms such as Amazon, Google and Facebook have done. Understanding Uber as a mode of platform intermediation can in this sense be useful for understanding the kinds of informational asymmetries and governance challenges provoked by a world of proliferating urban platforms – whether or not Uber itself continues to dominate the world of ride-sharing.

In the next section I want to explore some of the key concepts of platform intermediation, and how these apply to Uber. Spanning the literatures of microeconomics, internet and software studies, geography and sociology, the study of platforms evidences an increasingly widespread recognition that techniques of intermediation are now, centrally, techniques of data-driven governance, with destabilising impacts that extend beyond the reaches of any one given company.

Platform intermediation and the art of digital scale

As platform scholars Tarlton Gillespie and Mike Annany observed: "To understand the power and invisibility of platforms [...] we must study how they change." (Gillespie and Ananny 2016, p. para 4). Uber emerged as a company borne of the smartphone-era, an era when the majority of urban citizens were rapidly equipped with location-tracking, courtesy of the Global Positioning Satellite (GPS) receivers bundled into their smartphones. Location-tracking via smartphones provided the opportunity for Uber to more efficiently match drivers and riders than the traditional taxi service (Rayle et al. 2016). It served to become the perfect (digital) intermediary: to better match or "steer" the way different parties come into relationship with each other. Once a rider is matched with a driver via the Uber app on a user's smartphone, both parties can track each other, with the driver easily able to locate their rider and, via a Google map interface bundled within the Uber app, transport them to their chosen destination. No local knowledge is required: the app provides the requisite information to fulfil the transaction.

This "match making" service provided by the Uber app was relatively novel at the time of its launch in the transport domain, but ultimately it represented the extension of methods of digital intermediation into the domain of mobility services, made possible by virtue of ubiquitous smartphone use. As a platform, Uber also became one of many digital intermediaries whose service as match-makers was accompanied by a set of highly tactical design interventions that ensured its point-to-point digital service generated wider benefits for the company and its capacity to exert influence over the transactions it facilitated. This shift – from basic digital intermediation to "platform intermediation" – has been the focus for extensive scholarly and regulatory interest in recent years, as economists, regulators, internet scholars, geographers and sociologists grapple with the nature of platform influence in a society whose relations are increasingly structured and organised through methods of digital intermediation (Evans 2003; Olma 2014; Plantin et al. 2016; Langley and Leyshon 2017; van der Graaf and Ballon 2018).

A pivot towards "platforms" and "platform intermediation" as a focus of attention, away from more neutral conceptions of digitalisation (Barns 2019; Domurath 2019), has reflected growing recognition that platforms develop strategic ways of harvesting value through their digital intermediation services. As Langley and Leyshon (2017: 7) have written:

> It appears that the key for the platform is to intermediate the ever-expanding value created by user interactions across their market network. This is because continually increasing numbers of users – understood as producers

and creators of value and generators of data, and not as consumers – is crucial to a platform's capacity to cultivate and capture value, and to do so over time and on an ever greater scale.

Fundamentally, of course, the value harvested by platforms is in the data generated through digitally mediated transactions and relationships, which informs the terms dictated by platform companies for wider data-driven applications. Thus Facebook, initially an intermediary in the domain of college dating, has morphed into what John Lancaster called "the biggest surveillance-based enterprise in the history of mankind" (2017), by monetising the data generated by the digital transactions its platform facilitates. As has been underscored through the lens of the platform capitalism critique, the key advantage of the platform business model over others in an era of big data value chains is the capacity for the platform to occupy a privileged position as an intermediary and therefore "governor" of digital exchange and the data that result as a consequence (Srnicek 2016; Pasquale 2016).

Attention towards the tactics used by platforms to intermediate social relationships of value-sharing – whether via social media, ride-sharing, online retail or search platforms – has recognised the capacity for platform companies to broker ever-more diverse and expansive kinds of value-trading. In the lexicon of more business-oriented platform strategy, platforms orchestrate "ecosystems" of diverse market entities through methods of "open innovation" (Tiwana 2013), which ensures the functionality of the platform is extended by ostensibly outside actors. This means that rather than employing large numbers of people to work for a firm under traditional management structures as direct employees, platforms instead enlist their users to extend the reach of their proprietary technology (Simon 2011; Tiwana 2013), bringing more diverse entities into relationship with each other via code. Technology-based competition has thus seen to shift from that of a "battle of devices" to a "war of ecosystems" (Armstrong 2006).

The computational architecture put in place across a platform ecosystem hinges fundamentally on the role of the Application Programming Interface, or API, which enables different kinds of software to connect. APIs institute a system of protocols and standards that determine how data from user activity can be structured and used (Helmond 2015; Mackenzie 2018; Raetzsch et al. 2019). From the perspective of computational design, this architecture is designed to be "open" in the sense of allowing multi-sided markets of platform participants (like software makers, or Uber Drivers, riders or mapping software) to share services across a wider platform ecosystem. APIs operating at the "edge" of Apple and Facebook operating systems, for example, encouraged software developers to flock to their developer platforms, allowing the diversification of software services provided by these companies, not by employees but by ostensibly "open" ecosystems of developers (Helmond 2015). But these intermediated interactions also require conformance to centralised standards and protocols, which may be changed or altered in ways that are opaque to users.

Platform architecture can thus be read as highly *recombinatory*, ingesting and harvesting data from a highly porous, intermediated platform ecosystem of multi-sided relationalities and transactions, while underpinned by protocols and modes of value-exploitation that are closely controlled by platform owners. Social media theorists understand this condition as one of "proprietary opacity" (Mackenzie 2018), whereby the programmability of platforms, by virtue of their API, at once decentralises data production while simultaneously recentralising data collection (Helmond 2015). This leads to discursive and conceptual challenges to the study of platforms and their impacts. As Mackenzie has written, platforms can be *seen* to "engage the flexibility and mutability of programming and programmability to modulate interfaces, devices, protocols and, increasingly, infrastructures, in the interests of connectivity" (Mackenzie 2018: 4). These relatively porous boundaries between a platform company and its ecosystem of users have allowed platform companies to discursively position themselves as value-creating entities, through their ever-expanding network effects.

Such was the discursive logic underpinning celebrations of participatory media, the sharing economy and networked digital culture, which enabled internet platforms to assert their roles as neutral players, or "*just* intermediaries", that facilitated new kinds of content creation and sharing (Gillespie 2010). This positioning achieved by early internet platforms was enshrined in US legislation known as "Section 230", contained within the 1996 Communications Decency Act, which continues to protect "interactive computer services" from immunity against civil cases over the content posted on websites. It is also critical to the assertion by Uber that it does not employ its Drivers: it is *just the platform* on which others can choose to generate value (Domurath 2019). And yet, the mutability of programming interfaces and the relative "openness" of the platform to new value-creating users belies the control platform owners have over their ecosystems. In an increasingly data-driven economy, platforms own the infrastructure from which data as capital is generated, and they set the terms upon which others can leverage the value of that data for a wide variety of digital applications (Langley and Leyshon 2016: 7; Plantin, Lagoze et al. 2016). From neutral intermediaries to platform ecosystems, the world's largest platforms operate beyond sovereign boundaries, their value escalating as each of the billions of digital transactions they facilitate is utilised by platform companies as training data for new machine learning, or artificially intelligent (AI) applications.

A pivot in critical attention platform design and governance, which grapples with the unique natures of platform companies and their capacity to govern increasingly larger volumes of digitally intermediated social relationships, ultimately seeks to understand the nature of power and influence operating in an era of platform scale. As Domurath has noted, during 2019 there was hardly a topic that provoked greater regulatory attention than the contemporary nature of the platform economy (2019: 565). Platforms like Facebook, Google, Amazon and Uber are now cast as *rentiers* of the digital economy (Olma 2014; Sadowski 2019) and "infrastructures of capital accumulation" (Langley and Leyshon 2017: 18).

While attention has escalated recently, platform scholars can draw on extensive literature that has grappled for decades with the nature of competition in platform markets. Micro-economists studying the rise of platforms in the early 2000s were provoked by the market power exerted by Microsoft, to examine the relatively hidden incentives and forms of control, both computational and discursive, exerted by platform-based operating systems. Fundamentally operating as match-makers, platforms will always be seeking ways to maximise the nature and volume of transactions that take place, particularly when this can generate ever-greater volumes of data-as-capital. As the platform economist David Evans put it: "Businesses in these industries will devise entry strategies to get multiple sides of the market on board, and devise pricing, product and other competitive strategies to keep multiple customer groups on a common platform that internalises externalities across members of these groups" (Evans 2003). The tactics adopted by platform firms to entice users to choose one service over another are described as "steering tactics" (Rochet and Tirole 2003: 993). For economists, steering tactics are central to the operations of successful platforms, which devote significant attention towards and investment in strategies that can enhance the overall value of the platform to its users.

Increasingly, platforms draw on algorithms as steering tactics, whether to attract constant attention to their interfaces, or to undermine their competitors, or steer their users into particiular purchasing or other behavioural decisions. As van Dijck (2012: 171) has written in relation to social media platforms:

> What is important to understand [...] is how they activate relational impulses, which are in turn input for algorithmically configured connections—relationships wrapped in code—generating a kind of engineered sociality.

The "engineered sociality" of platform ecosystems provides a useful way to understand the nature of platform governance in urban settings. It extends from a purely transactional understanding of data capture and extraction to consider the different levels of embodiment, sociality, steering, transaction, competition and urban "nudging" at play. According to platform economists Rochet and Tirole, (2003: 1012) the use of steering strategies is particularly important in ride-sharing code-spaces, where customers can easily switch platforms (between, for example, Uber, Ola or Didi).

Platform economists call the ease of switching, where multiple apps and digital marketplaces are running successfully in a city, "multi-homing". For Rochet and Tirole (2003), the concept of multi-homing is a way of describing a process of "horizontal differentiation", which results in customers choosing to join and use several platforms. Multi-homing code-spaces of ridesharing like Uber demand steering strategies that essentially seek to limit or discourage users from switching platforms, through the use of price structures (such as surge pricing) and reduced wait-times (Bryan 2019). Surge pricing will essentially act to attract more drivers to the platform marketplace (through offers of higher prices for rides) while keeping wait times for users low.

These different insights underscore the nature of platform intermediation as inherently *processual* – always mutating, aggregating, scaling, match-making, harvesting and diversifying. Understanding the processual, and recombinatory, dynamics of platforms operating as ecosystems in turn highlights ever-present contingencies and relationalities operating across scales and beyond company boundaries.

Uber intermediation and recombinatory governance

Returning to Uber, the platform lens demarcates a somewhat different set of boundary conditions for this company. Rather than simply operating as a loss-making ride-hailing platform (which it clearly is), Uber Technologies Inc. can also be viewed as an integrated global data ecosystem whose diverse applications aim to intermediate more and more diverse forms of human mobility in contemporary cities. The platform lens also helps to situate the operations of Uber as a technology-based mobility company within a much broader ecosystem of users and contributors, which it seeks to govern and influence. Uber's ecosystem not only consists of its drivers and riders, but also incorporates a variety of open source licenses, proprietary platforms (Google Maps), data centres, cloud computing services, subsidiaries and broader supply chains, whether of vehicle service providers or hotels, airports and other industries integral to its capacity to deliver mobility services. As it instruments this ecosystem into action via the Uber API, Uber institutes common protocols that ensure diverse data is collected, stored and used in standardised ways. This infrastructure means the company can in turn capture the ever-expanding value created when it expands its global reach, harvesting and modulating globally scalar networked interactions in real-time.

Propelled by significant injections of capital by financial backers such as Softbank, Apple and Google, the faster Uber scales into new markets, the more advanced is its capacity to expand its data ecosystem. Ride-hailing may not be profitable in key markets, but continued expansion facilitates what early cybernetic thinkers described as the "intelligence amplification" of a computational network (Licklider and Taylor 1968). Data harvested through the greater volumes of transactions facilitates the expansion of Uber Technologies Inc. into new service domains, including its Autonomous Technology Services Division and other multi-modal transport services. Uber's capacity for data aggregation at scale has allowed it to leap frog other companies in the race to build self-driving cars (BBC 2018). This extension of the platform's capability also reflects significant investments made by the company, which reportedly has spent more than $1m per day expanding its capabilities in autonomous driving (BBC 2018).

Uber now aims to become, first and foremost, an urban data platform that supports integrated, multi-modal movement solutions (Hawkins 2019: para 7), the "operating system for everyday life in the city" (Hawkins 2019: para 4). Having aggressively sought to undermine the regulatory capabilities of city governments, Uber now seeks to leverage its data science to position itself as "trusted adviser" to governments. Oriented towards solving urban mobility challenges, Uber has launched the Uber Movement platform to share selected and anonymised data it harvests from its platform as a "tool" for regulators, planners and other transport users (Uber Movement, n.d.). The data delivered via this platform is sourced from the 10 billion trips taken on the platform, and features "Uber Movement Speeds" that provides average street speed data to assist with data-driven urban planning. Along with Lyft, Uber co-sponsored a study that found that TNCs were increasing traffic in key urban markets, including by up to 13% of traffic in San Francisco (Fehr and Peers 2019). As an urban data science platform, it is aiming to support its users to select a greater diversity of transport options, which may include public transportation, and not only private rides.

As Fran Bell, the head of data science at Uber has pointed out, "Uber tackles some of the world's most challenging data science problems at scale and in real time [...] We use data intelligently to build better experiences for our users and *solve problems* at scale" (Vorwerk 2019, italics added). As the world's most globally networked urban data platform, users of Uber's technology services will inadvertently participate in "upwards of hundreds" of experiments being run by Uber's data team on any one day, which seek to test out assumptions about how users respond to data provided, potential vulnerabilities present across the platform and also how trip information is used and adopted. As the Uber Engineering team has explained on its website: "Every engineer watching a dashboard tends to care about data in a particular location or region, around a set of experiments, or related to a certain product" (Lozinski 2017). Incoming metrics are then compared to predictive models based on historical data to determine whether current data is within the expected bounds (ibid).

The Uber user (driver, or rider, for example) acts here as something of data ranger, constantly harvesting city data on behalf of the platform, while at the same time benefitting from on-demand data delivered via the app interface, for example, in the form of ETAs (estimated time of arrival information). The way Uber manages and influences its users constitute dynamic integrations of code, commerce and corporeality. But in order to achieve continuous network orchestration at scale, a company like Uber cannot stand still. Multi-homing codespaces of ridesharing require constant steering strategies that discourage users from switching platforms, through the use of price structures (such as surge pricing) and reduced wait-times (Bryan 2019). Surge pricing will essentially act to attract more drivers to the platform marketplace (through offers of higher prices for rides) while keeping wait times for users low.

This sense of having to manage a state of constant flux pervades the company's S-1 filings. The company reports: "Our success in a given geographic market significantly depends on our ability to maintain or increase our network scale and liquidity in that geographic market by attracting Drivers, consumers, restaurants, shippers, and carriers to our platform" (2019b: 35). It reports constant risks associated with the exposure of its underlying software IP to competitors, via its reliance on third-party open-source licenses (56). And it must constantly invest, in order to, as put in its S-1: "increase the number of Drivers, consumers, restaurants, shippers, and carriers using our platform through incentives, discounts, and promotions; expand within existing or into new markets; increase our research and development expenses; invest in ATG and Other Technology Programs" (Uber 2019b: 34).

Conclusion

As I have discussed in this chapter, Uber's rapid advancement across global urban marketplaces has generated multiple spheres of conflict between the company and existing urban governance settings. Operating through a kind of "reverse innovation process", Uber has not sought permission to operate, but has scaled by relying on its discursive positioning as a technology platform that connects or intermediates the activities of diverse transport users and providers. While its controversial culture, and interventionist tactics, have attracted a great deal of public attention, leading to numerous court cases and rulings against the company across key jurisdictions, a highly public IPO in 2019 has also drawn attention to Uber's persistent operating losses, as it continues to invest in incentives and discounts and expansion into new markets.

While Uber attracts attention as a controversial company, in this chapter I have examined the way Uber makes manifest conditions of platform intermediation operating at scale. The processual lens of platform intermediation highlights the nature of Uber as a relatively porous and diverse ecosystem of users, which is, via methods of algorithmic steering, nevertheless subject to highly opaque conditions of platform governance. Through "upwards of hundreds experiments" conducted on the Uber platform each day, Uber's ecosystem evidences platform tactics of "engineered sociality" which incorporate both computational architecture and business strategy. Its mode of governance can be seen as recombinatory in nature, open to innovation and inputs from users acting as data-rangers, while also necessitating conformance to standardised protocols put in place through API infrastructure. Through conditions of recombinatory governance, platform ecosystems of the kind instrumented and engineered by Uber, through its "nifty ride-sharing app", establish not only altered code-space of everyday urbanity, but also represent new forms of contemporary urban governance managed through data science at scale and in real-time, in ways that seek to exceed the capacity of any single city government.

The scale of investment poured into Uber by Softbank demonstrates a clear link between conditions of platform intermediation and financial intermediation, with Uber acting as a vehicle for speculation on the conditions of AI-driven future infrastructure and asset management. To Softbank, driven by a focus on building the "foundational platforms" of our future cities, Uber's future growth potential rests on its capacity to constantly intermediate at scale, eventually become, to repeat again the hyperbolic pronouncements of its CEO, "the broker of all human movement". Platform intermediation, accelerated via financial intermediaries, thus sets out not only its globally networked ambitions but also its temporal reach, seeking to intervene in a way Tafuri (1976, in Jameson 2005: 228) once described as the "actuarial colonisation of the unknown".

These conditions of platform intermediation return us to the spectre of cybernetics, particularly when we consider how experiments within contemporary code/spaces are deployed by platform companies like Uber (Kitchin and Dodge 2011). Michael Gandy (2005) previously described the melding of the technological and the human as enabling a kind of "cyborg urbanisation", in order to draw attention to the role played by cities in facilitating the establishment of a "physical infrastructure that links the human body to vast technological networks" (Amin 2007). Though they lack the prosthetic limbs and dystopian lighting effects, platform ecosystems might be seen to enrol their users in a highly mobile, embodied "cyborg urbanity". Uber's platform intelligence operates through the palms of our hands, which, via the well-designed app interface activated via tapping on smartphones, links us to multi-scalar infrastructures that facilitate intimate, transient encounters with strangers (in a cab, in a house, on the street) via global networks of dataveillance and value extraction.

As the mounting numbers of court cases against Uber by disgruntled Driver-Partners and taxi drivers suggests, Uber will need to move quickly to recalibrate its platform ecosystem as a "broker" for integrated, multi-modal transport solutions, and not just a controversial ride-sharing platform. But its capacity to leverage the anonymised data harvested from its 10 billion trips to date suggests it remains well placed to pivot.

References

Amin, A. (2007). Re-thinking the urban social. *City, 11*(1), pp. 100–114. doi:10.1080/13604810701200961.

Armstrong, M. (2006). Competition in two-sided markets. *The RAND Journal of Economics, 37*(3), pp. 668–691.

BBC. (2018, 27 August). Toyota to invest $500m in Uber in driverless car deal. Retrieved from https://www.bbc.com/news/business-45324753.

Bainbridge, A. (2019). Uber 'came to our shores, illegally, like pirates', class action lead plaintiff says. *ABC News*. Accessed 22 September 2019 at https://www.abc.net.au/news/2019-05-03/uber-to-face-class-action-against-taxi-and-private-drivers/11073640

Barns, S. (2019). Negotiating the platform pivot: From participatory digital ecosystems to infrastructures of everyday life. *Geography Compass*. Published online July 2019. doi:10.1111/gec3.12464.

Barns, S. (2020; forthcoming). *Platform Urbanism: Negotiating Platform Ecosystems in Connected Cities*. London, Palgrave.

Bogost, N., & Montford, I. (2009). Platform studies: Frequently questioned answers. *Digital Arts and Culture*. Accessed 22 September at http://nickm.com/if/bogost_montfort_dac_2009.pdf.

Botsman, R., & Rodger, R. (2010). *What's Mine Is Yours: The Rise of Collaborative Consumption*. New York, HarperBusiness.

Browne, R. (2018, 19 December). Uber loses appeal against landmark UK workers' rights ruling. *CNBC*. Retrieved from https://www.cnbc.com/2018/10/31/uber-loses-appeal-against-landmark-uk-workers-rights-ruling.html.

Bryan, K. A. G., J. (2019). A theory of multihoming in rideshare competition. *Journal of Economics & Management Strategy, 8*(1). doi:10.1111/jems.12306.

Chesborough, H., & Alstyne, M. (2016). 'Permissionless innovation. *Communications of the ACM, 58*(8), August 2015, pp. 24–26. doi:10.1145/2790832.

Choksi, M., & Fujiu, R. (2016). Behind the wheel. Accessed 22 September 2019 at https://www.uber.com/en-AU/newsroom/behind-the-wheel/.

Choudary, S. P. (2015). *Platform Scale: How an Emerging Business Model Helps Startups Build Large Empires with Minimum Investment* (Kindle Edition). Boston, MA, Platform Thinking Labs.

Clark, J., Couldry, N., De Kosnik, A., Gillespie, T., Jenkins, H., Kelty, C., … van Dijck, J. (2014). Participations: Dialogues on the participatory promise of contemporary culture and politics. *International Journal of Communication, 8*(2014), pp. 1446–1473.

Cramer, K., & Krueger, A. (2016). *Disruptive Changes in the Taxi Business: The Case of Uber.* Retrieved from http://www.nber.org/papers/w22083.

Crunchbase. (n.d.). Softbank vision fund. Accessed 22 September 2019 at: https://www.crunchbase.com/organization/softbank-vision-fund#section-locked-charts

Crunchbase. (2019). Uber funding rounds. Accessed 22 September 2019 at: https://www.crunchbase.com/organization/uber/funding_rounds/funding_rounds_list

Domurath, I. (2018). Platforms as contract partners: Uber and beyond. *Maastricht Journal of European and Comparative Law, 25*(5), pp. 565–581.

Dupuis, N. (2018). Stories of the sharing economy: Policy narratives surrounding the entry of transportation network companies into four mid-sized American cities. *Critical Policy Studies*, pp. 1–22. doi:10.1080/19460171.2018.1437459.

Evans, D. S. (2003). Some empirical aspects of multi-sided platform industries. *Review of Network Economics, 2*(3), pp. 191–209.

Fehr and Peers. (2019). Estimated percent of driving by Lyft and Uber in six major U.S. regions, September 2018. Released September 2019. Accessed September 2019 at https://drive.google.com/file/d/1FIUskVkj9lsAnWJQ6kLhAhNoVLjfFdx3/view.

Fowler, S. (2017). *Reflecting On One Very, Very Strange Year at Uber.* Blog post at https://www.susanjfowler.com/blog/2017/2/19/reflecting-on-one-very-strange-year-at-uber. Accessed 22 September 2019.

Gandy, M. (2005). Cyborg urbanization: Complexity and monstrosity in the contemporary city. *International Journal of Urban and Regional Research, 29*(1), pp. 26–49.

Ghosh, S. (2018, 11 June 2018). SoftBank revealed the inside story of how it negotiated an $8 billion share deal with Uber. *Business Insider.* Retrieved from https://www.businessinsider.com.au/softbank-6-months-uber-8-billion-2018-6.

Gillespie, T. (2010). The politics of platforms. *New Media and Society, 12*(3), pp. 347–364.

Gillespie, T., & Ananny, M. (2016). *Exceptional Platforms.* Paper presented at the The Internet, Politics and Policy Conference, Oxford University, Oxford. http://blogs.oii.ox.ac.uk/ipp-conference/2016/programme-2016/track-b-governance/platform-studies/tarleton-gillespie-mike-ananny.html.

Gorwa, R. (2019). What is Platform Governance? *Information, Communication & Society, 22*, pp. 854–871. doi:10.1080/1369118X.2019.1573914.

Graham, M., Hjorth, L., & Lehdonvirta, V. (2017). Digital labour and development: impacts of global digital labour platforms and the gig economy on worker livelihoods. *Transfer: European Review of Labour and Research, 23*(2), pp. 135–162. Graham, M., & Shaw, J., Eds. (2017). *Towards a Fairer Gig Economy*. London, MeatSpace Press.

Griswald, A. (2018, 24 January 2018). Softbank - not Uber- is the real king of ride hailing. *Qz.com*. Retrieved from https://qz.com/1187144/softbank-not-uber-is-the-real-king-of-ride-hailing/.

Hartmans, A., & Leskin, P. (2019). The history of how Uber went from the most feared startup in the world to its massive IPO. *Business Insider*. May 19 2019. Accessed 22 September 2019 at https://www.businessinsider.com/ubers-history/?r=AU&IR=T.

Hawkins, A. (2019). Inside Uber's plan to take over city life with CEO Dara Khosrowshahi. *The Verge*. 26 September 2019. Accessed 19 October 2019 at https://www.theverge.com/2019/9/26/20885185/uber-ceo-dara-khosrowshahi interview-exclusive.

Helmond, A. (2015). The platformization of the web: Making web data platform ready. *Social Media + Society, 1*(2), 2056305115603080. doi:10.1177/2056305115603080.

Isaac, M. (2019). *Super-Pumped: The Battle for Uber*. New York, Norton.

Jameson, F. (2005). *Archaeologies of the Future: The Desire Called Utopia and Other Science Fictions*. London, Verso.

Kitchin, R., & Dodge, M. (2011). *Code / Space: Software and Everyday life*. Cambridge, MA, MIT.

Lancaster, J. (2017). You are the product. *London Review of Books, 39*(16), pp. 3–10.

Langley, P., & Leyshon, A. (2017). Platform capitalism: The intermediation and capitalization of digital economic circulation. *Finance and Society, 2*(1), pp. 11–31.

Lee, D. B. (1973). Requiem for large-scale models. *Journal of the American Institute of Planners, 39*(3), pp. 163–178. doi:10.1080/01944367308977851.

Lozinski, L. (2017). The Uber engineering tech stack, part I: The foundation [Press release]. Retrieved from https://eng.uber.com/tech-stack-part-one/.

Licklider, J., & Taylor, R. (1968). The computer as a communication device. *In Science and Technology: For the Technical Men in Management, 76*, pp. 21–31.

Mackenzie, A. (2018). From API to AI: Platforms and their opacities. *Information, Communication & Society*, pp. 1–18. doi:10.1080/1369118X.2018.1476569.

Mazzucato, M. 2011. *The Entrepreneurial State*. London, Demos.

Mohlmann, M., & Geissinger, A. (2018), "Trust in the sharing economy: Platform-mediated peer trust," In N. Davidson, M. Finck, and J. Unfranca (eds), *Cambridge Handbook of the Law of the Sharing Economy*. Cambridge: Cambridge University Press.

Olma, S. (2014). Never mind the sharing economy: Here's platform capitalism. Retrieved from http://networkcultures.org/mycreativity/2014/10/16/never-mind-the-sharing-economy-heres-platform-capitalism/.

Pasquale, F. (2016). Two narratives of platform capitalism. *Yale Law and Policy Review, 35*, p. 309.

Pelzer, P., Frenken, K., & Boon, W. (2019 November). Institutional entrepreneurship in the platform economy: How Uber tried (and failed) to change the Dutch taxi law. *Environmental Innovation and Societal Transitions, 33*, pp. 1–12. doi:10.1016/j.eist.2019.02.003.

Plantin, J.-C., Lagoze, C., Edwards, P. N., & Sandvig, C. (2016). Infrastructure studies meet platform studies in the age of Google and Facebook. *New Media & Society*, pp. 1–18. doi:10.1177/1461444816661553.

Plantin, J.-C., & Powell, A. (2016). *Open Maps, Closed Knowledge: What the Platformisation of Maps Means for Citizenship and Society*. Paper presented at the IPP2016: The Platform Society, Oxford Internet Institute.

Pollio, A. (2019 July). Forefronts of the sharing economy. Uber in Cape Town. *International Journal of Urban and Regional Research, 43*, p. 4. doi:10.1111/1468–2427.12788.

Raetzsch, C., Pereira, G., Vestergaard, L. S., & Brynskov, M. (2019). Weaving seams with data: Conceptualizing city APIs as elements of infrastructures. *Big Data & Society, 6*(1), 2053951719827619. doi:10.1177/2053951719827619.

Rayle, L., Dai, D., Chan, N., Cervero, R., & Shaheen, S. (2016). Just a better taxi? A survey-based comparison of taxis, transit, and ridesourcing services in San Francisco. *Transport Policy, 45*, pp. 168–178.

Rochet, J.-C., & Tirole, J. (2006 Autumn). Two-sided markets: A progress report. *RAND Journal of Economics, 37*, pp. 645–667.

Sadowski, J. (2019). Landlord 2.0: Tech's new rentier capitalism. *One Zero on Medium*.

Sadowski, J. (2019 June). When data is capital: Datafication, accumulation, and extraction. *Big Data and Society*, pp. 1–12. doi:10.1177/2053951718820549.

Sherman, L. (2017). Why can't Uber make money? *Forbes*. 14 December 2017. Accessed 22 September 2019 at https://www.forbes.com/sites/lensherman/2017/12/14/why-cant-uber-make-money/#3f1ffa8510ec.

Sherman, L. (2019). Can Uber Ever Be Profitable? *Forbes*. 2 June 2019. Accessed 22 September 2019 at https://www.forbes.com/sites/lensherman/2019/06/02/can-uber-ever-be-profitable/ -1ccbec9a5785

Simon, P. (2011). *The Age of the Platform: How Amazon, Apple, Facebook, and Google Have Redefined Business*. Henderson, NV, Motion Publishing.

Srnicek, N. 2016. *Platform Capitalism*. Cambridge, Polity.

Sundararajan, A. (2016). *The Sharing Economy: The End of Employment and the Rise of Crowd-Based Capitalism*. Cambridge, MA, MIT Press.

Swyngedouw, E. Governance innovation and the citizen: The Janus face of governance beyond-the-state. *Urban Studies, 42*, pp. 1991–2006.

Tiwana, A. (2013). *Platform Ecosystems: Aligning Architecture, Governance, and Strategy*. Waltham, MA, Morgan Kaufmann.

Uber. (n.d.). Uber guidelines for third party data requests and service of legal documents. Accessed online 22 September 2019 at https://www.uber.com/en-AU/legal/data-requests/guidelines-for-third-party-data-requests/en/.

Uber. (2019a). Facts and figures as at December 2018. Accessed 22 September 2019 at https://www.uber.com/en-PK/newsroom/company-info/.

Uber. (2019b). Amendment No. 1 to form S-1 registration statement under the securities act of 1933 for Uber Technologies, Inc (S-1 Filing). April 26 2019. Accesed 22 September 2019 at: https://www.sec.gov/Archives/edgar/data/1543151/000119312519120759/d647752ds1a.htm.

Uber Movement. (n.d.). Website accessed 22 September 2019 at https://movement.uber.com/.

Van Der Graaf, S., & Ballon, P. (2018). Navigating platform urbanism. *Technological Forecasting and Social Change, 142*(2019), pp. 364–372.

van Dijck, J. (2012). Facebook as a tool for producing sociality and connectivity. *Television & New Media, 13*(2), pp. 160–176. doi:10.1177/1527476411415291.

Van Dijck, J. (2013). *The Culture of Connectivity: A Critical History of Social Media*. New York: Oxford University Press.

van Dijck, J., Poell, J., & de Waal, M. (2018). *The Platform Society: Public Values in a Connective Society.* Oxford: Oxford University Press.
van Doorn, N. (2019). On platform urbanism and Airbnb citizenship. *New Media and Society.* Published online 29 October 2019. doi:10.1177%2F1461444819884377.
Vorwerk, M. (2019). Data science at scale: A conversation with Uber's Fran Bell. Retrieved from https://eng.uber.com/data-science-at-scale-a-conversation-with-ubers-fran-bell/.
Wong, J. (2017). Greyball: How Uber used secret software to dodge the law. *The Guardian.* 4 March 2017. Accessed 22 September 2019 at: https://www.theguardian.com/technology/2017/mar/03/uber-secret-program-greyball-resignation-ed-baker.

7

A NEW INSTITUTION ON THE BLOCK

On platform urbanism and Airbnb citizenship

Niels van Doorn

> Platforms are what platforms do. They pull things together into temporary higher-order aggregations and, in principle, add value both to what is brought into the platform and to the platform itself [...] As organizations, they can also take on a powerful institutional role, solidifying economies and cultures in their image over time.
>
> (Bratton 2016: 41)

Introduction

Taking a cue from Benjamin Bratton's reflections on the practical ontology of platforms, this article develops the argument that Airbnb should be understood as a new urban institution that is transforming relations between market, state, and civil society actors in post-welfare societies. I examine how this transnational "home sharing"/short-term rental platform accomplishes such profound transformations, which in turn requires an investigation into the specific nature of Airbnb as an institutional form. Institutions, in North's classic definition, are "the humanly devised constraints [and, I would add, affordances] that structure political, economic and social interaction" (1991: 97). Following Bratton, platforms can be understood as new institutional forms that deviate from conventional public and private institutions in "the apparently paradoxical way that they standardize and consolidate the terms of transaction through decentralized and undetermined interactions" (2016: 42). While undetermined, these interactions are nevertheless optimized insofar as they are "regularized by passage through the platform's established forms" and – when these forms are computational, as is the case for digital platforms – "that passage is the [centralized and] capitalized translation of interactions into data and data into interactions" (ibid.; cf. Helmond 2015). This ongoing, recursive process of data-driven translation and capitalization affords today's platform companies their growing institutional power, provided that they continue to offer value for their user base and manage to leverage this base, as well as their data assets, in interactions with stakeholders and shareholders.

As I will elaborate below, Airbnb's privileged avenue for gaining institutional power has increasingly been the world of policy and regulation. While Airbnb has been known to evade regulation and to litigate municipal governments aiming to restrict its operations (e.g. Kendall 2016), the company has more recently sought to become a partner in urban policy- and rule-making (e.g. Woolf 2016). This adjustment both reflects and further advances a broader transformation of international policy and regulatory landscapes over the past three decades, where top-down decision-making has been gradually losing ground to more experimental, "evidence-based" forms of (self-)regulation and policymaking, articulated in collaborative governance models that promote public-private partnerships and heterogeneous stakeholder networks (Peck et al. 2012). Assuming the more proactive and agenda-setting role of the urban "regulatory entrepreneur" (Pollman and Barry 2016), Airbnb aims to co-shape the terms of current and future policy debates pertaining not just to home sharing/short-term rental but also to the very fabric of city life, from tourism to housing and urban planning (Ferreri and Sanyal 2018; Gurran et al. 2018).

In the wake of neoliberal urbanism's gradual erosion of the "modern infrastructural ideal" that at least nominally promoted centrally governed universal services for city dwellers up until the late 1970s (Graham and Marvin 2001; Plantin et al. 2018: 300–301), the last decade has seen a "platformization of infrastructures" resulting in the formation of "complex platform-based ecosystem[s] encompassing private and public organizations and citizens" (Plantin et al. 2018; Van Der Graaf and Ballon 2019: 364). This, then, is where neoliberal urbanism reproduces itself as "platform urbanism", a condition "whereby platform-based business models ensure the generation of urban data largely takes place within proprietary data ecosystems" (Barns 2017: n.p.). It is also where Pasquale's (2017) distinction between territorial and functional sovereignty becomes instructive: by leveraging proprietary urban data and information asymmetries, companies like Airbnb exercise an infrastructural type of power that grants them increasing control over particular "functional arenas" (e.g. tourism, housing, urban planning) traditionally governed by state actors whose territorial sovereignty is punctuated (and might become supplanted) by border-crossing yet new boundary-setting platforms.[1] Consequently, this provokes a reassessment of the rights and responsibilities concomitant to a platform-mediated mode of urban citizenship in an era of ongoing welfare retrenchment. I argue that Airbnb pursues its own mode of platform urbanism by strategically mobilizing its valuable data assets as well as its "host community", which it frames as a collective of entrepreneurial citizens looking to supplement their income in a climate of economic insecurity and tech-enabled opportunity. I take the Airbnb Citizen initiative as paradigmatic of the company's efforts to establish itself as a thought leader in local as well as international public debates and policy circles. Part "advocacy channel", part public relations initiative, Airbnb Citizen has been responsible for organizing so-called Home Sharing Clubs around the world, producing Economic Impact reports, sharing Airbnb community data, and releasing a Policy Tool Chest as "a resource for governments to consider as they draft or amend rules for home sharing" (Airbnb Citizen 2016a).

The argument is elaborated in three parts. Part one addresses the rise of what has been termed "regulatory entrepreneurship", especially among tech companies that are taking advantage of the new collaborative opportunities provided by urban governance policies. It subsequently introduces Airbnb Citizen and the company's Policy Tool Chest as primary examples of this development. Part two investigates how Airbnb manages and frames its relationship to its user base, zooming in on the role and position of Airbnb hosts. Finally, part three examines the Airbnb Citizen figure promoted in the company's carefully orchestrated social imaginary, questioning the distribution of opportunities and risks related to platform-mediated citizenship in cities faced with the impacts of regulatory devolution and rising platform power.

The ascendency of regulatory entrepreneurship

Over the past three decades, global policy and regulatory milieus have experienced a broad shift from a paradigm of centralized government to one characterized by distributed *governance* (Jessop 1997, 2016). The ostensible strength of this new governance model is that it transcends the "false dilemma" of choosing between centralized regulation and deregulatory devolution, while also suggesting that – at least in principle – economic efficiency and democratic legitimacy can be mutually enforcing (Lobel 2016: 263–264). Governance blurs the lines between public and private, local and global, exposing a variety of new actors ("stakeholders") to the responsibilities and risks of government while streamlining its methods and outcomes so that they are translatable across time and space. Before I detail how Airbnb claims to make good on the promise of governance to establish economic efficiency and democratic legitimacy as two mutually enforcing policy goals, it is necessary to first have a brief look at the rise of what has come to be known as "regulatory entrepreneurship". According to Pollman and Barry, a regulatory entrepreneur is a company for which changing the law forms "a material part of its business plan" (2016: 387). Whereas corporate lobbying is generally a reactive venture, one that tries to protect existing interests against adverse regulations, regulatory entrepreneurship is a more proactive affair with higher stakes: without changing certain laws the company would not be able to operate – at least not legally (ibid.: 392–3). Furthermore, contemporary regulatory entrepreneurs generally "make an issue as salient as possible, rally the public to their cause, then use their popular support as leverage to win the change they want from resistant officials" (ibid.: 387).

Unsurprisingly, then, many of today's most prominent regulatory entrepreneurs hail from Silicon Valley. Although tech firms have become rather infamous for engaging in regulatory arbitrage, or the practice of taking advantage of regulatory loopholes, and even ignoring local laws altogether, they have more recently taken it upon themselves to engage lawmakers and other public officials with the express goal of *transforming* the regulatory landscape to meet their bottom lines. As such, Uber and Airbnb are spearheading a wave of so-called "reformer startups" that are taking advantage of the now hegemonic governance paradigm, not just by publicly challenging legislative institutions but also by pursuing collaborative partnerships with regulators and policymakers in need of expertise and data – thus becoming valued institutional actors in their own right (Pollman 2017). Regulators increasingly stimulate technology-driven innovation by allowing startups to experiment in so-called "regulatory sandboxes" where they can test their products/services under the (often informal) guidance of regulators that grant them temporary, conditional exemptions from standard regulations – often in exchange for concessions to share data (Pollman 2017: 19–20). From a reformer startup's perspective, the primary goal of such a "regulatory hack" is to become a dependable liaison in both local and global public policy networks, which in many cases requires the creation of a standardized and modular "toolkit" – or a set of policy-oriented strategies, activities, and documents that can be quickly mobilized in different regulatory situations and environments (ibid.).[2]

Airbnb's toolkit has been at least five years in the making. Its beginnings can be traced back to Airbnb's Community Compact, an initial set of public policy principles, released in 2015, proffering three broad commitments to city governments: paying a "fair share" of hotel and tourist taxes; building an "open and transparent community" by providing cities with anonymized data on listings that can inform policy decisions; and promoting "responsible home sharing to make cities stronger" by educating Airbnb's user community in order to develop and legitimize forms of self-regulation – frequently monitored in the context of local public-private partnerships and agreements (Airbnb Citizen 2015a). Released a full year before the official launch of Airbnb Citizen, the Community Compact is an early illustration of how Airbnb has been coming to terms with its rapid growth and expanding role as an institutional actor in policy and regulatory circles. Airbnb Citizen, previously called Airbnb Action, was inaugurated in November 2016 to function not just as "a tool for advocacy" but as "a more expansive site for the growing home sharing movement, offering resources for staying informed and taking action" (Airbnb Citizen 2016b). Presenting the improved initiative as "home sharing's new home", Airbnb emphasized its commitment to fostering this "movement" – ostensibly a grassroots phenomenon – in order to achieve positive change worldwide.

Since then, the company has followed through on its commitments by establishing (provisional) partnerships, memoranda of understanding (MOUs), and other agreements in all three areas of focus, which resulted in the publication of its Policy Tool Chest (in December 2016) as a resource that encapsulates "insights gained, lessons learned, and policy options developed through these hundreds of collaborations across five continents" (Airbnb Citizen 2016a). Airbnb's Policy Tool Chest offers four thematic sets of policy tools, also called "options", that elaborate on its earlier Community Compact and are intended to help officials "at every level of government" craft "fair and progressive" – or "modern" – rules for home sharing (ibid.). By formulating concrete, road-tested suggestions and recommendations with respect to tax collection, neighborhood responsibility, accountability and cooperative rule-making, and anonymized data-sharing, the Tool Chest essentially articulates a number of best practices which proactively shape home sharing as a practice in need of "smart policymaking" that can unlock its problem-solving potential. Importantly, Airbnb is careful to note that its Tool Chest should not be seen as "a one-tool-fits-all policy prescription or model legislation", but rather constitutes an "adaptable framework" that can be tailored to the needs of specific local jurisdictions (ibid.; cf. Peck et al. 2012).

The main wager that Airbnb's Policy Tool Chest aims to communicate is that *platform-facilitated home sharing markets form the solution to a plethora of problems faced by cities and their inhabitants.* Its story, perfectly summarized in the document's cover note written by Chris Lehane – Airbnb's Global Head of Policy and Public Affairs and former Obama Administration official – is one of economic *and* civic empowerment (whereby the former fuels the latter) through tech-enabled, decentralized market-making. Besides "economic opportunity", the key term here is "democratization": Airbnb ostensibly "democratizes capitalism" by "empowering people to use their homes to earn extra income" and thereby "fostering entrepreneurship"; it allegedly also "democratizes travel" by giving "more people and more communities the opportunity to benefit from tourism's growth"; finally, it "democratizes revenue" by generating "new tax revenue that governments can dedicate to existing critical services" or use to invest in Airbnb-assisted "new programs" that address local social challenges (ibid.). In other words, "everyone can win" – individuals, households, neighborhoods, communities, and cities are all empowered to leverage the Airbnb platform, which seeks to "align their interests in creating economic opportunity" (ibid.). For individuals and households, the platform provides easy – yet highly ordered and carefully delineated – access to home sharing markets by ostensibly lowering entry barriers and transaction costs. Meanwhile, cities can use the platform's data-rich compliance and enforcement tools not only to collect more taxes but also to expedite and automate zoning and land use regulations, while promoting "sustainable tourism" by spreading tourist flows outward from traditionally popular areas in city centers to peripheral neighborhoods whose communities are seeking economic revitalization. This, then, is the "social value proposition of home sharing as an economic solution" (Airbnb Citizen 2016a), and this is how Airbnb claims to make good on the promise of governance to establish economic efficiency and democratic legitimacy as two mutually enforcing policy goals: by reconciling everyday entrepreneurialism with a sense of civic purpose and conflating private and public interests, *both of which are premised on the marketplace.* As the next section will discuss, such a conflation works in Airbnb's strategic advantage, to the extent that it allows the company to identify itself with its users – the "demos" in whose name the public good has traditionally been defended.

A parallax view

Public-private partnerships are ideal arrangements in the governance paradigm, but when regulators are less inclined to cooperate and make exceptions (which tend to become new rules) because they have doubts regarding Airbnb's "everyone wins" narrative, the company has demonstrated that it is exceptionally well equipped to take a more antagonistic approach to its regulatory entrepreneurship.[3] Given that its main product is a digital platform that enables commercial transactions between users who in most cases derive value from their exchanges, Airbnb possesses an unprecedented capacity to mobilize this user base – particularly its "hosts" – as a scalable political force. Crucially, many Airbnb hosts are committed to changing legislation in order to maintain their business, which depends fully on the success and legality of Airbnb's platform business model. Moreover, this dependency is clearly mutual, given that Airbnb's business model can only succeed as long as hosts are willing and able to "share" their home on the platform. What I am getting at here is that the interests and objectives of Airbnb can be seen to converge with those of its entrepreneurial hosts: in the cosmology of the "sharing economy" it is markets all the way down. Consequently, it becomes trickier to discern what we are dealing with: is Airbnb a business engaging in regulatory entrepreneurship by instrumentalizing its user base to fight for its cause, or is it a platform facilitating a grassroots movement that fights for *its own* cause, which happens to be structurally aligned with Airbnb's cause? The short answer is that it depends on one's vantage point, or where one's analysis is situated. Put differently, we're confronted with a parallax view. To illustrate how this view's focus is trained by Airbnb, we should have a look at how it frames its relation to what it calls Home Sharing Clubs.

Still basking in the afterglow of the company's recent victory in San Francisco, where it successfully mobilized its user base to defeat the controversial "Proposition F" that would have restricted short-term home rental in the city, the then newly appointed Chris Lehane made the following boastful declaration in a blog post dated November 5, 2015 (six days before the publication of Airbnb's Community Compact):

> [...] this election may have been the first time that the broader public understood what we already knew – Airbnb hosts and guests are not just a community, they represent a people-to-people movement that is getting stronger as the days grow longer.
>
> *(Airbnb Citizen 2015b)*

To cultivate this budding movement, Airbnb had announced a day prior that it would "support the creation of 100 *independent* Home Sharing Clubs in 100 cities around the world in 2016" (ibid.; emphasis mine). Notice the tension here between the existence of Home Sharing Clubs as putatively "independent" organizations whose formation can be labeled "grassroots", and the extensive support structure the company is rolling out to organize, train, educate, advise, and support these Clubs. The platform idiom, however, helps to alleviate this tension: as a platform, Airbnb aims to become an increasingly central part of global urban infrastructures – a data-intensive operating system on which other services can run. One significant quality of infrastructures is that they tend to become invisible over time, as their invisibility is generally proportional to their ubiquity.

Applied to Home Sharing Clubs, this means that Airbnb's role in their expansion and development should be similarly understood as infrastructural, insofar as its support is – ideally – both ubiquitous and invisible.[4] This allows Airbnb to position itself as a "people-to-people platform – of the people, by the people and for the people" (Airbnb Citizen n.d.), while simultaneously minimizing its presence as a business corporation. The source of all the activist energy and agency is, from this perspective, located not in the company's San Francisco headquarters, in Silicon Valley, or even in Washington D.C., but originates from individuals and households assembled in local Home Sharing Clubs that "advocate for fair and clear home sharing regulations in their city, share best practices around hosting and hospitality, organize community service activities, and can serve as a forum to connect those who share a passion for home sharing".[5] In this sense, and again from this perspective, Home Sharing Clubs are vital examples of civil society's "institutional core constituted by voluntary associations outside of the state and the economy" (Flyvbjerg 1998: 210). Airbnb, in turn, operates as a platform that facilitates and optimizes such voluntary associations, while its own institutional logic likewise is "not reducible to those of states or markets" (Bratton 2016: 41).

Home Sharing Clubs share Airbnb's ambiguous political identity to the extent that they are, like civil society at large, neither fully public nor completely private entities. In the Habermasian view promulgated by Airbnb, they are rather composed of private individuals coming together to contribute to a public sphere that forms a counter to state and market powers. More specifically, Home Sharing Clubs agitate against repressive local regulations that stifle innovation and redistributive economic growth, while at the same time speaking truth to the power of industry incumbents (i.e. the hotel lobby). Yet what if we adopt a Foucaultian perspective here? For Foucault, as is well known, civil society is an arrangement of liberal governmentality in which *homo œconomicus* can be "appropriately managed" (Foucault 2008: 296). Even though a central feature of civil society is that it forms "a spontaneous synthesis within which the economic bond finds its place" (ibid.: 303), and it thus operates structurally like a marketplace that obviates the need for a social contract, it also exceeds the realm of merely economic interests and transcends the egoism of "economic men" that would otherwise threaten to fissure the social fabric. Consequently, then, "if it is true that civil society is already there, that it ensures its own synthesis, and that it has a sort of internal governmentality" guided by communitarian values and sentiments, it is reasonable to ask: "Does civil society really need a government?" (ibid.: 310). Can it not better govern itself, through a pragmatic mix of economic and moral calculations/relations?

I argue that this is exactly the political question posed by Airbnb's Home Sharing Clubs, at an opportune moment when neoliberal urbanism – through the rising governance paradigm – increasingly seeks "the recruitment of civil society to serve its objectives" (Lazzarato 2009: 111), which include the progressive downloading of public responsibilities and risks to private "partners". The so-called "disruptive innovation" introduced by Airbnb is to provide a digital platform that makes this process not only more "frictionless" but also turns it into a profitable enterprise for its users, whose interests are now represented locally Home Sharing Clubs. These Clubs represent a civil society movement that not just advocates for the *right* to rent out private homes – i.e., appealing to the "juridical structure of power" (Foucault 2008: 304) – but also supposedly democratizes capitalism from the inside out and thus obviates the need for government interference beyond ensuring the optimal conditions for platform-mediated market-making and self-regulation. The ambiguous political identity of Home Sharing Clubs – neither fully public nor completely private – hence does not so much derive from how these ostensibly autonomous associations form a counter to state and market powers, as the Habermasian view would have it, but is instead predicated on *the way they question the very legitimacy of the regulatory state through the gospel of redistributive market power.* This gospel, which posits the Airbnb platform as a formally neutral technical system that merely facilitates such redistribution, is carefully orchestrated by Airbnb itself. To properly grasp this, however, a gestalt switch is required – one that brings the company's institutional power back into clear focus.

After a decade of development and expansion, it is both timely and urgent to critically examine how Airbnb's institutional power is articulated into something that could be called a "political program". Importantly, following Bratton's insights, such a program "is not only to be found in the legal consensus (or dissensus) and policy admonitions of traditional 'politics' but also in machines directly" (Bratton 2016: 44). What this means is that Airbnb's politics are *intrinsic to and indissociable from* the operations of its platform, whose purportedly empowering and redistributive affordances reconfigure and expedite the political program of market-based risk- and responsibility-sharing central to neoliberal urbanism and its attendant governance paradigm. Airbnb's platform-orchestrated "deep capture"[6] campaigns are therefore deeply sociotechnical exploits, which is what gives them their extraordinary power.

By setting the terms of participation in areas such as short-term property rental, user-driven policy advocacy, and urban tourism markets according to protocols (from API documentation to commercial partnership and user agreements) that are less fixed than *fixing* insofar as they remain adjustable to meet changing needs and requirements, Airbnb actively contributes to the recomposition of urban polities. In cities around the world, the company participates in the ongoing reorganization of relations between market, state, and civil society actors, which have seen their traditional roles transform under the aegis of the governance paradigm. As the public good and private commercial interests become rhetorically conflated and materially entangled in Airbnb's political program, public officials are increasingly struggling with the question of who, exactly, benefits from (a lack of or a change in) regulation. Meanwhile, as more people join the platform and aspire to become Airbnb Citizens, and as this self-reliant figure comes to subtly express a kind of model citizenship in post-welfare times, we should ask, with Barns (2017: n.p.), "whether the urban spaces of technology-enabled citizenship today orient us toward risks associated with vertical integration, as much, if not more so than heralding the disruptive possibilities of a participatory public sphere?" So what *are* the risks and opportunities of platform-mediated and increasingly data-dependent citizenship in cities faced with the impact of regulatory devolution and responsibilization? This will be the topic of the final section.

A house is not a home

When, back in July 2014, Airbnb's CEO Brian Chesky revealed the company's new slogan and thereby inaugurated its rebranding strategy in a Medium blog post, he made the following distinction between houses and homes:

> For so long, people thought Airbnb was about renting houses. But really, we're about home. You see, a house is just a space, but a home is where you belong. And what makes this global community so special is that for the very first time, you can belong anywhere.
>
> *(Chesky 2014)*

In contrast to short-term housing rental, which consists of nothing more than discrete economic transactions, home sharing overflows the economic by creating a sense of global community where anyone (who can afford it) can belong anywhere. The transnational civil society imagined by Airbnb thus reimagines cosmopolitanism by countering its traditionally "thin conception of social life, commitment, and belonging", instead offering an account of how interactions between hosts and guests might become "the basis for active citizenship" (Calhoun 2002: 878–879; cf. Roelofsen and Minca 2018). Nevertheless, as civil society is an ensemble in which the economic bond always finds its place (Foucault 2008: 303), the forms of active citizenship and solidarity promoted by Airbnb Citizen are articulated with a celebration of economic empowerment and entrepreneurialism. So who is the quintessential Airbnb Citizen? Although it is essentially a dual figure composed of the "host" and the "guest" who together shape its community marketplace, the host's side of the equation has taken precedence ever since Airbnb's policy priorities have discursively shifted the pendulum from its guests' desire to belong anywhere to the needs of middle-class households leveraging its platform to make ends meet. The company's "economic empowerment agenda" is clear about how its platform serves as a de facto social safety net:

> Our people-for-people platform allows ordinary people to use their house – typically their greatest expense – to generate supplemental income to pay for costs like food, rent, and education for their children. [...] While governments are debating the best way to support groups such as seniors and the middle class, Airbnb is generating real money for families right now.
>
> *(Airbnb Citizen 2017)*

While a house is not a home, a home evidently still remains a house – a significant expense that, with prudent policy and regulation, can also be turned into an asset that finances the rising costs of social reproduction.[7] I therefore argue that Airbnb's model citizen is the middle-class homeowner seeking to provide for his/her family, a subject whose contours become sharply visible when positioned against a background of neoliberal welfare reform. What I have been calling "post-welfare societies" are the product of concerted efforts to restructure – and roll back – the welfare state as some used to know it and substitute a regime of privatized "asset-based welfare" for the public provision of social security (Doling and Ronald 2010). As a result of enduring state-mandated campaigns to increase homeownership rates in North America and Northwestern Europe and an attendant rise in housing property values that was only temporarily disrupted by the 2007–08 mortgage crisis and the subsequent recession, one favored speculative asset has been the home. Even in the aftermath of the crisis, during which so many households were forced to default on their mortgages, it remained a common practice to treat one's home as an investment that can generate returns in the form of access to welfare goods such as pension plans, education, and childcare facilities, which can be purchased by tapping into housing wealth.

Importantly, this is not only a matter of drawing on the home as a financial asset but also entails exploiting housing property as a consumption asset, for instance by sub-letting a room in order to pay the bills (Jarvis 2008). Airbnb has effectively scaled and standardized such practices, while retroactively positioning itself as a beacon of entrepreneurial opportunity amid a post-crisis climate of economic insecurity. In other words, the platform company derives its institutional power not just from how it takes advantage of the regulatory devolution accompanying the governance paradigm, but also from the way it manages to capitalize on emerging housing-based welfare regimes. Its professed public service to post-welfare societies is the provision of a platform on which home-owning households – that familial unit of "active welfare subjects" now rebranded as Airbnb Citizens (Jarvis 2008: 217) – can safely optimize the monetization of their "underutilized" domestic assets. In this way, it offers the operating system for "a fundamental reworking of social relations of property" (Stabrowski 2017: 328), by normalizing and intensifying household practices of financial calculation, competition, and (micro-)enterprise and thereby shoring up the popular appeal of asset-based welfare (cf. Allon 2011).

Yet, again, we should be careful not to take Airbnb's account of itself at face value. What disrupts the company's narrative of middle-class economic empowerment, which has increasingly been driven by its own Economic Impact reports, are recurring claims that a large part of the company's revenue is generated by a relatively small share of high-volume hosts – so-called "multi-listers" who own and market multiple properties (Cox and Slee 2016). Despite its repudiation of critical reports and its claim that its "One Host, One Home" policy should take care of this issue,[8] Airbnb has been confronted by independent studies that demonstrate the company's complicity in the ongoing gentrification of cities, by taking long-term housing stock off the market and consequently driving up house prices and rents (e.g. Wachsmuth and Weisler 2018; Wachsmuth et al. 2018). Multi-listers, often not your typical homeowner but rather so-called "enterprise customers" such as commercial property managers, are the disavowed but logical outcome not just of Airbnb's business model and platform evolution (cf. Helmond et al. 2019; Hein et al. 2018) but also of its entrepreneurial citizenship imaginary: after all, capital begets capital. While successful hosts could, in this imaginary, conceivably reinvest their earnings in a new property, it is more likely that multi-listers operate on behalf of real estate agents with access to (institutional) investors looking to purchase portfolios of housing properties. Professionally managed properties with high turnover rates – like those promoted through Airbnb Plus[9] – are economically much more appealing to Airbnb than homes that are intermittently and/or partly rented out, which is why the company has been expanding its platform infrastructure and API-based tools that allow businesses to "optimize listings at scale with professional features".[10] In terms of scalability and revenue, then, a house is most certainly not a home. By developing its platform-centric ecosystem to the advantage of enterprise customers who are thereby empowered to monetize their extensive housing assets, Airbnb can indeed be said to contribute to rising wealth inequalities in cities across the globe. In this sense, and in true neoliberal fashion, the actual Airbnb Citizen lifted up by the platform is not the middle-class homeowner but the business corporation.

Meanwhile, the risks of Airbnb's platform-mediated citizenship are largely borne by precisely those groups the company champions in its Airbnb Citizen campaigns: middle- and working-class families, women, and people of color, all seeking to – as Airbnb's Economic Empowerment Agenda phrases it – "stay afloat during tough times" (Airbnb Citizen 2017) In these tough times, which continue to be tough for many low-income families in spite of proclamations that we've entered a post-recession era of economic growth, Airbnb can indeed form a vital lifeline on which households grow to depend for supplemental income. Over time, this dependence is likely to increase due to the process of "generative entrenchment" that gives platforms their unique power to set norms and standards in an expanding array of settings. As Bratton explains, generative entrenchment is a mechanism

> by which one platform's early consolidation of systems (formats, protocols, and interfaces) decreases a *User*'s opportunity costs to invest more and more interactions into that particular platform, while it increases the costs to translate earlier investments into another platform's (at least partially) incompatible systems.
>
> *(2016: 47)*

It is a process that exposes both the opportunities and the risks of Airbnb's vertically integrated platform, to the extent that it does not only provide the means through which hosts can generate income, enhance their profile and reputation, and engage with their peers, but at the same time makes it increasingly costly to deviate from the platform's standards and rules or to transition out of the system altogether (Leoni and Parker 2018).

What happens when Airbnb decides to make (sudden) changes to its Terms of Service agreement, its hosting standards, or the way it algorithmically determines the visibility of its listings? The potential impact of such moves would surely be most severe among segments of the host community whose economic survival is structurally tied to the platform. Although a growing body of scholarship documents how Airbnb's interface design can facilitate and potentially counter discrimination as a peril affecting both hosts and guests on the platform (e.g. Edelman et al. 2017), there is a need for more critical research on the risks attendant to the structural power imbalance between the platform and its user base. Despite Airbnb's rhetoric of democratization, it offers its users no substantive mechanism for collective bargaining or decision-making: Airbnb Citizens may be economically supported by the platform (some more than others) but within its domain they have no political representation, just varying levels of privilege predicated on differential access to a set of data-driven tools. This ultimately makes Airbnb a risky platform for micro-entrepreneurial citizens, especially the more precarious among them. Will Airbnb use its technological and institutional power in the service of its most vulnerable citizens or will it further re-set the terms of participation to their disadvantage? As we debate emerging types of platform urbanism, it is of crucial importance to keep in mind that platforms like Airbnb are not just exploring the soft power exercised through urban policy and governance but also experiment with new and still dimly registered forms of sovereign power that secure them a lasting grip on the socio-material fabric of contemporary cities and their households.

Notes

1 According to Plantin et al. (2018: p. 294, 296, 306), some main qualities, or characteristics, of infrastructures include ubiquity, reliability, invisibility, operating as gateways, generating dependency, embeddedness, durability, and extensibility. After a full decade of global growth and extensive market capitalization (Sherwood 2019), Airbnb has managed to acquire many of these infrastructural qualities in its largest markets (including London, Paris, and New York city) and will likely develop these

features in many other urban regions where its presence still has room for growth. While Airbnb's ubiquity, reliability, embeddedness, extensibility, and its propensity to operate as a gateway that generates dependencies have so far primarily impacted urban tourism and hospitality markets, the company's strategy of "negotiating [its] 'programmability' toward specific stakeholder groups" (Helmond et al. 2019: p. 125) and deploying boundary-crossing partnership programs are also affecting housing and urban planning policy in many cities (e.g. Gurran et al. 2018). It is in these domains that, as this paper argues, Airbnb seeks to becoming increasingly infrastructural.

2 For a definition of "regulatory hacking", see the website of 1776, which presents itself as "the Northeast Corridor's largest network of entrepreneurial incubators": https://www.1776.vc/regulatory-hacking/ (accessed November 13, 2018).

3 Airbnb has faced fierce and tenacious public resistance against its operations in some crucial markets, such as New York City, Berlin, and Barcelona. While some of these initiatives were funded by the hotel lobby, many protests have been independently organized by housing rights advocacy groups and citizens concerned about the gentrification and livability of their neighborhoods. It should thus be noted that Airbnb's success as a regulatory entrepreneur is globally uneven and partial, also taking into account the fact that in many other cities around the world there has so far been little necessity for the company to proactively engage in public policy at all (given that these markets are still too marginal or because there are few available options for institutional engagement).

4 Airbnb currently hosts all its support resources and peer communication tools in its platform's Community Center: https://community.withairbnb.com/t5/Community-Center/ct-p/community-center.

5 This quote is taken from a Frequently Asked Questions document made available via Airbnb's Community Center: https://community.withairbnb.com/html/assets/ClubsFAQ.pdf (accessed November 13 2018).

6 Corporate deep capture campaigns are "replete with astro-turf organizing, the maintenance of front groups, and the sponsorship of knowledge production" (Yosifon 2006: p. 598), all with the objective of capturing the hearts and minds of the public at large – influencing collective dispositions, structures of feeling, and modes of reasoning.

7 As Foucault (2008: p. 148) – ventriloquizing the German Ordoliberals – asks: "What is a house if not an enterprise?"

8 Airbnb's "One Host, One Home" policy was initiated in 2016 as a token of good will toward local regulators and formalizes the company's commitment "to removing commercial operators from the platform", particularly multi-listers who signed up with the platform after November 1st 2016 (Airbnb Citizen 2016c).

9 See http://airbnb.com/plus.

10 See https://www.airbnb.com/b/host_pro.

References

Airbnb Citizen (2015a) The Airbnb community compact. *Blog post*, 11 November. Available at: https://web.archive.org/web/20180523053349/https://www.airbnbcitizen.com/the-airbnbcommunity-compact/.

Airbnb Citizen (2015b) Organizing in 100 cities: The Airbnb host movement. *Blog post*, 5 November. Available at: https://web.archive.org/web/20190202152622/https://www.airbnbcitizen.com/organizing-in-100-cities-the-airbnb-host-movement/.

Airbnb Citizen (2016a) Airbnb policy tool chest. Available at: https://www.airbnbcitizen.com/airbnb-policy-tool-chest/.

Airbnb Citizen (2016b) Meet Airbnb citizen. *Blog post*, 8 November.

Airbnb Citizen (2016c) One host, one home: San Francisco. *Blog post*, 24 October. Available at: https://web.archive.org/web/20190106181648/https://www.airbnbcitizen.com/one-host-onehome-san-francisco/.

Airbnb Citizen (2017) Introducing the Airbnb economic empowerment agenda. Blog post, 13 March. Available at: https://web.archive.org/web/20190622102015/https://www.airbnbcitizen.com/introducing-airbnb-economic-empowerment-agenda/.
Airbnb Citizen (no date) Community compact. Available at: https://www.airbnbcitizen.com/wp-content/uploads/2015/11/Airbnb-Community-Compact.pdf.
Allon F (2011) 'Home economics': The management of the household as an enterprise. *Journal of Australian Political Economy* 68: pp. 128–148.
Barns S (2017) Visions of urban informatics: From proximate futures to data-driven urbanism. *Fibreculture Journal* 29. https://doi.org/10.15307/fcj.29.204.2017
Bratton BH (2016) *The Stack: On Software and Sovereignty*. Cambridge, MA: MIT press.
Calhoun C (2002) The class consciousness of frequent travelers: Toward a critique of actually existing cosmopolitanism. *The South Atlantic Quarterly* 101(4): pp. 869–897.
Chesky B (2014) Belong anywhere. Blog post, 16 July. Available at: https://medium.com/@bchesky/belong-anywhere-ccf42702d010.
Doling J and Ronald R (2010) Home ownership and asset-based welfare. *Journal of Housing and the Built Environment* 25(2): pp. 165–173.
Edelman B, Luca M and Svirsky D (2017) Racial discrimination in the sharing economy: Evidence from a field experiment. *American Economic Journal: Applied Economics* 9(2): pp. 1–22.
Ferreri M and Sanyal R (2018) Platform economies and urban planning: Airbnb and regulated deregulation in London. *Urban Studies* 55(15): pp. 3353–3368.
Flyvbjerg B (1998) Habermas and Foucault: Thinkers for civil society? *British Journal of Sociology* 49(2): pp. 210–233.
Foucault M (2008) *The Birth of Biopolitics: Lectures at the Collège de France, 1978–1979*. Basingstoke, UK: Palgave Macmillan.
Graham S and Marvin S (2001) *Splintering Urbanism: Networked Infrastructures, Technological Mobilities and the Urban Condition*. London: Routledge.
Gurran N, Searle G and Phibbs B (2018) Urban planning in the age of Airbnb: Coase, property rights, and spatial regulation. *Urban Policy and Research* 36(4): pp. 399–416.
Hein A, Böhm M and Krcmar H (2018) Tight and loose coupling in evolving platform ecosystems: The cases of Airbnb and Uber. In Abramowicz W and Paschke A, *Proceedings from the 21st Business Information Systems International Conference*, pp. 295–306.
Helmond A (2015) The platformization of the web: Making web data platform ready. *Social Media + Society* 1(2). doi:10.1177/2056305115603080.
Helmond A, Nieborg DB and Van der Vlist FN (2019) Facebook's evolution: Development of a platform-as-infrastructure. *Internet Histories* 3(2): pp. 123–146.
Jarvis H (2008) 'Doing deals on the house' in a 'post-welfare' society: Evidence of micro-market practices from Britain and the USA. *Housing Studies* 23(2): pp. 213–231.
Jessop B (1997) Capitalism and its future: Remarks on regulation, government and governance. *Review of International Political Economy* 4(3): pp. 561–581.
Jessop B (2016) *The State: Past, Present, Future*. Cambridge: Polity Press.
Kendall M (2016) Airbnb fights unfriendly regulations with wave of lawsuits against San Francisco, other cities. *Mercury News*, September 18. Available at: https://www.mercurynews.com/2016/09/18/airbnb-fights-unfriendly-regulations-wave-lawsuits-san-francisco/.
Lazzarato M (2009) Neoliberalism in action: Inequality, insecurity and the reconstitution of the social. *Theory, Culture & Society* 26(6): pp. 109–133.
Leoni G and Parker LD (2018) Governance and control of sharing economy platforms: Hosting on Airbnb. *The British Accounting Review*. doi:10.1016/j.bar.2018.12.001.
Lobel O (2016) The law of the platform. *Minnesota Law Review* 101: pp. 87–166.

North DC (1991) Institutions. *Journal of Economic Perspectives* 5(1): pp. 97–112.
Pasquale F (2017) From territorial to functional sovereignty: The case of Amazon. Law and Political Economy, December 6. Available at: https://lpeblog.org/2017/12/06/from-territorial-to-functional-sovereignty-the-case-of-amazon/
Peck J, Theodore N and Brenner N (2012) Neoliberalism resurgent? Market rule after the great recession. *South Atlantic Quarterly* 111(2): pp. 265–288.
Plantin JC, Lagoze C, Edwards PN and Sandvig C (2018) Infrastructure studies meet platform studies in the age of Google and Facebook. *New Media & Society* 20(1): pp. 293–310.
Pollman E (2017) The rise of regulatory affairs in innovative startups. Loyola Law School Legal Studies Research Paper No. 2016–43. Available at: https://ssrn.com/abstract=2880818.
Pollman E and Barry JM (2016) Regulatory entrepreneurship. *Southern Californian Law Review* 90: pp. 383–448.
Roelofsen M and Minca C (2018) The superhost. Biopolitics, home and community in the Airbnb dream-world of global hospitality. *Geoforum* 91: pp. 170–181.
Sherwood H (2019) How Airbnb took over the world. *The Guardian*, May 5. Available at: https://www.theguardian.com/technology/2019/may/05/airbnb-homelessness-renting-housing-accommodation-social-policy-cities-travel-leisure.
Stabrowski F (2017) 'People as businesses': Airbnb and Urban micro-entrepreneurialism in New York City. *Cambridge Journal of Regions, Economy and Society* 10(2): pp. 327–347.
Van der Graaf S and Ballon P (2019) Navigating platform urbanism. *Technological Forecasting and Social Change* 142: pp. 364–372.
Wachsmuth D, Chaney D, Kerrigan D, Shillolo A and Basalaev-Binder R (2018) The high cost of short-term rentals in New York City. Report from the Urban Politics and Governance Research Group, McGill University, January 20. Available at: https://www.mcgill.ca/newsroom/channels/news/high-cost-short-term-rentals-new-york-city-284310.
Wachsmuth D and Weisler A (2018) Airbnb and the rent gap: Gentrification through the sharing economy. *Environment and Planning A: Economy and Space* 50(6): pp. 1147–1170.
Woolf N (2016) Airbnb regulation deal with London and Amsterdam marks dramatic policy shift. *The Guardian*, December 3. Available at: https://www.theguardian.com/technology/2016/dec/03/airbnb-regulation-london-amsterdam-housing.
Yosifon D (2006) Resisting deep capture: The commercial speech doctrine and junk-food advertising to children. *Loyola of Los Angeles Law Review* 39: pp. 507–601.

8

POLITICAL STRUGGLES IN THE PLATFORM ECONOMY

Understanding platform legitimation tactics

Luke Yates

Introduction

Platforms such as Airbnb, Uber and Deliveroo are packaged together using a rapidly shifting and energetically contested vocabulary, including the "sharing", "collaborative" and "platform" economy. Several of these terms, most notably the "sharing economy", proclaimed a new economic paradigm that linked these businesses with grassroots economic initiatives such as car-sharing apps, community energy cooperatives, Freecycle and alternative currencies. Combining the idealism of the grassroots initiatives and the proven scalability of the platform businesses, they would together address anxieties about the environment, the inequities of financialised capitalism, and stagnant growth. This position has gradually become untenable, but this change indicates that the ways in which society and academics understand urban platforms are developing as a result of conflicts and struggles among platforms, their competitors, states and social movements. These struggles have involved protests, subpoenas, occupations, lawsuits, petitions, blockades, fines from governments, strikes and boycotts, and there are also simultaneously discursive struggles over legitimacy.

This chapter explores the dynamics of these struggles that shape the transformation of the fields in which urban platforms operate. The focus of this piece is on introducing and theorising two key tactics through which platform businesses seek to legitimise their activity in public and to governments in a context of political pressures. In doing so the chapter makes a contribution towards the two fundamental debates in the literature: the problem of characterising the platform economy, and assessing its implications (Artioli 2018).

There are two sections. The first discusses directly the conflictual character of platforms. While discrete empirical case studies are hugely important, I argue for the importance of maintaining a view across different types of platforms in order to draw out common patterns in legitimacy and governance struggles that are affecting cities simultaneously. In particular, the tactics of platform businesses to legitimise themselves and place pressure on regulators are similar across businesses. The second section of the chapter then introduces two of these tactics of legitimation: one mainly discursive, the other mainly political. The first concerns how platform businesses deploy tropes about the future in their public communications. I show that platform businesses frame themselves as vectors of progress, a rhetorical tactic that suggests that criticism and regulation of them are reactionary. Platform businesses are invested in highly contentious visions of the future, but casting their current activity as monopolising "the future" suggests that existing problems with platform businesses should be overlooked because they represent progress and their consequences are still unknown. The second tactic of legitimation I discuss concerns grassroots lobbying approaches to public policy – where civil society pressures are created or co-opted to influence public opinion and regulation. I identify three broad approaches to this platform-based grassroots lobbying: the direct mobilisation of platform users, alliances with existing "grassroots" associations and the creation of front groups.

Political struggles in the platform economy

There are a range of obvious and less obvious conflicts and disputes in the platform economy. Presented as holding "significant potential to contribute to competitiveness and growth [...] to promote new employment opportunities, flexible working arrangements and new sources of income" (European Commission 2016: 2), the "collaborative economy" was supposed to address major social, economic and environmental challenges. Yet critics raise concerns about precarious forms of employment and labour rights abuses, non-regulated and untaxed consumption, the acceleration of problematic dynamics in cities such as gentrification and congestion, and the advancement of private economic interests at the expense of public (Slee 2016, Srnicek 2016, Rosenblat 2018).

Conflicts around platforms are now becoming well known and take various forms and trajectories. Protests, occupations, petitions and blockades by those affected by platform-mediated short-term rentals and platform-based taxi services, and strikes by food delivery drivers, are common (Aglionby and Davies 2014, Booth 2015, Coldwell 2017, Thelen 2018, Aguilera et al. 2019, Crisp 2018, Cant 2018). Alternative platforms or initiatives have emerged, including an "ethical" short-term rental service which seeks to involve local people in governing and benefiting from tourism (Fairbnb), and cooperative models of platform taxi or food delivery services (e.g. CoopCycle) who recognise working rights and distribute revenue among workers rather than shareholders. Several cities including Barcelona, Madrid, Amsterdam and Berlin have introduced legislation attempting to limit the impact of Airbnb listings on local residents (CEO 2018, Aguilera et al. 2019), and ride-sharing apps have had legislative pushback from administrations all over the world at the city, regional and national level (Collier et al. 2018, Thelen 2018, Rosenblat 2018). Flashpoints often obscure longer processes of contention, negotiation and lobbying in courts, city halls and conflicts in a range of other private and public arenas. The conflicts involve significant contributions by civil society actors as well as public institutions and businesses, and they have implications for how everyday life is organised in terms of housing, transportation, work and the provisioning of food and the way in which cities and urban space is governed. How should the conflicts be studied?

Conflicts in the platform economy are usually studied in relation to the fields that they concern. This is important because there are specific histories of labour struggles, movement organising around cities, and other areas where there is specific intellectual expertise. It also sidesteps the problem that the terms used to refer to the platform economy are difficult to pin down and, particularly in the case of the "sharing" or "collaborative" economy, are laden with morally charged assumptions and language which unduly influences how we understand them.

Yet the approach of this paper is to look across platform-based conflicts and to compare across contexts. The first reason for this is also the reason why I deliberately avoid using a definition of the platform economy in this chapter: that terminology is this sector is explicitly and unusually politicised, a phenomenon that must itself be studied. The positive case for platform businesses is often made specifically through reference to a catch-all term such as sharing or collaborative economy, for example that it is sustainable, is a source of new growth and jobs and value, it suits people's lifestyles, and it is more convivial than alternatives (see Botsman and Rogers 2010). Even though it makes no scientific sense to smooth over differences between for-profit and not-for-profit platforms, those with open versus closed data models, and across platforms where there are very different consequences and challenges, these terms and modes of categorising do political work often precisely through their obfuscating nature. Understanding the politics of the platform economy must acknowledge that the lumping together of disparate initiatives has an effect. This is demonstrated in part by the fact that despite widespread controversies surrounding businesses, nobody is against the sharing economy because it is so expansive, including, famously, things like the sharing of power drills among neighbours (Botsman and Rogers 2010). All platform economy businesses appeal to some core narratives, for example, perhaps most pervasively, about their providing a kind of solution, both for individuals and societies, to the fallout from the global economic crisis.

This relates to the second reason to look across platform conflicts. This is that the conflicts themselves, and the tactics of platforms and their opponents, exhibit similarities (see, for example, Collier et al. 2018, Thelen 2018, and Aguilera et al. 2019). Tactics of opponents to platform businesses opponents include challenging businesses' assertions about their benefits, setting up alternative urban platforms to those businesses, and using legal challenges (Collier et al. 2018: 929–930). Tactics of platform businesses also display similarities. These common approaches include discursive tactics, such as future-oriented tropes about innovation and progress; and political tactics, such as grassroots lobbying, which complements standard insider lobbying by creating civic pressure and influencing public debate in their favour. The article now moves to analyse these two specific tactics of platform businesses.

Tactics of platform businesses: platforms as self-styled vectors of progress, and platform-based grassroots lobbying

Platforms as self-declared vectors of progress

Much collective activity around platforms does not involve open conflict, but episodes of public disagreement emerge revealing competing economic and social visions, interests and possibilities that can be analysed and compared. One category among the several discursive tactics that are deployed by platforms publicly in media statements, speeches, websites, internal communications, and in public conversation with other actors, explicitly concerns the future.[1] A small literature looking at social movements, conflicts and the future highlights the strategic and performative significance of imagination in confrontations among collective actors (e.g. Brown 2016, Schultz 2016). This work and many key texts in social movement studies (e.g. Eyerman and Jamison 1991) highlight that movements produce and inspire new, insurgent and disruptive ideas, narratives about change that are strikingly ubiquitous. This particular rhetorical combination of disruption, insurgency and innovation has been borrowed by technology firms. Platform businesses associate their activity with stories of technological advance and tropes about innovation, inevitability and creative destruction that contrast their visionary approach with states and regulation, and which implicitly deny the existence of other possible futures. Shoshana Zuboff (2019) captures important dimensions of the implications of this tactic in her discussion of Silicon Valley's use of what she calls *messages of inevitabilism*. Zuboff points out that companies such as Google discuss their expansion as though there was no alternative, which "conveys the futility of opposition" (222), "protects power from challenge" (224) and "conceals the realpolitik of surveillance capitalism at work behind the scenes" (226). Platforms frame their activity using particular kinds of future visions that have a performative effect.

Appeals to the future by urban platforms such as Uber and Airbnb are also made in a pointedly confrontational register. Airbnb and Uber describe the legal, political and economic impediments to their operations as backward and outdated, with regulation and legal frameworks – implying the state per se – generally presented as unworkable, impracticable, fragmented, unresponsive and idiosyncratic. Arguments are made against regulation and legal frameworks in order to justify operating outside them. Uber's press statements, where it is threatened or restricted, regularly denounce the regulation or the local state initiating it as outdated or "backward" (e.g. Husser 2015). The CEO of Airbnb Brian Chesky (2014) explained controversy in New York around Airbnb in a way that demonstrated the important work that tropes around the future already played in legitimising the platform.

I understand why there are a lot of misconceptions. Airbnb is a really, really new idea. And I think we spent the last 15 or 20 years basically disrupting or changing the way content, bits, bytes, were suddenly consumed, interact, changed. But what happens when the internet started moving into your neighbourhood? With Airbnb, Uber, what happens when that actually happens?

> It starts to – well for some people they love it, but for other people they were very scared [...] One thing I try to tell people is before you try to regulate something, just try to learn about it. [...] Thomas Jefferson said, in 1812, "Laws and institutions must go hand in hand with the progress of the human mind. Institutions must advance also, and keep pace with the time..." If the government does not continue to regulate and update their law, and they're regulating 21st-century businesses with 20th-century laws, we could be in trouble.

Chesky explains criticism of Airbnb as arising from an irrational and emotional reaction. He equates criticisms with reactionary attitudes towards the internet, which is typically seen as having had an irreversible and largely positive impact on society. Platforms are businesses that are not just innovative, but they also represent a rising tide of social and intellectual progress. Chesky frames regulation as being antithetical to progress, and even suggests, in choosing to quote Thomas Jefferson, that opposition to Airbnb is un-American or unconstitutional. Yet the quotation also highlights a paradox in platform businesses' depiction of the future: that platforms rarely invoke a future with any content to it beyond the further expansion of their own business. They frame their platforms as vectors of progress in a world that is otherwise old-fashioned and backward. It is the state and its opponents that are out of date, and the platform and its users, "community" or "movement" who are the present and future.

Platform economy businesses invoke vacuous or controversial future visions, but their future visions are still potent because states have largely abandoned this discursive terrain. Under the economic mode of neoliberalism, states and public bodies tend not to develop or elaborate future visions. Visions of society are privatised or outsourced at the same time and for similar reasons as public services, innovation and societal transformation. This, and the continuation of the cliché, variously attributed to Frederic Jameson, Slavoj Zizek or Mark Fisher that claims it is easier to imagine the end of the world than the end of capitalism, leaves Airbnb, Uber and other Silicon Valley corporations and their backers with a near-monopoly over the discursive power of remotely desirable future visions for the economy. This is becoming recognised as significant politically, given that the discussion of future arrangements has a performative effect, drawing certain possibilities closer and discounting others (Mische 2009).

Debates about platforms' current activity as being about the future demands an attitude of indulgence and agnosticism: after all, their actual implications are still unknown and their potential might bring benefits we need to "learn about". This, in turn, allows the technologies to expand and scale-up, rapidly becoming everyday, ordinary, convenient and imbricated in people's lives and social practices. It starts to appear anachronistic and backward to threaten this, which allows a potential political constituency who will defend the business to develop (see below, and Thelen 2018). In other words, the ambivalence about their effects that platform businesses claim if accepted creates a self-fulfilling prophecy where the business builds up its potential political power vis-a-vis the state, which can be framed as reactionary because it literally offers no alternative future. This lack of competition in imaginaries allows platform businesses to misleadingly brand the conflict between their corporate power versus the state and social movements as a conflict between "the people" and the future (represented by the users of the platform and the technology) versus the state, the establishment and the past.

This is not a foregone conclusion. States, through legal injunctions normally proposed by strong unions, have regulated or prohibited platform businesses at the outset such as with Uber in France and Germany, and this approach appears most effective in maintaining democratic control over employment relations and public space that platform businesses may threaten (Thelen 2018). Secondly, social movements and alternative economic initiatives produce alternative visions of the future that compete with those of platform businessees, despite, drawing from similar tropes mobilised by platform businesses. These tropes include critiques of entrenched corporate power and consumer culture, celebrations of community, cosmopolitanism, sustainability and reciprocity – as well as values which are usually ignored by platform businesses such as solidarity and equality. The idea of platform cooperatives is a device for breaking the monopoly on imagining the future economy and helps avoid opposition to platform businesses being framed as reactionary (Scholz and Schneider 2016). Thirdly, the most plausible alternative "sharing" or "platform" visions to those of platform businesses are often supported or articulated by public institutions and governments. Barcelona's city hall, controlled until May 2019 by Ada Colau's radical "platform" Barcelona en Comú and now run in coalition with the Catalan Socialist party (PSC), had a yearly budget of several million euros to spend on "social and solidarity" economic alternatives, many of which comprise some digital or online component (e.g. Ajuntament de Barcelona 2017). It defined these as "a broad (formal and informal, individual and collective) range of socio-economic initiatives that prioritise the satisfying of needs – whether of its members or of other people – above financial profit" (Ajuntament de Barcelona 2017: 5). This involved sponsoring and supporting new cooperatives, attempts to integrate existing worker-owned enterprises, cooperatives and other entities to cultivate alternative models of ownership and the production of "social value", the open-sourcing and public availability of urban data (Open Data Barcelona), and new procurement rules governing the outsourcing of public services that evaluate bids on broader criteria than cost, but also employment conditions and gender equity. Interventions from movements or states can weaken the monopoly that platform businesses attempt to assert over the future. By imagining and articulating alternative futures, and investing in, promoting and demonstrating the technology and social relations that might lead to them, states – and to a lesser extent social movement organisations and projects – can subvert the powerful discursive monopoly over the future that platform businesses otherwise claim.

The manufacture of civil society support: platform-based grassroots lobbying

The second tactic common across various platform businesses that I present here is platform-based grassroots lobbying. This refers to the various ways in which platform businesses have blurred the lines between the corporate and civic through the adoption or co-optation of repertoires of civil society in their public policy strategies. I will describe what I mean by this here and offer a preliminary typology of platform-based grassroots lobbying. Several authors have noted that firms in the platform economy are engaged in public policy work involving civil society practices or associations, but existing academic and journalistic work is mainly descriptive or limited to one business at a time (Collier et al. 2018, Thelen 2018, Aguilera et al. 2019).

Platform businesses engage various existing civil society campaigning techniques complementing standard lobbying, campaign donations and PR strategies. They take three forms: politically mobilising their "user" base and supporters; making alliances with existing civil society actors through donations, sponsorships or joint campaigns; or developing front groups and activist stories which hold claim to be authentically "grassroots", who play a role in platform businesses" public policy. These categories – user mobilisation, grassroots alliances and front groups – help demonstrate the breadth and diversity of extant approaches to platform-based grassroots lobbying. This shows how and where platforms build upon existing corporate grassroots strategies (Walker 2014) and help identify the controversies that these techniques throw up.

The first form of platform-based grassroots lobbying is *user mobilisation*. User mobilisation refers to short-term initiatives where the everyday users of platforms are encouraged to support a corporation in response to a specific and local regulatory threat by signing petitions, contacting representatives and responding to consultations.[2] Edward Walker (2014) describes how corporate public affairs agencies have, since the 1970s, been used to mobilise businesses' customers or workers, usually responding to a specific threat. App-based platforms have introduced innovations to these repertoires, for example by using push-button notifications and splash screens to target their large number of app users, who are urged to sign and share petitions or send form letters to representatives. Uber and Lyft both mobilise their users regularly in the context of regulatory struggles. In an early campaign in California against two bills which aimed to toughen up requirements on insurance, background checks of drivers and drug and alcohol testing, Lyft employed specialist digital advocacy firm Phone2Action to help it contest the legislation (Said 2015). A similar example was Uber London's #SaveYourUber campaign petition (Elvidge 2017), initiated by UK head of Uber Tom Elvidge in November 2017 when their licence to operate was not renewed by Transport for London after the company refused to comply with safety concerns (Andrews 2017).[3] This has followed similar heavily mediatised petition campaigns all over the world, in at least nine US states, and countries as diverse as Denmark, Spain and India (ibid). In New York, threatened by a cap on provider numbers, Uber added a "De Blasio" button on their app that illustrated how much longer users would have to wait for a taxi if the mayor's proposals went ahead and provided a template email for users to register their opposition to them (Collier et al. 2018). De Blasio's proposal was not implemented. Another important example is Airbnb's legislative struggles with cities, which frequently morph into the third tactic of creating new grassroots actors or front groups (see below). In 2015 concerns around the effect of Airbnb on housing in San Francisco, already among the most expensive cities in the world to live in, led to a referendum asking residents to vote for or against various measures including the limiting of short-term rentals to 75 days per year. Airbnb outspent the "yes" campaign by 30 times (over $8m). The campaign hired new staff including 11 full-time political campaigners with experience in the Obama campaign, who made 32,000 phone calls to the 6,500 hosts in the city, several hundred of whom subsequently attended protests and court hearings (Atkin 2016). Altogether, the press releases claimed to have enrolled 2,000 volunteers and knocked on 285,000 doors (Somerville 2015), probably the most spectacular example of platform-based grassroots lobbying to date.

Like all grassroots lobbying tactics, there is controversy in mobilising the base because these efforts are not expressions of spontaneous support from civil society. Rather, they are resourced and professionally executed by companies without public transparency or oversight beyond that which the platforms choose to share publicly (with the exception of certain states in the US, where money spent on grassroots lobbying must now be added to the public lobbying register). A more specific controversy is that users appear quite easily to be mobilised against their own interests due to companies being able to provide partial information via apps and short-circuiting traditional news reporting or other sources of information. In the case of Uber, Collier et al. (2018: 927) point out, both consumers and drivers have been "mobilized against regulations that may have benefited both groups: more extensive consumer protection for consumers and Uber-provided commercial insurance for drivers". A third and related controversy is that while some interpretations of grassroots lobbying may consider the opportunities that platform businesses offer their users in democratic processes as empowering and deepening democracy (see some passages in Walker 2014), platforms may also use their user-oriented resources – specifically their apps – to support their objectives by user *demobilisation*. In Seattle, where the city council voted to allow a union for ride-hailing apps, Uber was widely reported as obliging their drivers to listen to misleading company-run podcasts opposing unionisation, as well as sending them text messages and invitations to meetings to try and persuade them that the union is a threat to their livelihoods (e.g. Ghoshal 2017, see also Rosenblat 2018).

The second form of platform-based grassroots lobbying is *grassroots alliances*, whereby platform businesses create public-facing relationships with existing grassroots campaigns or groups, normally by funding them. This may have the effect of "amplifying existing voices" which may be sympathetic to the business (Walker 2014: 6). In most cases the alliance involves seeking public expressions of support, usually focused on proposed legislation that appears to be the grassroots actors' part of the exchange. During the 2015 legislative struggle in California (mentioned above in relation to Lyft's employment of the Phone2Action agency), Uber partnered with Mothers Against Drunk Driving (MADD) (Kalanick and Withers 2015, cited in Rosenblat 2018) by offering special promotions to the group and initiating a joint media campaign. Uber's donations to MADD led the group to organise a letter-writing campaign to the California state governor and against legislature which proposed Uber requiring commercial insurance (Collier et al. 2018). More recently, Lyft partnered with the National Center for Transgender Equity (NCTE) to offer legal support and resources to gender-transitioning drivers (Lyft 2019), a move unlikely to affect significant numbers of people but both an important symbolic legitimation of transgender identity in the current political context, and one that further burnishes the company's liberal credentials in defining itself against its chief competitor Uber.[4] In New Jersey Uber attempted to recruit 3,000 new drivers from low-income minority neighbourhoods to drive for the company, drawing praise from the local branch of the NAACP (Rosenblat 2018: 183). A fourth interesting example is Airbnb's sponsorship of Ouishare, a sharing economy advocacy association who organised yearly conferences bringing together activists, tech and innovation communities and businesses in the sector. Ouishare's corporate partners and funders of conferences have regularly included Airbnb, as well as Blablacar, the French insurance company MAIF, and many others.

With grassroots alliances there are again concerns about the transparency of the funding and support given to the grassroots actors. In addition there is the risk of co-optation. Where existing civil society organisations are trusted for their expertise and the links they have to the grassroots of people affected by particular policies, issues and identities, these alliances confer legitimacy by proxy onto the business. Yet these groups do not or cannot readily disclose the resources and influence they receive from the business, or what effect the partnership has on their normal activity. Even though these organisations may fiercely protect their reputation for independence, the financial or in-kind support is clearly contingent on an open expression of support.

The third form of platform-based grassroots lobbying can be referred to as *front groups*: creating actors which, with varying plausibility, claim grassroots legitimacy, who play a role that is complementary to platform businesses but which may or may not be directly controlled by them. So far, these have taken two main forms. The most common is the creation of largely hollow lobbying organisations that attempt to recruit support from ordinary people in addition to directly influencing public debate and government on behalf of the sector or the business who has funded them. Examples include local tech lobby associations such as San Francisco Citizens Initiative for Technology and Innovation, and San Francisco Planning and Urban Research Association, both of whom are sponsored by industry actors (Meronek 2014). The most interesting example of these largely hollow lobbying organisations was Airbnb's organisation Peers, who called itself "a grassroots organisation to support the sharing economy movement", and was co-founded by the Airbnb CEO Brian Chesky, former Airbnb Global Head of Community for Airbnb Doug Atkin, and Natalie Foster, a former Obama campaign organiser. The organisation launched in 2013, engaged in various activities including coordinating and marketing petitions to support Airbnb and Lyft in regulatory struggles, and attempting to crowdfund donations. They closed down in 2016 after significant public criticism, but a new organisation with links to sharing economy businesses Purpose was immediately founded with overlapping personnel (Slee 2016). Several thousand people were reported to have joined Peers, which might appear to bear out the group's claims to grassroots membership. Yet Peers' funding came from "mission-aligned independent donors" and foundations, who prominently included investors and executives of sharing economy businesses, and whose listed partners were almost entirely for-profit companies (Kamenetz 2013, Slee 2016).

The second form of front groups appears so far to be unique to Airbnb and emerged out of its practices of user mobilisation. User mobilisation took on a more proactive and ambitious character since prominent struggles around 2014–2015 in San Francisco, New York and Barcelona. The company now runs ongoing public policy campaigns using community organising tools, hiring people with experience in grassroots organising and election campaigning, as Community Organisers and Campaign Managers (Steinmetz 2016 and personal research). They claim to have established over 4,000 groups of rentiers that they call "home sharing clubs" (Airbnb 2019) who engage in civil society on its behalf – organising protests, talking to press and attending meetings as civil society representatives in decision-making forums around short term rentals. Home-sharing clubs are, the company maintains, "independent", in some cases setting their own agendas and organising events themselves. Nevertheless they are initially curated and leaders selected by Airbnb staff, and they enjoy support oriented towards the specific regulatory campaigns of the city in question – usually involving training, convening and logistical support after exhaustive phone-banking to find potentially interested hosts with appropriate profiles (i.e. rentiers without multiple listings) (personal research).

Controversies associated with front groups also revolve around co-optation and transparency. Peers did not disclose openly its industry backing, and home-sharing clubs' protest behaviour is regularly reported by the media as though they are traditional grassroots campaigns. Front groups use mobilising tools that are distinctive for being modes of collective action that allow people without resources to place pressure on elites, appearing to allow elites their own access to the power of collective action in addition to traditional sources of corporate political power held through traditional lobbying and PR work. The distinction that front groups have with user mobilising is that front groups tend to be more sustained resourcing efforts and are further distanced from the business that funds them or their activity. Sometimes, however, the distinction is difficult to maintain or it appears that platform businesses are pursuing both forms of grassroots lobbying simultaneously. For example, food delivery service Doordash in summer 2019 offered a form letter email campaign protesting a California Supreme Court decision to stiffen regulations around who can be classified as an independent contractor, a standard case of user mobilisation (Doordash 2019). But they also linked explicitly to a front group, a campaign platform called the "I'm Independent Coalition" (California Chamber of Commerce 2019) where contractors or supporters are encouraged to change their Facebook profile photographs to support the campaign, sign form letters and inform themselves over the legislation. The platform's corporate supporters include Doordash and most gig economy businesses operating locally, including Lyft, Uber, Instacart, Postmates, Handy and Glamsquad. As well as demonstrating the slippages between user mobilisation and front groups, this example also shows that platform economy and technology businesses sometimes work together on creating some of the grassroots actors or third-party agencies who subsequently mobilise them, and the example of Peers shows that sometimes actors are set up who have an effect on the sector that is broader than the business which funded them. This coordination between businesses, and the effects of these initiatives on the entire platform sector, might have been lost through exclusively focusing on businesses as case studies.

Conclusions

This chapter contributes to understandings of the politics of the platform economy (see also Collier et al. 2018, Rosenblat 2018, Thelen 2018, Aguilera et al. 2019). It summarises and begins to theorise the political struggles surrounding platforms, from rhetorical struggles over the meaning of the concepts used, to flashpoints of open conflict among civil society, businesses and government, to backstage negotiations and trajectories of transformation in which collective actors change the practices of everyday life around platforms. It argues that looking across urban platforms as well as at specific businesses is important for understanding these dynamics. It demonstrates this by focusing on two tactics of legitimation of platform businesses that cut across several platforms: the invocation of tropes about the future I have called *platforms as self-declared vectors of progress,* and the manufacture of grassroots support I have called *platform-based grassroots lobbying.*

There are some implications for future research. Although there is a growing literature on the political struggles around future visions and the performative character of debating the future (Mische 2009, Schultz 2016, Brown 2016), little work yet has discussed the relationships, conflicts and confrontations involved.

The example of platform businesses and the way in which states and regulation are depicted also suggest how neoliberal economic trends towards privatisation not only outsource key services of the state to companies and avoid the planning or execution of large socio-economic projects or infrastructure. They also rhetorically neuter states in their attempts to regulate platforms, in disputes where they are cast as reactionary in the face of technological progress. This masks the ideologically extreme and economically extractive character of the businesses involved. Examples show how the alternative visions of city councils, states and social movement initiatives can subvert this dynamic.

Another set of implications surrounds the development of platform-based grassroots lobbying, which in various ways extends our understanding of grassroots lobbying practices reviewed to date (see Walker 2014). The above debate categorises the most common tactics of platform-based grassroots lobbying into three, with illustrative examples. Research should explore these examples further to describe and explain the emergence of platform-based grassroots lobbying, its specific practices and its outcomes, where measurable. This should be done, where possible, using comparison across contexts, practices and businesses – allowing for an assessment of these tactics in a wider understanding of how the political struggles in the platform economy are playing out, and their implications for policy and movements that are attempting to establish better practice and outcomes in the sectors affected.

Notes

1 The two mottos "Move fast, break things" and "Don't ask permission, ask forgiveness", and the frequent substitution of references to the business in favour of ventriloquising users as a community suggest further facets to the discursive legitimation tactics that are linked to platforms' wider strategic approaches.
2 Less often, it has entailed requests and support for higher-cost forms of activity such as attending protests, for example Uber have offered free Uber taxi rides to locations of demonstrations in its favour (Pasick 2015) and Uber and Lyft supported protests in San Francisco in 2015 (Said 2015).
3 The petition received 858,001 signatures by the time that it closed. Andrews (2017) notes that in addition to press releases, users were prompted to sign via their app, there was promotion by Google ads, and, possibly, promotional marketing paid for on Change.org.
4 As Lyft president John Zimmer puts it "We're woke. Our community is woke, and the U.S. population is woke" (Time 2017).

References

Aguilera, T., Artioli, F. and Colomb, C. (2019) 'Explaining the diversity of policy responses to platform-mediated short-term rentals in European cities: A comparison of Barcelona, Paris and Milan', *Environment and Planning A*, pp. 1–24. doi: 10.1177/0308518X19862286.

Airbnb (2019) 'Home sharing clubs'. *Airbnb Citizen*. Available at: https://www.airbnbcitizen.com/clubs/ (last accessed 16/12/19)

Ajuntament de Barcelona (2017) 'The impetus plan for the social and solidarity economy'. Available at: https://ajuntament.barcelona.cat/economia-social-solidaria/en (last accessed 12/2019)

Andrews, S. (2017) Uber and the corporate capture of e-petitions. *Red Pepper.* Retrieved from https://www.redpepper.org.uk/uber-and-the-corporate-capture-of-e-petitions/ (last accessed 10/2018)

Artioli, F. (2018) 'Digital platforms and cities: A literature review for urban research', *Cities Are Back in Town Working Papers*, 208 (1). Available at: http://spire.sciencespo.fr/hdl:/2441/5ktrkc24c69j5bdrfoofk99nek/resources/wp01-2018-artioli.pdf (last accessed 10/2018).

Booth, K. (2015) 'Protesters occupy Airbnb HQ ahead of housing affordability vote', *The Guardian*, 2 November. Available at: https://www.theguardian.com/us-news/2015/nov/02/airbnb-san-francisco-headquarters-occupied-housing-protesters (last accessed 10/2018)

Botsman, R., & Rogers, R. (2010) *What's Mine Is Yours: How Collaborative Consumption Is Changing the Way We Live.* New York: Harper Business.

Brown, K. P. (2016) 'The prospectus of activism: Discerning and delimiting imagined possibility', *Social Movement Studies.* Routledge, 15(6), pp. 547–560. doi: 10.1080/14742837.2016.1191338.

California Chamber of Commerce. (2019) I'm independent coalition. Available at: https://imindependent.co/ (last accessed 16/12/19)

Cant, C. (2018) 'The wave of worker resistance in European food platforms 2016–17', *Notes from Below*, 29 January. Available at: https://notesfrombelow.org/article/european-food-platform-strike-wave (last accessed 10/2018)

Chakrabortty, A (6th March 2019) 'In an era of brutal cuts, one ordinary place has the imagination to fight back'. *The Guardian.* 6th March. https://www.theguardian.com/commentisfree/2019/mar/06/brutal-cuts-fight-back-preston-dragons-den (last accessed 16/12/19)

Chesky, B. (2014) 'Sharing the future'. Keynote address at Personal Democracy Forum 2014. 19 June. Available at: https://personaldemocracy.com/conferences/nyc/2014 (last accessed 16/12/2019)

Coldwell, W. (2017) 'First Venice and Barcelona: Now anti-tourism marches spread across Europe', *The Guardian*, 10 August. Available at: https://www.theguardian.com/travel/2017/aug/10/anti-tourism-marches-spread-across-europe-venice-barcelona (last accessed 10/2018)

Collier, R. B., Dubal, V. B. and Carter, C. L. (2018) 'Disrupting regulation, regulating disruption: The politics of Uber in the United States', *American Political Science Association*, 16(4). doi: 10.1017/S1537592718001093.

Corporate Europe Observatory (2018) UnFairbnb: How online rental platforms use the EU to defeat cities' affordable housing measures. Available at: https://corporateeurope.org/power-lobbies/2018/05/unfairbnb (last accessed 10/2018)

Crisp, J. (2018) 'Deliveroo riders go on strike in Belgium and Netherlandst'. *Daily Telegraph*, 20 January. Available at: https://www.telegraph.co.uk/news/2018/01/20/deliveroo-riders-go-strike-belgium-netherlands/ (last accessed 10/2018)

Doordash (2019) 'Tell California lawmakers: Protect our freedom to work independently!' Available at: https://p2a.co/fxdVXjs (last accessed 16/12/19)

European Commission (2016) *A European Agenda for the Collaborative Economy.* Brussels.

Elvidge, T (2017) 'Save our Uber in London'. Change.org. Available at: https://www.change.org/p/save-your-uber-in-london-saveyouruber (last accessed 12/2019)

Eyerman, R. and Jamison, A. (1991) *Social Movements: A Cognitive Approach*. University Park: Penn State University Press.

Ghoshal, A. (2017) 'Uber is forcing drivers in Seattle to listen to anti-union propaganda'. *Business Insider*, 13 March. Available at: https://thenextweb.com/us/2017/03/13/uber-is-forcing-drivers-in-seattle-to-listen-to-anti-union-propaganda/#.tnw_W2w4tbFx#.tnw_BMC5Ia4L#.tnw_X60FlcJc#.tnw_J74yZ1Ke#.tnw_RaNFJ1HJ (last accessed 12/19)

Husser, A. (2015) 'Taxis likely hurting themselves, helping Uber with protests', *CBC News*, 13 December. Available at: https://www.cbc.ca/news/business/uber-taxi-toronto-protest-1.3362193 (last accessed 16/12/2019)

Lyft (2019) 'Supporting transgender drivers'. *Lyft Blog*, May 28. Available at: https://blog.lyft.com/posts/nationalcenterfortransgenderequality (last accessed 16/12/19)

Meronek, T. (2014) 'Tech companies adopt Astroturf to Get Their (Wicked) way'. *Truthout*. April 20. Available at: https://truthout.org/articles/tech-companies-adopt-astroturf-to-get-their-wicked-way/ (last accessed 7/2019)

Mische, A. (2009) 'Projects and possibilities: Researching futures in action', *Sociological Forum*, 24(3), pp. 694–704. doi: 10.1111/j.1573–7861.2009.01127.x.

Rosenblat, A. (2018) *Uberland: How Algorithms Are Rewriting the Rules of Work*. Berkeley: University of California Press.

Scholz, T. and Schneider, N. (2017) *Ours to Hack and to Own: The Rise of Platform Cooperativism, A New Vision for the Future of Work and a Fairer Internet*. New York: OR Books.

Schor, J. B. (2017) 'Does the sharing economy increase inequality within the eighty percent?' *Cambridge Journal of Regions, Economy and Society*, 10(2), pp. 263–279. doi: 10.1093/cjres/rsw047.

Schulz, M. S. (2016) 'Social movements and futures research', *World Futures Review*, 8(2), pp. 98–107. doi: 10.1177/1946756716634310.

Slee, T. (2015) *What's Yours Is Mine: Against the Sharing Economy*. New York: OR Books.

Somerville, H. (2015) 'Airbnb to create 100 clubs to advocate for home-sharing'. *Reuters*. November 5. Available at: http://www.reuters.com/article/us-airbnb-sanfrancisco-idUSKCN0ST2RL20151105 (last accessed 6/2018)

Srnicek, N. (2016) *Platform Capitalism*. Cambridge: Polity.

Steinmetz, K (2017) 'How Lyft is capitalizing on Uber's scandals'. *Time*. 28 March. Available at: https://time.com/4712746/lyft-interview-uber/ (last accessed 16/12/19)

Thelen, K. (2018) 'Regulating Uber: The politics of the platform economy in Europe and the United States', *Perspectives on Politics*, 16(4), pp. 938–953. doi: 10.1017/S1537592718001081.

Walker, E. T. (2014) *Grassroots for Hire*. Cambridge: Cambridge University Press.

Zuboff, S. (2019) *The Age of Surveillance Capitalism*. London: Profile Books.

9

ANALYSING URBAN PLATFORMS AND INEQUALITY THROUGH A "PLATFORM JUSTICE" LENS[1]

Richard Heeks and Satyarupa Shekhar

Introduction

It is a given that the intersection between the urban and the digital is fast-growing. Among the multiple strands and terminologies – smart city, digital city, city 2.0 – a recent trend has been the emergence of urban platforms. We can define these as a set of digital urban resources – including services and content – that enable value-creating interactions between external producers and consumers (adapted from Constantinides et al. 2018). Examples include urban labour platforms (e.g. Uber, Deliveroo), urban innovation platforms (e.g. Change by Us, Sidewalk Labs) and urban data platforms (e.g. those openly sharing either government or community data).

Like other digital technologies more generally, urban platforms have brought with them many aspirations for benefits of efficiency and effectiveness: greater productivity and income, faster and better innovations, greater accountability of city governments, etc (Schaffers et al. 2011, Desouza & Bhagwatwar 2014). However, alongside these benefits have come a set of growing concerns. These are sometimes expressed fairly specifically, for example, in relation to loss of privacy or around the capture of urban development resources or agendas by private interests (Coletta et al. 2017, Crommelin et al. 2018). Overarching these, though, is the apprehension that platforms may be associated with growing urban inequality (Verrest & Pfeffer 2018).

This, in turn, is problematic because – alongside climate change – inequality is the central factor that may lead to urban unsustainability (Romero-Lankao et al. 2016). This is particularly an issue for cities of the global South given their already high levels of inequality and the negative impacts thereof: reduced economic growth rates, higher urban crime and violence, social and political discord, etc (Parnell & Oldfield 2014). However, due to the recency of urban platforms, little research has yet been undertaken on the links between platforms and inequality. It is the purpose of this chapter to address that lacuna.

The selected focus is the platformisation of slums in developing countries. Slums are a physical manifestation of urban inequality and marginalisation, and a key current question is whether the move to the digital city is intensifying or ameliorating inequalities for those who are already marginalised within the physical city. In this chapter, we focus on urban data platforms that provide digital data produced by one urban group as a basis for decision-making by a consumer group (Barns 2018). The specific example chosen is community data platforms that provide data about low-income groups or settlements, with the platform typically displaying this via a map-based interface for consumption by urban decision-makers such as local government or international NGO staff. These will be analysed through the lens of "platform justice" and the contribution of this chapter is thus two-fold. First, that it specifically addresses the question of the relationship between urban platforms and inequality; second, that it provides a conceptual framework through which to analyse that relationship.

There follows a brief review of urban platforms, leading into the presentation of the platform justice model. An explanation of research methods and cases leads to a set of findings and then conclusions.

Urban digitality, inequality and justice

As noted in the Introduction, digital platforms are an increasing component of the economic, social and political urban landscape, and they are also increasingly central to future visions of the city. Platforms undoubtedly bring benefits to that landscape. For example, urban labour platforms (e.g. Uber, Upwork) are enabling the formation or greater efficiency of labour markets in cities, creating new jobs for workers and reducing costs for clients (Drouillard 2017). Urban governance platforms are enabling greater citizen participation in city decision-making (Anttiroiko 2016).

Yet we can also identify from the literature two sets of concerns. Some focus on constraints: barriers that prevent the beneficial impacts of platforms from being realised. For example, "digital divide issues, including the inability to access the platforms and/or use the platforms effectively, can … curtail effective civic engagement" (Desouza & Bhagwatwar 2014: 46). Others, as noted above, focus on disbenefits: negative impacts of platformisation such as loss of privacy, the capture of gains by private corporations and, particularly, growing urban inequality (Taylor 2014, Verrest & Pfeffer 2018). However, analysis of real-world experiences is lacking: in part due to the recency of platformisation but also due to a lack of clear analytical frameworks.

So how should urban platforms be understood and evaluated? One approach would follow descriptors used for the platformisation concerns expressed above: "ethical concerns" (Taylor 2014), "ethical implications" (Coletta et al. 2017), "injustices" (Verrest & Pfeffer 2018). This would suggest analysis using ideas that are emerging at the intersection of critical digital studies and social justice; most strongly represented to date in the recent literature on "data justice" (Heeks & Renken 2018). In this chapter, we adapt this material to form the notion of "platform justice", defined as the specification and pursuit of ethical standards for platform-related resources, processes and structures.

Platform injustice

For the idea of platform justice to have meaning and relevance, there must be some injustice that the ethical standards of justice would seek to remedy. We illustrate this through reference to the particular focus for this paper – digital platforms and slum communities – and selecting just one illustrative issue: the right for those communities to be represented. This could be understood from a consumption perspective: the digital divide that prevents most slum area residents from being represented among users of digital platforms (Heeks 2018). And it can be understood from a productive perspective; for example that residents of slum areas either do not figure among the workers employed through labour platforms or are significantly under-represented (Graham et al. 2017).

Most concretely, this can be understood in terms of urban data platforms: not merely that slum areas themselves are unmapped as locations but that data on assets, services, voices and livelihoods of citizens is absent. An illustrative comparison (Figures 9.1a and b) is the appearance on the Google Maps platform of two different areas of Kenya's capital, Nairobi: the slum area of Kibera vs. the wealthy central district. Despite Kibera being full of hundreds of businesses, schools, clinics, etc of the type shown in Figure 9.1b, these are almost entirely absent on the platform. (Kibera also has no Street View while the wealthier area does; all adding to the sense of Kibera as invisible and as a "no go" area.)

FIGURE 9.1A Platform representation of low-income area. Kibera on Google Maps.

FIGURE 9.1B Platform representation of high-income area. Nairobi Central District on Google Maps.

Thus the physical marginalisation of low-income communities is mirrored by a virtual marginalisation within most urban platforms. In the case of data platforms, this injustice has knock-on effects on infrastructural and political marginalisation. The lack and poor quality of data about slums "directly results in poor planning and maintenance of public infrastructure, and poor provision of public services" (Shekhar & Padmanabhan 2015: 3). Lack of data also marginalises communities politically, ceding and skewing power within decision-making and service provision to political elites and their interests (Menon 2013, Feruglio & Rifai 2017). These platform injustices must be understood and countered through platform justice but, to date, no model for this exists.

A model of platform justice

Platform justice could be derived from a number of different perspectives building, for example, on the ideas of various different philosophers of social justice such as Amartya Sen or Iris Marion Young (Heeks & Renken 2018). These ideas suggest five different understandings or dimensions of social justice: procedural, instrumental, rights-based, structural and distributive. In the initial application of platform justice undertaken in this paper, we will use all five; thus allowing the reader to understand each in more detail and to evaluate which would be relevant to their own particular applications and interests.

In brief, these five dimensions of platform justice are:

- *Procedural*: fairness in the way in which the platform operates. This evaluates the ethicality of the everyday functioning of the platform; specifically its value chain practices which, in the case of urban data platforms, capture data that is stored, processed and visualised via the platform, and then leads to urban development decisions and actions.

- *Instrumental*: fairness in the results of the platform's operation. This evaluates the ethical implications of the results that emerge from the enacted value chain when it is fully operational and has an impact.
- *Rights-based*: adherence to basic digital rights. An increasing number of such rights is emerging but typical examples would include a right to be represented on a platform, a right to privacy (e.g. not to be represented), a right to access the platform and a right to ownership of and/or control over data that is being used on a platform.
- *Structural*: the degree to which interests and power in wider society support fair outcomes in other forms of platform justice. Platforms shape but are also shaped by the key constituents of social structure: formal and informal institutions, social and technical resources, social relations and epistemics (knowledge and discourse). The justice or injustice of those wider constituents will thus partly determine the justice or injustice of platform usage.
- *Distributive*: an overarching dimension relating to the (in)equality of outcomes that can be applied to each of the other dimensions of platform justice.

For further understanding of these dimensions, readers are referred to earlier papers (Heeks 2017, Heeks & Renken 2018) and to the application of the dimensions in the Findings below.

These dimensions fit well with the concerns of literature on platforms. They incorporate a focus on praxis, given the importance of understanding the real experiences of those creating, populating and using urban platforms (Coletta et al. 2017). They incorporate the argument of critical platform studies literature that broader rights and social structure must be incorporated into any analysis because of their role in shaping outcomes associated with platforms (Leszczynski 2016, Verrest & Pfeffer 2018). They incorporate the argument that platforms are increasingly important determinants of equality or inequality in today's world (Kenney & Zysman 2015, Srnicek 2017). The five dimensions therefore allow understanding of a range of potential platform justices and injustices, spanning the quotidian use of platforms, through the procedures necessary to make them work, to their immediate and broader impact, to the structural conditions of their context.

Putting all of these elements together, the platform justice model is summarised in Figure 9.2 (adapted from Heeks 2017). The centre of the model is the platform and its value chain leading to the instrumental value of the system in terms of its results. This value chain is shaped, even partly determined, by the affordances of digital platforms.[2] Surrounding the platform is its context. The platform is driven to function by some particular utility or value. It is shaped by a set of digital rights and, in turn, shapes those rights as it functions. And it has a similar interrelation – shaping and being shaped by – the elements of social structure.

A further illustration of the model will follow below. The model is generic and should thus be applicable to analyse all types of digital platforms. But, as noted above, we will use it to evaluate four urban data platform in developing

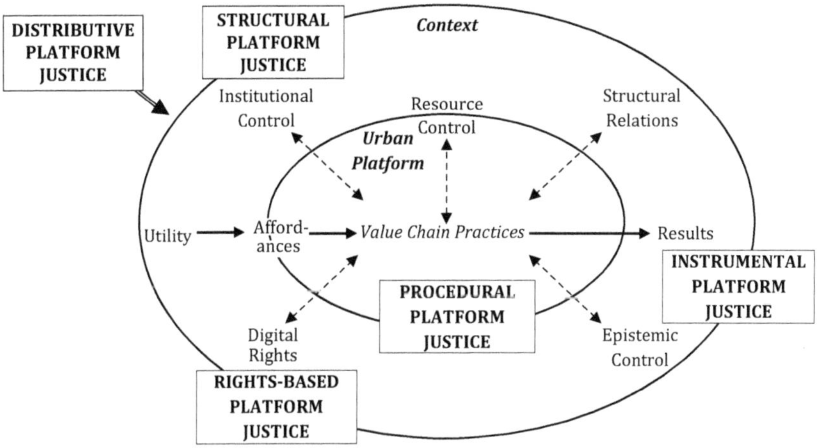

FIGURE 9.2 Platform justice model.

countries. Before doing that, we outline the initiatives and methods used for evidence-gathering.

Case background and methods

The cases used here are four urban data platforms from the global South that map data generated from within urban communities. These were chosen because the use of such platforms is increasing and because these mapping applications are of direct relevance to urban inequality.

In one case, the map is restricted to a slum area only. This is Map Kibera (http://mapkibera.org/) which was set up in 2009 by two US development activists. It uses OpenStreetMap as the platform through which to display data about the Kibera community of Nairobi (e.g. https://www.openstreetmap.org/#map=16/-1.3119/36.7876).

The three other platforms take a city-wide approach. They have pro-equity intentions but decided that the best way to expose inequalities was to provide data about the whole city. They are:

- Our Pune, Our Budget (http://ourpuneourbudget.in/pune-wise/) from India, which provides a data platform mapped as wards rather than the building-level detail provided by Map Kibera, and includes various urban infrastructure and services.
- Solo Kota Kita (https://solokotakita.org/) from the city of Solo in Indonesia has more limited data displayed and has tended to rely more on "mini atlases" of individual communities printed from the platform.
- Transparent Chennai (http://www.transparentchennai.com/) from India is similar to the Pune case in mapping various forms of urban infrastructure and services on its platform.

Findings are drawn from four sources that were combined and then analysed through the thematic lens of the platform justice model presented above. First, nine interviews, structured around the concepts of platform justice, with senior figures in the platform intermediary organisations associated with each of the four projects: Map Kibera, Kota Kita and Transparent Chennai were also the names of the organisations in three cases; for Our Pune, Our Budget, the platform was managed by the local NGO, the Centre for Environment Education. Second, evaluation documents from the projects that are reflective, to some degree self-critical, and sometimes based on primary fieldwork (e.g. Hagen 2017). Third, independent secondary sources of research specifically on these projects, most of which are based on primary fieldwork (e.g. Grillos 2017). Fourth, broader independent sources that incorporate analysis of these projects among a number of others (e.g. Baud 2016).

Findings

As per the model shown in Figure 9.2, the findings here will be presented in terms of each of the dimensions of platform justice in turn, but with each of procedural, rights-based, instrumental and structural considered through the lens of distributive platform justice given the over-arching focus on inequality in this chapter.

Procedural

Procedural platform justice studies the practices that operationalise the platform. Those associated with urban data platforms can be divided into three main parts: capture, processing and output. Data for the case study platforms was captured on the ground; typically by groups of local students walking through a slum area and mapping what they saw. These mappers gained technical skills, confidence and a network of contacts within the community and within the platform intermediary organisation. Direct interaction with community members did occur but only with a small minority sample.

Processing activities – recording and processing the data on the platform and then visualising it typically as some form of the map – were generally undertaken by voluntary or paid professionals from outside the community; professionals who were able to gain higher-level technical skills and experience as a result. The map itself was accessible via the online platform, sometimes in association with additional tools allowing a user to filter, for example, to show only health assets or only community facilities. However, this form was rarely accessible by the community because of their lack of access to the Internet-connected PCs necessary to engage with the online platform. Instead, paper-based versions of the platform such as those shown in Figure 9.3 (a translation of the actual "mini-atlas" provided in Bahasa Indonesia as part of the Solo Kota Kita project) were circulated at community meetings or as posters in parts of the community.

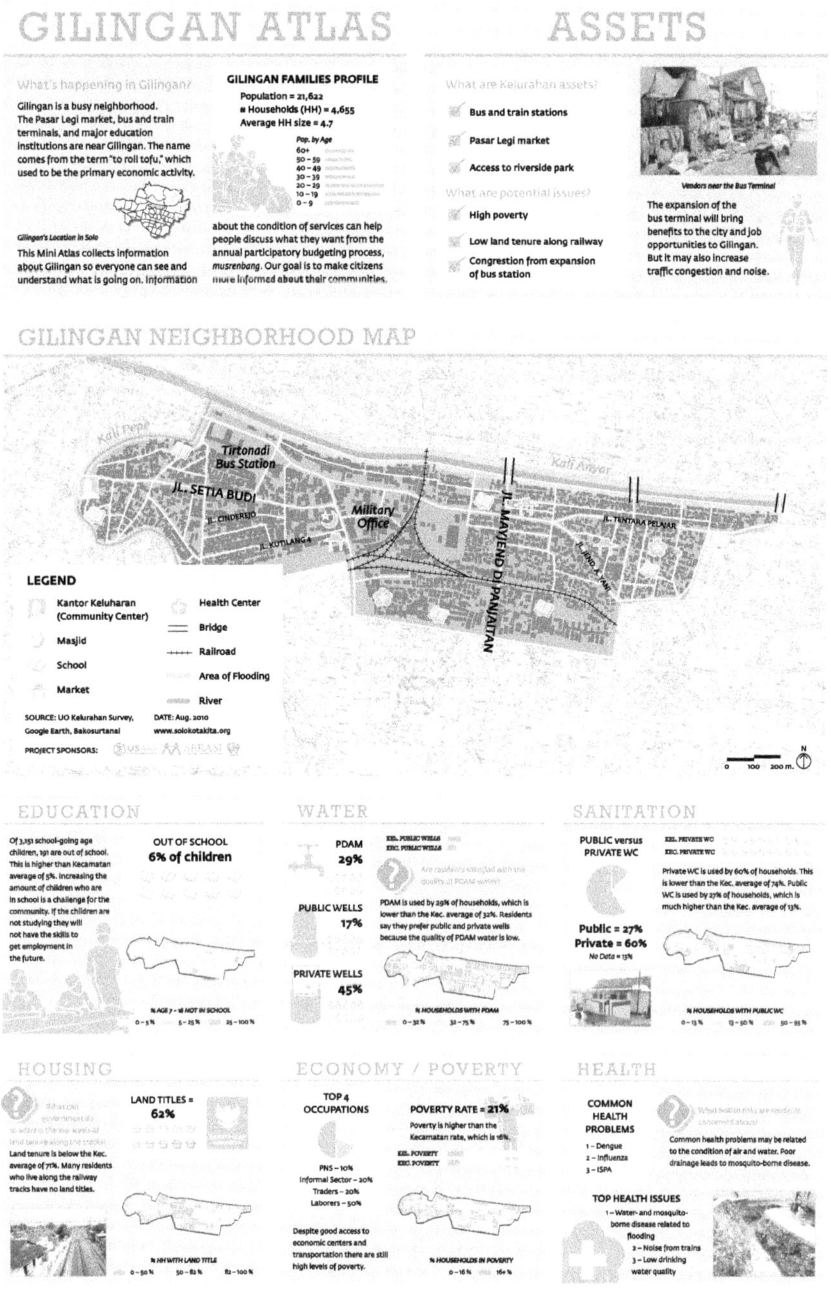

FIGURE 9.3 Example paper-based "Mini-Atlas" from Solo Kota Kita (2010).

Even with these outreach methods, it was only a small minority of the community that ever used the data.

Procedurally, then, the distributive impact of these urban data platforms can be seen as a pyramid. At the base, the majority of community members are not involved in any processes of the platform: data is not captured from or by them and they make no use of the platform, even in paper-based form. The mappers who capture data for the platform are far fewer in number but they do gain capabilities and contacts. The data processing/visualisation professionals are fewer still but gain greater capabilities. The platform intermediary organisations who own the platform benefit most of all because it is their staff who internalise the capabilities of platform project design and implementation; and who build social capital downwards into the community, outwards with the various professionals they engage and upwards into government and other organisations that can make use of the online data platforms.

Rights

Rights-based platform justice can deal with the adherence or otherwise to a series of digital and other rights, but here the main focus is the right to representation which would make previously "invisible" communities visible via the platform. As with procedural justice, the application of a distributive lens shows that a pyramidal outcome can be seen in relation to rights. The most marginalised – those without homes, those without identification, those residing on the physical margins who come in to the city to make their living – are rarely made visible by urban data platforms. They therefore cannot benefit in any way; indeed one might argue that the spotlighting of slum areas via the platforms throws those not represented further into shadow.

The slum communities are made much more visible: compare, for example, the representation of Kibera in Figure 9.1a with that in Figure 9.4 – a blank space on the map now becomes filled with buildings, roads, paths and multiple assets including schools, businesses, markets, religious and community buildings, etc.

As discussed next, the communities that are made visible do reap some benefit directly. However, they lose control of their representation; becoming legible to others who can make use of the community's representation on the platform for their own purposes. From this, we can see the ambivalence of legibility that urban data platforms offer. Slums must be legible to government, NGOs, development agencies, etc if they are to benefit from the resources, services, support, etc that these external organisations offer. But that same legibility exposes slum residents to any other agendas these organisations may hold. Those agendas may be orthogonal to slum interests or even counter to those interests: extractive, persecutory or predatory.

Even if direct evidence of the latter was limited, the fear of it led some platform intermediaries to avoid gathering certain types of data. For example, Transparent Chennai gathered data on issues facing informal waste-pickers in the city. But

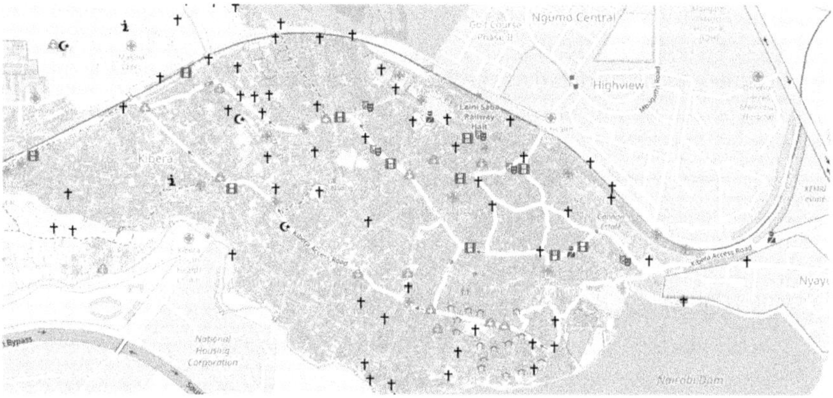

FIGURE 9.4 OpenStreetMap view of Kibera.

the waste-pickers' location was not recorded on the platform in order to protect them from state action. This issue also led some groups to prioritise their right to privacy and either resist or refuse to engage with data gathering. For example, some schools and pharmacies in Kibera did not wish to be mapped. They feared visibility to the state might lead to closure if their location became known and their informal status or activities (e.g. sales of stolen drugs) were then discovered.

Regarding the top of the "pyramid", government officials also remained largely invisible with low-income communities granted no right to access: the process by which these officials made decisions about the communities; the final decisions; or the extent of implementation of those decisions. As a result, community members were unable to either participate in or monitor the decisions being made about them unless the platform intermediary organisations undertook specific actions to counteract this.

Instrumental

Instrumental platform justice focuses on the direct impact of use of the platform. The first instrumental finding was an absence of tangible impact of these platforms in some situations. The data provided on the platform was not always used; either because no relevant decisions existed to make use of it, or because decision-makers such as government officials did not know about the data or did not want to use it. Even where it was used, it was rare for this to be systematically tracked, and so the evidence base is thin:

- In Solo, there were community development projects funded from the wider participatory budgeting that the platforms were intended to support. However, the use of data from the platforms – even in paper-based form – appears to have been somewhat limited. Whether used or not, wealthier residents were more effective in getting budget decisions shaped to their concerns.

As a result, the analysis showed spending flowing disproportionately to the least-poor areas of the city.

- In Pune, the platform's data was used to set an objective, needs-based budget that took account of the prevalence of poverty and inadequate housing across the city. Unfortunately that budget was then "unset" by local politicians, and investment flows remained skewed: 40% of Pune's population lives in slums but only 10% of the budget was allocated for slum improvement. Instead, funding flowed more into "middle-class" issues such as improvements for car drivers.
- Map Kibera data was used to improve government resource flows into local schools though mainly to the 25% of schools that were government-run. There was some evidence of international development agencies and NGOs using the platform's data for improved planning of education resource flows, or for the siting of water and sanitation facilities.
- Transparent Chennai engaged more with government from the start of its projects but the reported outcomes were rather circumscribed. The city built 15 homeless shelters not the 75 that the data had shown to be required, or laid on a ceremony to present a few waste-pickers with entry cards to the local waste site but not the ID cards they had been seeking.

We can say that urban data platforms have served the slums: these initiatives bring better-planned and more resources and services. But alongside the glass half-full is a glass half-empty. These absolute improvements are always less than intended or needed, and sometimes merely symbolic or temporary. Marginalised groups find some greater voice and place in urban decision-making than was previously the case thanks to these platforms. But, in distributive terms, relative inequality still grows: as seen most clearly in the Solo and Pune cases, more formalised areas and wealthier residents retain a stronger voice and benefit more from the decisions being made on the basis of urban data.

Structural

Structural platform justice examines the extent of (in)justice within the broader context that shapes the operation and impact of a platform. The overall picture outlined above is both explained by and reflected in the structural platform justice evidence. Path dependency is the main distributive narrative: the wider structural inequalities of power–interest significantly shape the data platforms and their use. For example, a consistent theme across all initiatives was that impact depended on whether or not the platforms had utility for powerful local actors. In Solo, the platform's data became an officially mandated part of participatory budgeting because Mayor Widodo wanted his urban development innovations to work. By contrast, government officials in Pune were resistant to participatory budgeting, making it much harder for new data to be utilised and for the platform to therefore have value.

In turn, the use of platforms largely reproduces wider structural configurations and constrains pro-equity aspirations. For example, the availability of data via platforms favoured those with the digital technologies and other resources – skills, knowledge, time, confidence, money – necessary to access the data and take and implement decisions with that data. These are all resources more available to those outside rather than within the marginalised communities. The use of data by external actors could benefit the community. Examples included the platform intermediaries empowered to better advocacy on behalf of communities, or the international NGOs empowered to planning better interventions. But control and empowerment still resided outside the community. In particular, the platform intermediary organisations are empowered. Interrogating openness, one finds that it is often sub-sets or summaries or visualisations of data that are being circulated. The foundational dataset was typically not accessible via the platform. Instead, control over that dataset including ownership and rights to update the data typically rested with the platform intermediaries.

This is not a completely static picture: platforms do reshape – subtly, incrementally – the structural determinants of power and utility. There is an incremental empowerment in terms of new data access, external perceptions of self-interest and of slums and their residents, and the landscape of organisations and trust. But there is no evidence of a wider transformation of the substrates of urban inequality, with communities potentially more dependent on external forces because of their heightened legibility via the data platforms.

Conclusions and recommendations

Platformisation is a key trend within urban development; with effects growing every day across an array of urban sectors. Yet analysis of real-world experiences and their breadth of impact has to date been limited; in part due to a lack of analytical frameworks.

This chapter therefore makes two significant contributions. First, it exposes a full picture of the impact of one type of urban platform – urban data platforms – on those who are marginalised within the city. Platforms do have a rights-based impact in counteracting the injustice of invisibility, but from a structural justice perspective they are shown to disproportionately serve those with the motivation and power to use the data provided by the platforms. Some instrumental results are certainly beneficial for slum communities and other marginalised citizens, and these platforms can be justified on that basis. However, though there can be no exact calibration from qualitative research, it is likely in distributive terms that these urban data platforms actually increase relative inequalities. Ordinary community members have seen some benefits but external actors who use the platforms to match their agenda and capabilities, benefit more. It is the latter who are more empowered to access, use and control the platforms.

Second, and in order to perform this evaluation, the chapter presents and demonstrates an explicit, systematic and comprehensive framework for the analysis of urban platforms. The framework is platform-oriented: acknowledging the functionalities and processes of platforms and the central role of the data that urban platforms handle. But it simultaneously decentres the platform by guiding attention to wider processes, impacts and structures. There was some overlap in the dimensions used; with the rights discussion particularly touching on procedural, instrumental and structural issues, and – in this particular application – with the distributive justice lens being subsumed within the other four dimensions. However, in general the perspectives were additive. They offer new insights and particularly substantiate the need to incorporate and understand context – rights, structures, power, interests – in order to fully understand the implications of urban platformisation.

The framework was applied here to analyse just one type of urban platform. However, its generic nature means it should be equally applicable to all types of platform: those relating to labour, enterprise, health, education, etc. It could be used to interrogate digital labour platforms active in urban areas such as Uber and Upwork; for example, exposing the fairness or otherwise of their operation, and their impact on inequality. It could be used to examine the role of social media platforms in city government elections; for example, analysing their support for or undermining of digital and democratic rights. Or it could be used to investigate other urban data platforms such as those being created by government or the private sector; for example, analysing who and what is represented, and whether there are gender or race biases in their use.

This expansion in the scope of application is one part of the future research agenda. This application could make use of the whole model as presented or could extract particular elements and dimensions that are seen to be especially relevant to the platform being studied. Even individual elements could be the basis for further research. For example, to explore the specific relation between platform affordances (such as those mentioned: datafication, digitalisation, generativity) and the dimensions of justice. Or to understand how formal and informal institutional forces shape the justice or injustice of the use of platforms.

A second strand of the future research agenda will be action research: working with platform-using organisations and with platform-owning organisations in using the framework to guide design and implementation. The findings here support the value of imbuing platforms with the values of platform justice; ensuring their design and implementation follow principles of procedural, rights-based, instrumental, structural and distributive justice. These might be summarised as a "Manifesto for Platform Justice":

1. Demand just and legal uses of digital platforms.
2. Demand consent of those providing data/content for platforms that is truly informed.
3. Build platform-related capabilities (capture-, processing-, output- and use-related) among those who lack them.
4. Promote rights of digital access, privacy, ownership and representation.

5. Advocate sustainable use of platforms.
6. Create a social movement for the "platform subalterns" who are marginalised or excluded by platforms.
7. Stimulate an alternative discourse around platforms and socio-economic development that places issues of justice at its heart.
8. Develop new organisational forms such as platform cooperatives.
9. Lobby for new platform justice-based laws and policies (including action on platform monopolies).
10. Open up, challenge and provide alternatives to the technical platform structures (code, algorithms, standards, etc) that increasingly control socio-economic development.

Notes

1 Acknowledgements: This chapter is developed from an earlier paper – Heeks, R. & Shekhar, S. (2019) Datafication, development and marginalised urban communities: an applied data justice framework, Information, *Communication & Society*, 22(7), pp. 992–1011 – and from a presentation made at the international workshop on Urban Platforms and the Future City held at the University of Manchester on 28 February–1 March 2019.

2 These include datafication (greater breadth, depth and granularity of data about urban phenomena that platforms hold), digitalisation (the conversion of all parts of the value chain from analogue to digital versions that can enable, for example, the shift from human to algorithmic management of urban processes, or the shift from human to machine-based learning about urban processes) and generativity (the use of platform data and development of digital processes in ways not planned at the origination of the platform).

References

Anttiroiko, A. V. (2016). City-as-a-platform: The rise of participatory innovation platforms in Finnish cities. *Sustainability*, 8(9), p. 922.

Barns, S. (2018). Smart cities and urban data platforms: Designing interfaces for smart governance. *City, Culture and Society*, 12, pp. 5–12.

Baud, I. (2016). Digitisation and participation in urban governance. In *Local Governance, Economic Development and Institutions*, G.M. Gomez & P. Knorringa (eds), Palgrave Macmillan, London, pp. 86–97.

Constantinides, P., Henfridsson, O., & Parker, G. G. (2018). Introduction—platforms and infrastructures in the digital age. *Information Systems Research*, 29(2), pp. 381–400.

Coletta, C., Heaphy, L., Perng, S. Y., & Waller, L. (2017). Data-driven cities? Digital urbanism and its proxies. *Tecnoscienza: Italian Journal of Science & Technology Studies*, 8(2), pp. 5–18.

Crommelin, L., Troy, L., Martin, C., & Pettit, C. (2018). Is Airbnb a sharing economy superstar? Evidence from five global cities. *Urban Policy and Research*, 36(4), pp. 429–444.

Desouza, K. C., & Bhagwatwar, A. (2014). Technology-enabled participatory platforms for civic engagement. *Journal of Urban Technology*, 21(4), pp. 25–50.

Drouillard, M. (2017). Addressing voids: How digital start-ups in Kenya create market infrastructure. In Digital Kenya, B. Ndemo and T. Weiss (eds), Palgrave Macmillan, London, pp. 97–131.

Feruglio, F., & Rifai, A. (2017). *Participatory budgeting in Indonesia*. Making All Voices Count.

Graham, M., Hjorth, I., & Lehdonvirta, V. (2017). Digital labour and development: impacts of global digital labour platforms and the gig economy on worker livelihoods. *Transfer: European Review of Labour and Research*, 23(2), pp. 135–162.

Grillos, T. (2017). Participatory budgeting and the poor. *World Development*, 96, pp. 343–358.

Hagen, E. (2017). *Open Mapping from the ground up*. Making All Voices Count.

Heeks, R. (2017). *A Structural Model and Manifesto for Data Justice for International Development*, GDI Development Informatics Working Paper no.69. University of Manchester, UK.

Heeks, R. (2018). *Information and Communication Technology for Development*. Routledge, Abingdon, UK.

Heeks, R., & Renken, J. (2018). Data justice for development: What would it mean? *Information Development*, 34(1), pp. 90–102.

Kenney, M., & Zysman, J. (2015). Choosing a future in the platform economy: the implications and consequences of digital platforms. *Paper presented at Kauffman Foundation New Entrepreneurial Growth Conference,* Amelia Island, FL, 18–19 Jun.

Leszczynski, A. (2016). Speculative futures: Cities, data, and governance beyond smart urbanism. *Environment and Planning A: Economy and Space*, 48(9), pp. 1691–1708.

Menon, S. (2013). *Participatory Budgeting in Pune*. Centre for Environment Education, Pune.

Parnell, S., & Oldfield, S. (eds). (2014). *The Routledge Handbook on Cities of the Global South*. Routledge, Abingdon, UK.

Romero-Lankao, P., Gnatz, D., Wilhelmi, O., & Hayden, M. (2016). Urban sustainability and resilience: From theory to practice. *Sustainability*, 8(12), 1224.

Schaffers, H., Komninos, N., Pallot, M., Trousse, B., Nilsson, M., & Oliveira, A. (2011). Smart cities and the future internet: Towards cooperation frameworks for open innovation. In *Future Internet Assembly 2011*, J. Domingue et al. (eds), Springer, Berlin, pp. 431–446.

Shekhar, S., & Padmanabhan, V. (2015). *The Quality of Civic Data in India and the Implications on the Push for Open Data*. Transparent Chennai, Institute for Financial Management and Research, Chennai, India.

SKK (2010). *Gilingan Neighbourhood Map*. Solo Kota Kita, Solo.

Srnicek, N. (2017). *Platform Capitalism*. Polity Press, Cambridge, UK.

Taylor, A. (2014). *The People's Platform: Taking Back Power and Culture in the Digital Age*. Fourth Estate, London.

Verrest, H., & Pfeffer, K. (2018). Elaborating the urbanism in smart urbanism. *Information, Communication & Society*, 22(9), pp. 1–15.

SECTION 3

What kinds of urban knowledge are generated, legitimised, and valued through platforms?

10
WHEN DATA IS CAPITAL
Datafication, accumulation, extraction

Jathan Sadowski

Introduction

Data has become central and essential for increasingly more sectors of contemporary capitalism. Industries focused on technology, infrastructure, finance, manufacturing, insurance, and energy are now treating data as a form of capital. Until recently, it was commonplace for companies to delete or not collect data because paying for storage did not seem like a good investment (Oracle and MIT 2016). But things have changed. Just as we expect corporations to be profit-driven, we should now expect organisations to be data-driven. The drive to accumulate data now propels new ways of doing business and governance. It is a key factor in major corporate decisions, such as Amazon's acquisition of Whole Foods for $13.7 billion (Stevens and Haddon 2017), and of government policies such as investment in urban sensor networks (Heinzmann 2014). Indeed, as *The Economist* (2017b) has noted, "Industrial giants such as GE and Siemens now sell themselves as data firms." In short, data—and the accumulation of data—is a core component of political economy in the 21st century.

As a paradigm and logic, the idea of data-as-capital affects and transforms many spaces and sectors. Thanks to technologies like the Internet of Things, online platforms, and data analytics the list of things that now count as "digital products and services"—and hence what counts as part of the digital economy—is growing at a rapid pace (Srnicek 2016). This, in turn, means that data is a foundational form of capital for everything from the "smart home" to the "smart city," finance to governance, production to distribution, consumer devices to enterprise systems, and much more (Kitchin 2014). Without data, many of these technologies and organisations would not be able to operate, let alone be able to generate value.

This chapter contributes to the study of data within contemporary capitalism by analysing data as a form of *capital*. The existing literature on the social, political, and economic dimensions of data treats data as a *commodity*. Yet, the distinction between capital and commodity is important. Framing data as a form of capital casts new light on the imperatives motivating contemporary organisations, the ways value can be derived from data, and the normative importance of data extraction. Datafication is not just passive collection; this process takes shape as a political-economic regime driven by the logic of perpetual (data) capital accumulation and circulation.

Data-driven capitalism

There are now a variety of labels that refer to the political-economic relationship between data and capitalism, such as "surveillance capitalism" (Foster and McChesney 2014; Zuboff 2015), "informational capitalism" (Fuchs 2010), and "platform capitalism" (Srnicek 2016). These different labels are not interchangeable, but they do share common themes and conclusions. I build on three broad insights from the growing literature on the critical political economy of data: (1) data is valuable and value-creating (Roderick 2014; Arvidsson 2016); (2) data collection has a pervasive, powerful influence over how businesses and governments behave (Bouk 2017; Wu and Taneja 2019); and (3) data systems are rife with relations of inequity, extraction, and exploitation (Fourcade and Healy 2013; Poon 2016; Aitken 2017).

Fourcade and Healy (2017) have argued that "modern organisations" are now driven by a "data imperative" that demands the extraction of all data, from all sources, by any means possible. "Storing and studying people's everyday activities, even the seemingly mundane, has become the default rather than the exception" (Angwin and Valentino-Devries 2012: np). Fulfilling the data imperative involves more than just passively collecting data; it means actively creating data (IBM 2014). This entails the (total) datafication and surveillance of people, places, processes, things, and relationships among them (van Dijck 2014). As IBM states that, "Everything is made of data these days" (IBM 2014).

Such statements do not merely *reveal* or *reflect* the world. They *order* and *construct* the world (boyd and Crawford 2012; Kitchin et al. 2015). They put those with data capital in a position of access and authority. They establish the context through which accumulation and use of data not only occurs but becomes a driving logic that influences behaviour. They perform the power/knowledge relationship by examining its features and characteristics, sorting it into categories and norms, rending it legible and observable, and excluding other metrics and methods of knowing it (Bowker and Star 2000).

Data is a recorded abstraction of the world created and valorised by people using technology. The framing of data as a natural resource that is everywhere and free for the taking reinforces regimes of data accumulation. Thus, data mining is a misleading name; a more apt term would be data manufacturing. Data is not out there waiting to be discovered as if it already exists in the world like crude oil and raw ore (Gitelman 2013).

It is not a coincidence that data is treated as a universal substance right at the time when there is so much to gain for whoever can lay claim to that data and extract it from every source. Indeed, there is a feedback loop: many control systems rely on the constant gathering and processing of data, and in turn those control systems enable more data to be generated (Sadowski and Pasquale 2015). Flows of data correspond to flows of power and profit, thus the alchemy of datafication promises to produce infinite reserves of both. At the same time, the rhetoric of universality reframes everything as within the domain of surveillance/platform/digital capitalism

The goal of transforming everything into data and the search for new sources of data echoes imperialist modes of accumulation (Luxemburg 1951; Thatcher et al. 2016). In short, as capitalism faces crises of accumulation, there is a need to find new sources of value and new places to offload goods. This could mean subjecting previously non-commodified and non-monetised parts of life to the logic of capitalism or colonising new territories so they are brought into the global capitalist web as sites of extraction (Moore 2015). We can see this dynamic of "data colonialism" when technology corporations like Facebook and Google move into territories like India and Africa (Thatcher et al. 2016). These new places with new people provide new opportunities for data accumulation—imperialism updated for the digital age.

A growing body of research on critical data studies (Dalton et al. 2016) has shown how the production, distribution, and use of data is situated within an emerging political economy that has wide-ranging implications across society: from the restructuring of cities and the state (Kitchin et al. 2015; Leszczynski 2012), to the (re)development of electrical and computational infrastructure (Levenda et al. 2015; Pickren 2018). But this chapter returns to foundational questions in the political economy of data: What is the economic form of data? How can value be derived from data? Why does data collection matter? Opening back up these questions, I argue, productively reframes how we understand the form and dynamics of data.

Data capital

The "big data strategist" for Oracle, one of the largest software companies in the world, has said, "Data is in fact a new kind of capital on par with financial capital for creating new products and services. And it's not just a metaphor; data fulfils the literal textbook definition of capital." (OracleANZ 2015). As businesses and government bodies begin treating data as capital, there is a need for examining the characteristics and dynamics of "data capital" (Oracle and MIT 2016). This section aims to do so by first reviewing two theories of capital, from Karl Marx and Pierre Bourdieu, then using them to analyse data.

In *Capital, Volume 1*, Marx (1990) describes capital as a relationship between money (M) and commodities (C); namely, the ways they circulate and transform, which he simplifies into two general formulas. The first formula represents consumption, C-M-C: a commodity is sold for money which is then used to buy another commodity. Therefore, C-M-C is the cycle of using money to turn one qualitatively different thing (e.g. labour power) into another qualitatively different thing (e.g. coffee).

The second formula represents capital, M-C-M′: money is used to buy a commodity which is then sold for more money. The cycle of capital is motivated by exchange-value and the cycle does not complete because capital requires continuous circulation. In order to expand, money capital is converted into two forms of "real capital" employed in the creation of surplus value. Constant capital is the means of production for commodities (i.e., factories, machinery, raw materials, etc.). Variable capital is the means of subsistence for labour power (i.e., the costs of hiring workers).

Expanding on Marx's foundational analysis, Bourdieu theorised two new forms of capital that are distinct from economic capital: cultural capital and social capital stand alone in their own right while also being "convertible, in certain conditions, into economic capital" (1986: 242). They are, at their root, "transformed, disguised forms of economic capital" (Bourdieu 1986: 251).

Cultural capital contributes to a person's status and success in ways that go beyond the idea of "human capital," which focuses on monetary investment in education and skills. Cultural capital is a representation of class and tends to be invested by a person's family and transmitted from a person's domestic environment. Bourdieu (1986) identifies three types of cultural capital: embodied (e.g. character traits), objectified (e.g. art collection), institutionalised (university degree). Social capital, according to Bourdieu (1986: 248), "is the aggregate of the actual or potential resources which are linked to possession of a durable network of more or less institutionalized relationships of mutual acquaintance and recognition." Think of networking at a conference. This form of capital accrues by being included in privileged groups, whether that means being inducted through rites of passage (e.g. fraternal orders) or through rites of inheritance (e.g. noble lineage).

Building from Marx, we can now frame two common analyses of data in terms of a debate about what economic form data represents. On one hand, data is cast as a digital raw material—constant capital—necessary in the production of commodities. The cover of a 2017 issue of *The Economist* (2017b) proclaims "The World's Most Valuable Resource" above an illustration of offshore oil platforms labelled with the names of major digital platforms like Facebook, Google, and Uber presumably drilling into an ocean of data. On the other hand, data is cast as a commodity produced by digital labour of people posting on Facebook, clicking on Google, exercising with Fitbits, and all the other things we do that create data and that data is created about (Fuchs 2014, Till 2012). Through the work of using platforms and devices, people are turned into commodities that take the form of personal data, which is sold to advertisers and data brokers. Therefore, at the risk of oversimplification, these two ways of analysing data—as raw material and as product of digital labour—can be recast as a debate about the relationship between real capital and commodities in the digital economy.

Building from Bourdieu, I suggest a better framing of data is as a form of capital that is distinct from, but has its roots in, the economic capital. Data capital is more than knowledge about the world, it is discrete bits of information that are digitally recorded, machine-processable, easily agglomerated, and highly mobile. Like social and cultural capital, data capital is convertible, in certain conditions, to economic capital. But not all value derived from data is necessarily or primarily monetary. Data capital is institutionalised in the information infrastructure of collecting, storing, and processing data; that is, the smart devices, online platforms, data analytics, network cables, and server farms.

Importantly, these characteristics of data capital mean it can be continually captured and circulated, thus data collection is driven by the logic of capital accumulation as described by Marx. This unending accumulation of capital, represented by M-C-M′-C-M″-C-M‴…, is a defining feature of capitalism. In digital capitalism, the imperative is to constantly collect and circulate data by producing commodities that create more data and building infrastructure to manage data. The stream of data must keep flowing and growing.

Ultimately, continuing the cycle of data capital becomes an intrinsic motivation, a driving force, for firms. The capitalist is not concerned with the immediate use of a data point or with any single collection, but rather the unceasing flow of data-creating. This point is illustrated by the fact that data is very often collected without specific uses in mind. Indeed, the practice of collecting data first and figuring it out later is increasingly a core part of how businesses and government bodies operate (Fourcade and Healy 2017). At a public talk in early 2017, Andrew Ng, an artificial intelligence researcher who has held top positions at Google, Baidu, and Coursera, was candid about this prevailing logic of data accumulation: "At large companies, sometimes we launch products not for the revenue, but for the data. We actually do that quite often... and we monetize the data through a different product" (Stanford 2017). The conditions needed to convert data capital into economic capital may never arrive, but that does not stop the cycle of accumulation.

The shift towards data capital takes advantage of the ideological and regulatory groundwork that has been laid since at least the 1980s to create a political-economic landscape conducive to finance capitalism (Konczal and Abernathy 2015). Under neoliberal governance, financial capital is treated as if it exists in transnational space beyond borders and governance (Major 2012). The same attitudes are directly applied to data capital. This means, for example, that a company could collect the personal information of Americans, store the data in Taiwan, and sell it in Europe (Rossiter 2017). Any restraints on this flow of data are said to hinder economic growth and technological innovation (Morozov 2015). This view was crystallised by Carl Bildt (2015), the former Prime Minister of Sweden and chair of the Global Commission on Internet Governance, in an op-ed for the *Financial Times*: "Barriers against the free flow of data are, in effect, barriers against trade." Bildt was rebuking proponents of "digital sovereignty" rules in Europe, which would require non-EU companies to keep data about EU citizens in servers that are geographically based in Europe.

The focus on data and datafication should not be seen as usurping financialisation, but rather as adding new sources of value and new tools of accumulation. There is a long history of crossover between innovations in information technology and innovations in finance (MacKenzie 2018). Far from being in competition with each other, Wall Street and Silicon Valley are converging around data capital as the new frontier of accumulation and circulation.

Data extraction

Analysing this process in terms of extraction emphasises the people targeted by, and the exploitative nature of, dataveillance.

Much of the valuable data capital extracted from the world is about people—their identities, beliefs, behaviours, and other personal information. This means that accumulating data often goes hand-in-hand with increasingly invasive systems for probing, monitoring, and tracking people (Schneier 2016). Surveillance—or, "dataveillance"—capabilities are integrated into everything ranging from consumer goods to civic infrastructure. Not only purchasing but interacting with smart technologies—especially ones integrated into your everyday, personal life—generates reams of data that would otherwise be out of reach to the companies that want it. And, it seems, to the governments that want that data: In February 2016, the then US director of national intelligence, James Clapper, admitted to a Senate panel that government agencies may treat networked smart technologies as a portal into people's homes and lives (Ackerman and Thielman 2016: np).

A typical example of a smart update to an everyday technology is the refrigerator. A regular refrigerator is a passive object: it just keeps food cold. A smart refrigerator is an active object: it keeps food cold, but it also keeps track of things like your favourite brands, what foods you eat at what times, and when your food is almost out or expired. The smart refrigerator can then take that data and use it, for example, to send targeted advertisements, recommend sponsored recipes, monitor your dietary intake, and purchase replacement food from the grocery store. The smart refrigerator can also be used for other purposes that are far from fridge-like, such as a surveillance device remotely accessed by police who wish to peek into the owner's house (Butler 2017).

This is how the logic of accumulation works: it transforms the refrigerator into a data producing, collecting, and transmitting machine. The same logic is behind the growing stable of smart technologies that are increasingly embedded with sensors, processors, and network connections. Rather than existing only as a commodity to be sold, a smart device becomes (perhaps primarily) a means of producing data. This logic influences the design of systems ranging from robotic vacuum cleaners secretly mapping users' homes so the manufacturer can exploit that data (Deahl 2017) to the methods of urban planning deployed to manage cities (Barns 2017). Data accumulation drives many key decisions about technological development, political governance, and business models.

When data is treated as a form of capital, the imperative to collect as much data, from as many sources, by any means possible intensifies existing practices of accumulation and leads to the creation of new ones. Indeed, following in the footsteps of other extractive enterprises through capitalism's history (Mezzadra and Neilson 2017), many of the now-common practices of data accumulation should actually be understood in terms of the more forceful practice of data *extraction,* wherein data is taken without *meaningful consent* and *fair compensation* for the producers and sources of that data. The terminology used to describe the ways data is accumulated—especially data about people—elides the fact that this data is often acquired in hidden ways for purposes unknown to the targets of dataveillance (Andrejevic 2014).

The question of consent is relatively straightforward. The problematic way technology firms treat consent is no secret; it's an issue raised often by journalists and academics. When companies seek consent to record, use, and/or sell a person's data, it is typically done with an end-user licensing agreements (EULA). They are a hallmark of digital technology and are the predominant form of the contract we enter into—almost on a daily basis for many Internet and software users (Thatcher et al. 2016). These are the pages on websites and applications that make you click "agree" or "accept" before you can use the service. EULAs are known as "standard-form" or "boilerplate" contracts because they are generically applied to all users (Zamir 2014). They are one-sided, non-negotiated, and non-negotiable; you either agree or you are denied access (Birch 2016). Companies are routinely caught smuggling dubious clauses into their EULAs; like, for example, requiring users to give up rights to control their data or to restrict how data is used. Moreover, EULAs are designed to prevent even the most enterprising person from being informed of the binding terms and conditions. They are long, dense legal documents. One study concluded it would take 76 days, working for 8 hours a day, to read the privacy policies a person typically encounters in a year (Madrigal 2012).

EULAs are the ideal-type of *pro forma* "consent," which may be better-termed acquiescence (Pasquale 2015). That is, EULAs are less a method of consent in any meaningful and more a form of compliance. Few, if any, alternatives are offered. Thus, even in many cases where people must actively agree to their data being accumulated, this agreement bears little resemblance to common meanings of consent—let alone robust forms of informed consent. When a thing is taken without consent we call it "theft." Just because the thing taken here is information about a person, rather than some material object, the ethical relevance should not be nullified. It is extraction nonetheless.

The question of fair compensation is more complicated, in large part because it can be difficult to put a fair price on personal information. Different types of data are valued differently by different businesses. The value of data also rises non-linearly in relation to the scale of data. The larger and more diverse a data bank, the more information and uses can be derived from it. So one individual's data may not be readily converted to economic capital, but the aggregated data of hundreds, thousands, millions of individuals can be immensely valuable. Even though it is difficult to price data, we can judge the fairness of compensation in at least two ways: (1) what kind of compensation, if any, is offered for data and (2) what is the difference between the compensation for data producers and the value obtained by data capitalists?

First, compensation most often comes in the form of access to services like Facebook's platform and Google's search engine. Rather than charging money to use the service, the owner collects data as payment. Even if we concede that some people think this is perfectly fair compensation, these service providers are outnumbered by the countless companies that collect, use, and sell personal data often without the knowledge of—let alone compensation for—those whose data they possess (Bouk 2017; Crain 2016). Many companies fail the first test right away: receiving nothing can hardly be seen as fair.

Second, the value of data capital is massive. Some of the wealthiest companies in the world, like Facebook and Google, are built on data capital. The data broker industry is estimated to generate $200bn in annual revenue (Crain 2016). The three biggest data brokers alone— Experian, Equifax, and Transunion— each bring in billions of dollars annually. Even for relatively small data brokers, the difference between the value of data and the compensation provided for it is striking (Roderick 2014). Additionally, other major sectors like finance, insurance, and manufacturing are increasingly relying on data capital to generate value. For many of these companies the data they use is primarily about people and created by those people doing things. These companies are accumulating billions of dollars in surplus value from the "digital labour" done by people (Scholz 2012), while paying little to nothing in return. Thatcher et al. (2016: 994) argue that these extractive practices go so far as to "mirror processes of primitive accumulation or accumulation by dispossession that occur as capitalism colonises previously noncommodified, private times and places." When a person doesn't receive a fair offer for the work they've done or thing they've sold, we call it "exploitation"—and this level of exploitation and inequity is indicative of extraction.

Before concluding, it is important to note that not all data extraction is equal. There are crucial issues related to the ways identity and class affect how, what, and why data is extracted. At times, data is disproportionally extracted from certain groups, such as when poor people of colour are subjected to systematic tracking by government agencies and financial institutions (Eubanks 2018). At other times, certain groups are missing from data sets, such as when facial recognition systems inaccurately identify people of colour because they've been trained with data composed of mostly white male faces—people who look like their programmers (Lohr 2018). There is a need for further analysis of the unevenness of data extraction.

Conclusion

This chapter has centred data and *data capital* as a core component of political economy in the 21st century. By applying the theories of Marx and Bourdieu, data is analysed as a form of capital that is distinct from, but has its roots in, the economic capital. Data collection is thus driven by the perpetual cycle of capital accumulation, which in turn drives capital to construct and rely upon a world in which everything is made of data. The supposed universality of data reframes everything as falling under the domain of data capitalism. All spaces must be subjected to datafication. If the universe is conceived of as a potentially infinite reserve of data, then that means the accumulation and circulation of data can be sustained forever. The imperative to capture all data, from all sources, by any means possible influences many key decisions about business models, political governance, and technological development. Following this imperative leads to accumulation by extraction in which personal data is taken with little regard for consent and compensation. Reframing surveillance technology and the data economy in terms of extraction helps us move beyond focusing (almost exclusively) on privacy and security. As important as these issues are, they elide the systemic issues of inequity and exploitation that are endemic to the contemporary political economy of data (Coll 2014).

Moreover, conceiving of many common practices of data collection as extraction helps lay the normative groundwork for political and legal responses to rampant, invasive data accumulation. Such responses could include regulations—essentially capital controls—on what types of data companies can collect, how they can collect it, where they can send and store it, and how much data a company can possess, both in aggregate and about individuals. It could also include new models of data ownership and governance like, for example, "managing crucial parts of the data economy as public infrastructure" and breaking up monopolistic firms like Amazon (*The Economist* 2017a: np). The mainstreaming of such ideas should be seen as a bellwether for how powerful Big Data (as in Big Oil and Big Finance) has become. This illustrates the need for further critical thought about the political economy of data, as well as reforms and alternatives to data capitalism.

The analysis here is not meant to mark a new epoch in political economy wherein—as executives and engineers in Silicon Valley are fond of saying—everything has changed and nothing will ever be the same. Instead, data capitalism is more of a shift in focus; it's a transition towards conceptualising a new kind of capital and new methods of accumulation. This transition follows from one of the dominant socio-economic regimes of the past few decades: finance capitalism (Krippner 2005; Konczal and Abernathy 2015; Davis and Walsh 2017). There are similarities between financialisation and datafication. Both have significant "implications for the production of space, corporate governance, accumulation regimes, and everyday life" (Fields 2017: 1). Both seek to maximise value extraction by using innovative methods of capital creation and circulation, whether through complex financial instruments or complex information technologies. Both use technically opaque systems that shield them from oversight (Pasquale 2015), use their political influence to skirt regulation (Roderick 2014), and use their powerful capabilities to engage in exploitative and predatory practices (Taylor and Sadowski 2015). In addition to these similarities, there is direct overlap between the two regimes, such as credit agencies using large sets of personal and demographic data to create hyper-individualised policies and scores (Hurley and Adebayo 2017) and Wall Street traders using "high frequency trading" algorithms to circulate capital at hyper-speed (Arnoldi 2016).

The institutions leading the way in data capitalism are explicit about the connections between financial capital and data capital. They are not calling for one to replace the other, rather they are arguing that finance and data should be seen as different but equal forms of capital, which supercharge each other. Datafication, like financialisation before it, is a new frontier of accumulation and next step in capitalism. Compared to financialisation, datafication is still in its early days, but the level of wealth and power wielded by data capitalists is already massive and still growing. The theories and methods used to analyse finance capitalism and information technology must now be synthesised and applied to studying the meaning, practices, and implications of datafication as a political-economic regime.

References

Aitken, R. (2017). 'All Data is Credit Data': Constituting the Unbanked. *Competition & Change* 21 (4): pp. 274–300.

Andrejevic, M. (2014). The Big Data Divide. *International Journal of Communication* 8: pp. 1673–1689.

Angwin, J. and Valentino-Devries, J. (2012). New Tracking Frontier: Your License Plates. *The Wall Street Journal* (September 29). Accessed February 28, 2016. http://www.wsj.com/articles/SB10000872396390443995604578004723603576296

Arnoldi, J. (2016). Computer Algorithms, Market Manipulation and the Institutionalization of High Frequency Trading. *Theory, Culture & Society* 33 (1): pp. 29–52.

Arvidsson, A. (2016). Facebook and Finance: On the Social Logic of the Derivative. *Theory, Culture & Society* 33 (6): pp. 3–23.

Barns, S. (2017). Visions of Urban Informatics: From Proximate Futures to Data-Driven Urbanism. *Fibreculture Journal* 29. doi: 10.15307/fcj.29.214.2017.

Bildt, C. (2015). EU Should Resist Urge to Rig the Rules of Cyber Space. Accessed 9 October 2017: https://www.ft.com/content/5d626a4e-f182-11e4-88b0-00144feab7de.

Birch, K. (2016). Market vs. Contract? The Implications of Contractual Theories of Corporate Governance to the Analysis of Neoliberalism. *Ephemera: Theory & Politics in Organizations* 16 (1): pp. 107–133.

Bouk, D. (2017). The History and Political Economy of Personal Data over the Last Two Centuries in Three Acts. *Osiris*, forthcoming.

Bourdieu, P. (1986). The Forms of Capital. In J. Richardson (ed.), *Handbook of Theory and Research for the Sociology of Education*, pp. 241–258. Westport, CT: Greenwood.

Bowker, G. and Star, S.L. (2000). *Sorting Things Out: Classification and Its Consequences*. Cambridge, MA: The MIT Press.

boyd, d. and Crawford, K. (2012). Critical Questions for Big Data: Provocations for a Cultural, Technological, and Scholarly Phenomenon. *Information, Communication & Society* 15 (5): pp. 662–679.

Butler, J. (2017). QLD Passes Laws to Turn Your Fridge into Police Surveillance Device. *Huffington Post,* 6 September. Accessed 9 October 2017: http://www.huffingtonpost.com.au/2017/09/06/qld-passes-laws-to-turn-your-fridge-into-police-surveillance-device_a_23198327/.

Coll, S. (2014). Power, Knowledge, and the Subjects of Privacy: Understanding Privacy as the Ally of Surveillance. *Information, Communication & Society* 17 (10): pp. 1250–1263

Crain, M. (2016). The Limits of Transparency: Data Brokers and Commodification. *New Media & Society*. doi: 10.1177/1461444816657096.

Dalton, C.M., Taylor, L., and Thatcher, J. (2016). Critical Data Studies: A Dialog on Data and Space. *Big Data & Society* 3 (1): pp. 1–9.

Davidson, A. (2017). A Washing Machine that Tells the Future. *The New Yorker,* October 23. Accessed 23 October 2017: https://www.newyorker.com/magazine/2017/10/23/a-washing-machine-that-tells-the-future.

Davis, A. and Walsh, C. (2017). Distinguishing Financialization from Neoliberalism. *Theory, Culture & Society* 34 (5–6): pp. 27–51.

Deahl, D. (2017). Roombas Have Been Busy Mapping Our Homes, and Now that Data Could Be Shared. *The Verge,* 24 July. Accessed 9 October 2017: https://www.theverge.com/2017/7/24/16021610/irobot-roomba-homa-map-data-sale.

Eubanks, V. (2018). *Automating Inequality: How High-Tech Tools Profile, Police, and Punish the Poor*. New York: St. Martin's Press.

Fields, D. (2017). Urban Struggles with Financialization. *Geography Compass*. doi: 10.1111/gec3.12334.

Foster, J.B. and McChesney, R.W. (2014). Surveillance Capitalism: Monopoly-Finance Capital, the Military-Industrial Complex, and the Digital Age. *Monthly Review,* 1 July. Accessed 9 October 2017: http://monthlyreview.org/2014/07/01/surveillance-capitalism.

Fourcade, M. and Healy, K. (2013). Classification Situations: Life-Chances in the Neoliberal Era. *Accounting, Organizations and Society* 38: pp. 559–572.

Fourcade, M. and Healy, K. (2017). Seeing Like a Market. *Socio-Economic Review* 15 (1): pp. 9–29.

Fuchs, C. (2010). Labor in Informational Capitalism and on the Internet. *The Information Society* 26 (3): pp. 179–196.

Fuchs, C. (2014). *Digital Labour and Karl Marx*. New York: Routledge

Gitelman, L. (ed.) (2013). *'Raw Data' is an Oxymoron*. Cambridge, MA: MIT Press

Heinzmann, D. (2014). New Sensors Will Scoop Up 'Big Data' on Chicago. *Chicago Tribue*, 23 June. Accessed 9 October 2017: www.chicagotribune.com/news/local/breaking/ct-big-data-chicago-20140621,0,3075626.story.

Hurley, M. and Adebayo, J. (2017). Credit Scoring the Era of Big Data. *Yale Journal of Law and Technology* 18 (1): pp. 148–216.

IBM (2014). A World Made with Data. Made with IBM. *YouTube*, 27 May. Accessed 9 October 2017: https://www.youtube.com/watch?v=QCgzrOUd_Dc.

Kitchin, R. (2014). *The Data Revolution: Big Data, Open Data, Data Infrastructures and Their Consequences*. London: Sage.

Kitchin, R., Lauriault, T.P., and McArdle, G. (2015). Knowing and Governing Cities Through Urban Indicators, City Benchmarking, and Real-Time Dashboards. *Regional Studies, Regional Science* 2 (1): pp. 6–28.

Krippner, G.R. (2005). The Financialization of the American Economy. *Socio-Economic Review* 3: pp. 173–208.

Konczal, M. and Abernathy, N. (2015). *Defining Financialization*. New York: Roosevelt Institute.

Leszczynski, A. (2012). Situating the Geoweb in Political Economy. *Progress in Human Geography* 36 (1): pp. 72–89.

Levenda, A., Mahmoudi, D., and Sussman, G. (2016). The Neoliberal Politics of 'Smart': Electricity Consumption, Household Monitoring, and the Enterprise Form. *Canadian Journal of Communication* 40 (4): pp. 615–636.

Lohr, S. (2018). Facial Recognition Is Accurate, if You're a White Guy. *The New York Times*, 9 February. Accessed 12 September 2018: https://www.nytimes.com/2018/02/09/technology/facial-recognition-race-artificial-intelligence.html.

Luxemburg, R. (1951). *The Accumulation of Capital*. New York: Monthly Review Press.

MacKenzie, D. (2018). Material Signals: A Historical Sociology of High-Frequency Trading. *American Journal of Sociology* 123 (6): pp. 1635–1683.

Madrigal, A.C. (2012). Reading the Privacy Policies You Encounter in a Year Would Take 76 Work Days. *The Atlantic*, 1 March. Accessed April 22, 2016 http://www.theatlantic.com/technology/archive/2012/03/reading-the-privacy-policies-you-encounter-in-a-year-would-take-76-work-days/253851/.

Mahmoudi, D. and Levenda, A. (2016). Beyond the Screen: Uneven Geographies, Digital Labour, and the City of Cognitive-Cultural Capitalism. *TripleC* 14 (1): pp. 99–120.

Major, A. (2012). Neoliberalism and the New International Financial Architecture. *Review of International Political Economy* 19 (4): pp. 536–561.

Marx, K. (1990). *Capital, Volume 1*. Translated by B. Fowkes. London: Penguin Classics.

Mezzadra, S. and Neilson, B. (2017). On the Multiple Frontiers of Extraction: Excavating Contemporary Capitalism. *Cultural Studies* 31 (2/3): pp. 185–204.

Moore, J.W. (2015). *Capitalism in the Web of Life: Ecology and the Accumulation of Capital*. London, UK: Verso Books.

Morozov, E. (2015). What Happens When Policy is Made by Corporations? Your Privacy is Seen as a Barrier to Economic Growth. *The Guardian*, 11 July. Accessed 9 October 2017: https://www.theguardian.com/commentisfree/2015/jul/12/ttip-your-data-privacy-is-a-barrier-to-economic-growth.

Mueller, G. (2016). Piracy as Labour Struggle. *TripleC* 14 (1): pp. 333–345.

OracleANZ (2015). What is Data Capital. *YouTube*, 24 September. Accessed 9 October 2017: https://www.youtube.com/watch?v=ffMi-JwotiM.

Oracle and MIT Technology Review Custom. (2016). *The Rise of Data Capital*. Accessed 9 October 2017: http://files.technologyreview.com/whitepapers/MIT_Oracle+-Report- The_Rise_of_Data_Capital.pdf.

Pasquale, F. (2015). *The Black Box Society: The Secret Algorithms that Control Money and Information*. Cambridge, MA: Harvard University Press.

Pickren, G. (2018). 'The Global Assemblage of Digital Flow': Critical Data Studies and the Infrastructure of Computing. *Progress in Human Geography* 42 (2): pp. 225–243.

Poon, M. (2016). Corporate Capitalism and the Growing Power of Big Data: Review Essay. *Science, Technology & Human Values* 41 (6): pp. 1088–1108.

Roderick, L. (2014). Discipline and Power in the Digital Age: The Case of the US Consumer Data Broker Industry. *Critical Sociology* 40 (5): pp. 729–746.

Rossiter, N. (2017). Imperial Infrastructures and Asia Beyond Asia: Data Centres, State Formation and the Territoriality of Logistical Media. *The Fibreculture Journal* 29: pp. 1–20.

Sadowski, J. and Pasquale, F. (2015). The Spectrum of Control: A Social Theory of the Smart City. *First Monday* 20 (7): online.

Schneier, B. (2016). *Data and Goliath: The Hidden Battles to Collect Your Data and Control Your World*. New York: W.W. Norton & Company.

Scholz, T. (2012). *Digital Labor: The Internet as Playground and Factory*. New York: Routledge.

Srnicek, N. (2016). *Platform Capitalism*. Cambridge, UK: Polity Press.

Stanford Graduate School of Business. (2017). Andrew Ng: Artificial Intelligence is the New Electricity. *YouTube*, 2 February. Accessed 21 January 2018: https://www.youtube.com/watch?time_continue=2041&v=21EiKfQYZXc.

Stevens, L. and Haddon, H. (2017). Big Prize in Amazon-Whole Foods Deal: Data. *The Wall Street Journal,* 20 June. Accessed 9 October 2017: https://www.wsj.com/articles/big-prize-in-amazon-whole-foods-deal-data-1497951004

Taylor, A. and Sadowski, J. (2015). How Companies Turn Your Facebook Activity into a Credit Score. *The Nation*, 27 May. Accessed 9 October 2017: https://www.thenation.com/article/how-companies-turn-your-facebook-activity-credit-score/

Thatcher, J., O'Sullivan, D., and Mahmoudi, D. (2016). Data Colonialism Through Accumulation by Dispossession: New Metaphors for Daily Data. *Environment and Planning D* 34 (6): pp. 990–1006.

The Economist (2017a). Data is Giving Rise to a New Economy. *The Economist*, 6 May. Accessed 9 October 2017: https://www.economist.com/news/briefing/21721634-how-it-shaping-up-data-giving-rise-new-economy.

The Economist (2017b). The World's Most Valuable Resource is No Longer Oil, But Data. *The Economist*, 6 May. Accessed 9 October 2017: https://www.economist.com/news/leaders/21721656-data-economy-demands-new-approach-antitrust-rules-worlds-most-valuable-resource.

Till, C. (2012). Exercise as Labour: Quantified Self and the Transformation of Exercise into Labour. *Societies* 4 (3): pp. 446–462.

van Dijck, J. (2014). Datafication, Dataism and Dataveillance: Big Data Between Scientific Paradigm and Ideology. *Surveillance & Society* 12 (2): pp. 197–208.

Wu, A.X. and Taneja, H. (2019). How did the data extraction business model come to dominate? Changes in the web use ecosystem before mobiles surpassed personal computers. *The Information Society* 35 (5): pp. 272–285.

Zamir, E. (2014). Contract Law and Theory – Three Views of the Cathedral. *University of Chicago Law Review* 81: pp. 2077–2123.

Zuboff, S. (2015). Big other: Surveillance Capitalism and the Prospects of an Information Civilization. *Journal of Information Technology* 30: pp. 75–89.

11

PLATFORM URBANISM AND KNOWLEDGE-POWER

Maroš Krivý

"Platform urbanism invokes for us a twist on the notion of 'platform capitalism'," write Scott Rodgers and Susan Moore (2018), setting out an agenda for the critical study of digital platforms and the urban. Drawing on Nick Srnicek's (2016) study of data and digital infrastructures under capitalism, Rodgers and Moore propose a speculative Henri Lefebvre-inspired substitution of "urbanism" for "capitalism." In a similar fashion, Sarah Barns (2017) identifies platform urbanism with the rise of business and policy models that pivot on urban data. She asks, what are the governance and citizenship implications of late capitalist, data-driven urban platforms? Adding to and departing from these debates, this chapter explores cultural fantasies surrounding the metaphor of the "platform" and asks, what forms of power does it sustain? It also argues for a more nuanced understanding of "urbanism" in relation to platform capitalism.

Urbanism as a way of life and knowledge-power

Since the publication of the classic essay by Louis Wirth (1938), the term "urbanism" has generally referred to a distinctively urban way of life engendered by modern urbanization and industrialization. While Wirth's environmental and technological determinisms permeate the "urbanology" of the likes of Richard Florida and Edward Glaeser, critical scholars have challenged Wirth's emphasis on population size, density and diversity, and instead examined the urban through the political and institutional lenses (Peck 2016). In doing so, they have either historicized but otherwise retained Wirth's sense of the term (e.g. "neoliberal urbanism"), or used the term in a different sense, as an investigation into the diversity of urban life forms ("comparative urbanism") (Peck, Theodore and Brenner 2009; Robinson 2016). As a result, in current urban scholarship, the term describes either an urban way of life (albeit one that has now been "politicized") or a study thereof. The puzzle, in other words, is that *urbanism* appears to be identical with *a theory of urbanism* (for contrast, imagine that the term "capitalism" would also mean a theory of capitalism or even Marxism). In putting the term "platform urbanism" into circulation, we would therefore do well to avoid confusing critical theory with its object.

Avoiding this conundrum, in this chapter "urbanism" is understood neither as a way of life nor as its critical study, but as an institutionalized disciplinary knowledge, a meaning of the term derived from French *urbanisme*. Lefebvre (1991: p. 414, 12) writes that the term is "borrowed from official pronouncements" and that "towns, cities—urban space—are the bailiwick of the discipline of urbanism." *Urbanisme* is what Lefebvre has in mind when he theorizes the practices of "conceiving" and "representing" space as intrinsic to spatial production. "Urbanism" was for Lefebvre not a placeholder term for, nor a theory of, urban way of life, but a historically specific and geographically situated institutional apparatus for manipulating, regulating and intervening into the urban (Pinder 2003). Lefebvre's collaborators from the French *Utopie* collective further expanded on this definition by asking, how is the disciplinary—or rather interdisciplinary—knowledge of urbanism implicated within the welfare-capitalist mode of social production? According to sociologist Hubert Tonka (2011: 156f1, 176), for example, the interdisciplinary nature of urbanism—which he situated at the intersections of "the organization of space, urban development, urban planning, regional development..., various urbanisms: organic, semiological, spatial, subterranean..., urbanology, ekistics"—masked its contradictory character of "a science of synthesis [... that] attempts to totalize separations." Urbanism, according to *Utopie*, was an exemplary case of what Louis Althusser's called the ideological state apparatus: the discipline's claim "to change the city in order to change society," members of the collective wrote in a joint statement, have effectively concealed "the fact that it participates in this existing order" (Utopie 2011: p. 113).

Given that Lefebvre and *Utopie* identified urbanism with various modernisms *à la* Bauhaus and the post-war welfare state, how is their work significant to platform urbanism today? If it is true that, on one hand, the ideal of synthesis mystifies social conflicts today as much as in the 1960s, contemporary urbanism, on the other, aims at this synthesis less by securing coherence and repression, than by unlocking, governing, and capitalizing on complexity (Chandler 2014; Grove et al. 2019). Rather than asking whether and how it is ideological, we might therefore consider urbanism as a form of expert power, and investigate ways in which it derives legitimacy from the metaphor-program of the platform. In this endeavour, urbanism could be examined as an interdisciplinary and transnational culture of expertise, drawing on the work of anthropologists who study experts as "actor[s] who ha[ve] developed skills in, semiotic-epistemic competence for, and attentional concern with, some sphere of practical activity" (Boyer 2008: p. 39). The question would then concern the gravitational pull of "platforms" on the skills, epistemologies, and attention of urbanists, who might be also considered as a transnational, loosely integrated thought collective, sharing in knowledge exchange, circulation of ideas and epistemic culture (Knorr-Cetina 1999; Plehwe 2009; Lobsinger 2013; Peck and Theodore 2015). We might further draw on Michel Foucault's investigation of power and knowledge, but also on various attempts to think knowledge and property relations as complementary, such as ones by Jacques Bidet on the rise of organizational "competent elites" or by William Davies on "elite power." Bidet's development of Foucault's (1982) concept of *pouvoir-savoir*, translated to English as "knowledge-power" (rather than the customary "power-knowledge"), emphasizes "a supposed knowledge that is doubled with a conferred authority," while Davies places at centre stage the power and competency of "expert intermediaries" to interpret data and algorithms for the political arena (Bidet 2016: p. 68; Davies 2017).[1]

Platforms: science, solutions, socialization

This chapter is not about how digital business platforms such as Airbnb, Facebook or Uber transform and upset urbanism (in Wirth's sense). Rather it is about the knowledge-power of urbanists-experts, and varieties of expertise that centre on platforms as a pivot of good urban governance. The chapter draws on examples from contemporary urbanism spanning the fields of urban science, urban policy and urban design. There is of course no single way in which to delineate the discipline of urbanism—its disciplinary authority seems to correspond with the normalization of particular programs, techniques or slogans as incontrovertible common sense beyond the discipline itself (for example, the need for cities to become "creative")—but it is possible to identify three strands to urbanistic expertise centred on platforms.

First, it is the so-called urban science, associated with research centres such as Santa Fe Institute, degree-awarding programs such as MIT's newest major in Urban Science and Planning with Computer Science, and institutions that combine research and graduate education such as Bartlett's Centre for Advanced Spatial Analysis (CASA) and New York University's Center for Urban Science And Progress (CUSP). Urban science combines the insights of complexity theory, neo-classical urban economics and anti-modernist strands of urban thought. The city is understood as a non-linear aggregation of individual behaviours apprehensible as data, and modelled as a self-organizing, emergent and unpredictable system. What sets urban science apart from the broader field of spatial sciences, as well as from the long history of using natural science to explain urban change, is the premise that thanks to the availability of "Big Data," urban analysis could be scientific without being deterministic. Despite the bias-free mantle, urban science has been applied in contested contexts ranging from the policing of urban riots in London to offering services to the developer of Hudson Yards in New York (Batty 2005, 2013; Bettencourt and West 2010; Baudains, Johnson and Braithwaite 2013; Bettencourt 2015; Kontokosta 2016; West 2017).

The second strand—let's call it "solutionist" following Evgeny Morozov (2013)—includes various "laboratories" encompassing applied research, urban design and contemporary start-up and maker cultures. The MIT Senseable City Lab is a prominent representative, as is a number of other MIT labs that collaborate with the Senseable City Lab and tinker with the urban, but also consultancies such as the New York-based Bits and Atoms. The notion of the urban platform, in this case, places emphasis on the ever-untapped potential of the city as a substrate of potentially disruptive innovations, the source of which are technologically mediated but also face-to-face human interactions. Wildly participatory and mildly counter-cultural, solutionism is at home in fab-labs, hackathons, urban living labs and unlikely interdisciplinary collaborations. It could be described as a form of institutionalized hacking, an aggressive techno-utopianism that thrives on bricolage and design thinking. This strand intervenes into the urban by means of all sorts of widgets and contraptions that are modular, mass-customizable and scalable, such as the dynamic curb invented for the widely touted Sidewalk Labs project in Toronto (Ratti and Townsend 2011; Kloeckl 2013; Townsend 2013; cf. Evans, Karvonen and Raven 2016; Ratti and Claudel 2016; Marshall 2018).

Third, the umbrella term "platform urbanism" could and should also consider various part architecture part urban policy initiatives that derive from "non-digital" participatory culture (it is also more European in provenance), and, as such, have been largely ignored by scholars of platforms. This strand is centred around enlivening public spaces, pedestrianization and other forms of more or less temporary urban interventions that aim to bring people together for shared and ostensibly non-instrumental experiences of the city, on sidewalks and streets, but also in (former) wastelands and other unconventional settings. Its proponents range from Danish urban design consultancy Jan Gehl Associates, the champions of "cities for people" and "livability," to Berlin-based art/architecture collective raumlabor, whose statement of purpose is "bye bye utopia," complete with ways in which their and others' work has been implicated and co-opted in municipal policies across the globe. In this context, "platforms" have been considered in terms of physical space and as a trigger for social change, typically DIY pavilions and regeneration projects aiming to foster communal experience. The platform urbanism of this third kind pivots on "socialization," encouraging the practice of mixing socially with others, but also normalizing cultural patterns of mixing with some rather than others, and remains oblivious to gentrification to which this urbanism not a little contributes, as epitomized by New York's High Line elevated park, a community initiative turned global urbanistic template. (Gehl 2010, 2011; David and Hammond 2011; Speck 2013; Liesegang 2014; SenStadt... 2007).

While these strands naturally overlap (and should be therefore read analytically), they highlight a range of meanings to "platform," from a database on urban behaviour patterns, through an environment of innovation, to a physical space where different bodies assert their liveliness. Posed in this way, a question of platform urbanism is encapsulated in catchlines such as "smart cities that put people not technology first." The mandate of the platform appears robust because it functions as a "legitimacy exchange" between disparate strands to urbanism revolving around participatory ethos and computational infrastructures (Bowker 1993: p. 108; Turner 2006: p. 25). What various urbanistic vision-programs centred around platforms have in common is that they claim an elusive mandate to govern that appears as its opposite: optimizing yet tranquilizing, ubiquitous and soothing. Amongst other reasons, this chapter argues, the platform has become compelling to urbanists of different kinds because it seemingly resolves the conundrum of governing but not determining cities. It provides a persuasive answer to questions such as, how to know the city in its unknowability? How to plan, design and otherwise govern the urban, while appearing to increase individual choice? How to embrace the urban as a self-organizing, distributed and complex system, while preserving the expertise of urbanists to identify points of intervention, their competency to steer the system towards preferred vectors of change and their knowledge-power to govern urban complexity?

How do platforms govern?

In his work on platform capitalism, Srnicek (2016: p. 48) defined platforms as a networked environment of sensing infrastructures that engenders, extracts, refines and circulates data, "a designed core architecture that governs the interaction possibilities." Tarleton Gillespie's (2010: p. 350) definition of the platform as a "'raised level surface' designed to facilitate some activity that will subsequently take place" underlines a similar meaning. Gillespie (2014, 2015, 2017) cautions against technological determinism and highlights the metaphor's performative force: by way of foregrounding a capacity to survey the terrain and act powerfully from an elevated, vantage point, the "platform" downplays if not entirely hides hierarchies, services and labour necessary to maintain platform infrastructures, but also and foremost the invisible power to design, intervene into and algorithmically operate them.

Platform urbanism is, this chapter suggests, a focal point for environmental and algorithmic forms of governing rationality (Foucault 2008, 2009; Rouvroy 2012; Rouvroy and Berns 2013; Gabrys 2014; Yeung 2017, 2018; Hörl 2018). While the former concerns governing from a distance, bearing on "social environment," on "the rules of the game" rather than on "the players," the latter preserves these tenets while placing "the players" at centre stage—not, however, as reasoning individuals or political subjects but as "infra-individual data and supra-individual profiles" (Foucault 2008: p. 146, 260; Rouvroy 2012: p. 145). While environmental and algorithmic techniques have long captivated the imagination of architects and urban planners (Scott 2016), it could be argued that they have consolidated into a distinctive governing rationality, and have become increasingly hegemonic, only on the heels of the Third Way mutation within late capitalism in the 1990s (Peck and Tickell 2002; Boltanski and Chiapello 2005; Mudge 2018: pp. 260–364). That historically and the geo-politically significant moment is consistent with the reinvention of urbanism along the lines of so-called New Public Management, an approach to public intervention that favours, as cultural theorist Stuart Hall (2003: p. 18) put it, not changing what people believe but "changing their practices" (which also explains why "knowledge-power" is for us more useful than "ideology").

The condition of possibility for the environmental and algorithmic types of power is a historically specific conundrum faced by the urbanistic competent elites: how do you intervene, as an urbanist, in order to maximize individuals' freedom from that intervention? Freedoms asserted by platform urbanists tend to be of a particular type: individual freedoms *from* politics, planning and societal intervention upheld in contrast and opposition to collective freedoms *to* (affordable public housing, free education), associated with post-war socialist, social democratic and other welfare-type urban programs, typically grouped under the umbrella term of "modernist" urbanism—in fact, and not without irony, co-opting some aspects of the critique of urbanism by Lefebvre and *Utopie* (cf. Kaminer 2011: pp. 39–40). It is significant that not capitalism but the so-called modernist urbanism centred around master planning, functional zoning and extensive public infrastructural investments into council or public housing have been consistently scapegoated for urban failures by platform urbanists of different shades (Swenarton, Avermaete and Van Den Heuvel 2015). The trope of modernism remains compelling to them as a kind of zombie allegory of totalizing urbanism, one that has outlived the specifically post-modernist critique as a ghostly foil against which a barrage of catchwords, programs and sub-themes of platform urbanism, such as smart, liveable, participatory or resilient cities, stands out as commonsensical (Martin 2010).

By highlighting the significance of environmental and algorithmic power to platforms, I do not want to provide a totalizing description of the contemporary city. This caveat is in place because of a tendency among critical scholars to describe urban effects of digital technologies as totalizing: take sociologist Richard Sennett's (2012) argument that smart cities "repress informality in the name of coherent control," urbanist Adam Greenfield's (2013) interpretation of the smart city as a neo-modernist update to Le Corbusier's Plan Voisin and cities like Brasília (which is intended as a knee-jerk insult), or even media anthropologist Shannon Mattern's (2017) otherwise imaginative critique of "the totalizing idea of the city as computer."[2] The "platform" is compelling, I suggest by contrast, because it holds a promise of squaring the circle of being in control without controlling, of engaging the totality of the urban—from its planetary infrastructure to the hive mind of its denizens—in some kind of non-totalizing way.

This is in no way to deny the continued relevance of class, ethnic and gender analysis to urban change (Harvey 1982; Lees, Slater and Wyly 2008). On the contrary, it is to draw attention to the role of platform thinking in perpetuating and amplifying existing inequalities. As media scholar Wendy Chun (2016: p. 58) writes,

> by finding seemingly unrelated correlations, Big Data can aggravate existing inequalities and lead to racist and discriminatory practices, justified

> through the use of seemingly innocuous proxies. Through these proxies, the allegedly 'coarse' and 'out-dated' categories of race, class, sexuality, and gender are accounted for in unaccounted ways.

Forms of urban science and policy that extol data as invariably beneficial are dangerous not because they are totalizing and normalizing but because they are neutralizing and pre-emptive, disappearing into the background of benevolent algorithms and unobtrusive environments.[3]

From environmental to algorithmic power?

This brings me to "nudge," a much-talked-about concept in behavioural economics that aims to improve individual decisions by intervening at the level not of subjects but their incentive structure (Thaler and Sunstein 2008). The theory and its remarkable popularity amongst public and urban policy makers (Jones, Pykett and Whitehead 2014; BIT 2016) gives a sense of prescience and almost tangible quality to Foucault's notion of environmentality. Its authors have presented it as a "libertarian paternalist" solution that pivots on an "improved choice architecture" (and is outright hostile to socialism) (Thaler and Sunstein 2008: p. 183; Sunstein 2019a). Put otherwise, the power of "nudge" derives from what legal scholar Karen Yeung (2017: p. 130), following psychologist Shoshana Zuboff (2015), refers to as the "ownership of the means of behavioural modification."

However, the popularization of "nudge," complete with its neo-behavioural challenge to the *homo oeconomicus* concept covered in a humanistic mantle, highlights also the limits to the argument that environmental power doesn't bear on individuals. In his lectures on ordo- and neo-liberalism, Foucault (2008, 2009) addressed the principle of the "rule of law," and more broadly the biopolitics of security predicated on statistically normative reason. "Nudge" is compelling because it extends from what Yeung (2017, 2018) calls "regulation by design" to what she calls "hypernudge," that is to say, from types of environmental interventions through standard-setting that indiscriminately target entire populations, to techniques of influence that are algorithmically tailored to particular individuals and groups, being predicated on a machinic capacity to identify behavioural correlations indiscernible to humans (cf. Sunstein 2019b). It also poses questions about the algorithmic operations of financial and real estate capitals (Morozov and Bria 2018).

Antoinette Rouvroy and Berns (2013: p. 173), also a legal scholar, refers to algorithmic governmentality as an "(a)normative or (a)political rationality" that is anticipatory, pre-emptive and neutralizing. The subject of algorithmic governmentality is not population but (in)dividuals that are simultaneously hyper-subjectified and de-subjectified. Rouvroy calls "data-behaviourism" a mode of governance that "circumvents and avoids reflexive human subjects, feeding on infra-individual data which are meaningless on their own, to build supra-individual models of behaviours" (Rouvroy 2012; Rouvroy and Berns

2013: pp. 173–174). If neoliberalism is usually conceived in substantive terms—for Wendy Brown (2015: p. 118, my highlight), for example, it is a "specific *normative* form of reason"—platform urbanism opens the question of neoliberal "governing without norms": an algorithmic power grafted onto an environmental one (Rouvroy 2018).

Urbanization and the "City"

It is useful to remind ourselves here of another, even if widely known, critical distinction addressed by Lefebvre between urbanization and the "city." For Lefebvre, the urban exceeds the "city": urbanization was not born out of industrialization but rather will have reached its apotheosis with de-industrialization.[4] What Lefebvre calls "urban revolution" is then tantamount to the process through which the urban has become a primary, critical site of capital accumulation, rather than its secondary circuit (Harvey 1982; Lefebvre 2003; Gotham 2006; Haila 2016). But urban revolution, especially at its current moment of planetary "explosion," is coextensive with a paradoxical "implosion" that fixes the "city" in time and space. It is accompanied by a tenacious ideology of the city, which breaths into it an air of timelessness and universality (Wachsmuth and Weisler, 2018). Networked infrastructures have proliferated across the planet hand in hand with the ideology of the pre-modern, pre-industrial and pre-capitalist "city," propagated by a torrent of intellectual precepts such as New Urbanism and the liveable city (Harvey 1997; Krivý and Ma 2018).

The urbanization-city dialectics has overlapped with a tension within the field of urbanism between the factions of "futurists" and "nostalgics," as Lefebvre (2014: p. 766) calls them. This tension has manifested in a dispute between techno-utopians of global circulatory networks and starry-eyed guardians of the classical city, which has been endemic to modern urbanism since the 19th century (Tafuri 1976; Schorske 1981: pp. 24–115; Vidler 1998). The flurry of cultural fantasies around urban platforms might then signal a perplexing convergence between the two factions. It may be that platform urbanism is situated, historically and intellectually, at the intersection of, on one hand, the planetary-wide smartness mandate (cf. Brenner & Schmid 2012; Halpern, Mitchell and Geoghegan 2017) and, on the other hand, the abstractly humanist mandate to remake "cities for people" (Gehl 2010; Speck 2013). Futurists and nostalgics are no longer strange bedfellows: this double-edged agenda lends platform urbanism an "apolitical" mantle that appears commonsensical and compelling.

Conclusion

The critical study of platform urbanism could build on but also re-examine the limits to the theory of platform capitalism: take Srnicek's (2017) argument that companies such as Airbnb, which operate on a "lean platform" model disproportionally dependent on surplus capital, are likely to fall apart soon, but considers neither the real state capital nor the experiential and ideological aspects to the "city" (cf. Wachsmuth and Weisler 2018; Stehlin 2018). Further research should better illuminate how platform urbanism "preserves" the false opposition between digital techno-utopianism and neo-conservative determinism, and contribute to building alliances around, beyond and outside of it (note that the Sidewalk Labs in Toronto, met with legal resistance and citizen initiatives such as #BlockSidewalk, has been touted as both smart and liveable city project).

Urbanists have been entranced by the potentials of environmental and algorithmic techniques to model and intervene into the urban circulation of humans and non-humans, while giving a new lease of life to behaviourally and cognitively deterministic notion of the "city." They have embraced and disseminated a type of expertise that doesn't go against the grain of urban life and doesn't tame it, but rather instigates and surrenders to its contingency and complexity: according to one exemplary statement of this position, the challenge is to "steer the system in ways that mesh with the way the system functions routinely" (Batty and Marshall 2009: p. 570). One reason for being cautious not to confuse different meanings of "urbanism," which I discussed in the opening of this chapter, and reassert the dated, dialectical character of the relationship between theory and practice, is that the cunning of platform urbanism-as-expertise might consist precisely in that it now appears continuous with urbanism-as-way-of-life.

Rather than adding to such obfuscation, we need to study and critique forms of expertise, competence and knowledge that have made it possible for urbanists to cast themselves as the instigators and governors of the spontaneous city. To make sense of this apparently ironic predicament, we might consider that the knowledge-power of platform urbanism has centred not only around "making bodies docile in relation to norms," but increasingly around "making norms docile in relation to bodies" (Rouvroy, in Rouvroy and Stiegler 2015: p. 128). Within the purview of the platform, the sheer matter-of-factness of urbanization *and* the city has become a principle of veridiction for governing the urban. It is urgent to address how algorithmic regulation of urban platforms—while grafted onto the environmental power and disciplinary normalizations—pre-empts and neutralizes the capacity for democratic government.

Notes

1 Translator Steven Corcoran (in Bidet 2016: p. xi) further explains that the rendering of "knowledge-power" better resonates with Bidet's other term "property-power" (*pouvoir-propriété*).

2 Mattern (2017) further claims that "the *city as computer* model ... hinder[s] the development of healthy, just, and resilient cities," despite the idea of resilient cities being a descendant of this model (Adams 2014; Grove 2018).

3 By "dangerous" I evoke Foucault's (2000: p. 256) methodological aside: "I would like to do the genealogy of problems, of *problematiques*. My point is not that everything is bad, but that everything is dangerous, which is not exactly the same as bad. If everything is dangerous, then we always have something to do."

4 It could be argued that the aforementioned sense of "urbanism" as a study or theory of the urban developed out of the crisis of the industrial city.

References

Adams, R. E. (2014) "Notes from the Resilient City" *Log* 32: pp. 126–139.

Barns, S. (2017) "Visions of Urban Informatics: From Proximate Futures to Data-Driven Urbanism" *The Fibreculture Journal* 29, online at http://fibreculturejournal.org/wp-content/pdfs/FCJ-214SarahBarns.pdf.

Batty, M. (2005) *Cities and Complexity* (Cambridge, MA: MIT Press).

Batty, M. (2013) *The New Science of Cities* (Cambridge, MA: MIT Press).

Batty, M. and Marshall, S. (2009) "The Evolution of Cities: Geddes, Abercrombie and the New Physicalism" *Town Planning Review* 80 (6): pp. 551–574.

Baudains, P., Johnson, S. and Braithwaite, A. (2013) "Geographic Patterns of Diffusion in the 2011 London Riots" *Applied Geography* 45: pp. 211–219.

Bettencourt, L. (2015) "Cities as Complex Systems" In Furtado, B. A., Sakowski, P. and Tóvolli, M. (Eds.) *Modeling Complex Systems for Public Policies*, pp. 217–238 (Brasília: IPEA).

Bettencourt, L. and West, G. (2010) "A Unified Theory of Urban Living" *Nature* 467 (7318): pp. 912–913.

Bidet, J. (2016) *Foucault with Marx* (London: Zed Books).

Boltanski, L. and Chiapello, E. (2005) *The New Spirit of Capitalism* (London: Verso).

Bowker, G. (1993) "How to Be Universal: Some Cybernetic Strategies, 1943–70" *Social Studies of Science* 23 (1): pp. 107–127.

Boyer, D. (2008) "Thinking through the Anthropology of Experts" *Anthropology in Action* 15 (2): pp. 38–46.

Brenner, N. and Schmid, C. (2012) "Planetary Urbanization" In Gandy, M. (Ed.) *Urban Constellations*, pp. 10–13 (Berlin: Jovis).

Brown, W. (2015) *Undoing the Demos* (New York: Zone Books).

Chandler, D. (2014) "Beyond Neoliberalism: Resilience, the New Art of Governing Complexity" *Resilience: International Policies. Practices and Discourses* 2(1): pp. 47–63.

Chun, W. (2016) *Updating to Remain the Same: Habitual New Media* (Cambridge, MA: MIT Press).

David, J. and Hammond, R. (2011) *High Line: The Inside Story of New York City's Park in the Sky* (New York: Farrar Straus Giroux).

Davies, W. (2017) "Elite Power under Advanced Neoliberalism" *Theory, Culture & Society* 34 (5–6): pp. 227–250.

Evans, J., Karvonen, A. and Raven, R. (2016) *The Experimental City* (Abingdon: Routledge).

Foucault, M. (1978) *The History of Sexuality. Volume 1: An Introduction* (New York: Pantheon Books).

Foucault, M. (2000 [1983]) "On the Genealogy of Ethics: An Overview of Work in Progress" In Rabinow, P. (Ed.). *Ethics, Subjectivity and Truth*, pp. 253–280 (London: Penguin Books).

Foucault, M. (2008 [1979]) *The Birth of Biopolitics: Lectures at the Collège de France 1978–1979* (New York: Palgrave Macmillan).

Foucault, M. (2009 [1978]) *Security, Territory, Population: Lectures at the Collège de France 1977–1978* (New York: Palgrave Macmillan).

Gabrys, J. (2014) "Programming Environments: Environmentality and Citizen Sensing in the Smart City" *Environment and Planning D: Society and Space* 32 (1): pp. 30–48.
Gehl, J. (2010) *Cities for People* (Washington: Island Press).
Gehl, J. (2011) *Life between Buildings* (Washington: Island Press).
Gillespie, T. (2010) "The Politics of Platforms" *New Media & Society* 12 (3): pp. 347–364.
Gillespie, T. (2014) "The Relevance of Algorithms" In Gillespie T., Boczkowski P. J. and Foot K. A. (Eds.) *Media Technologies: Essays on Communication, Materiality and Society*, pp. 167–194 (Cambridge, MA: MIT Press).
Gillespie, T. (2015) "Platforms Intervene" *Social Media + Society* 1 (1): pp. 1–2.
Gillespie, T. (2017) "The Platform Metaphor, Revisited" *Culture Digitally* (24 August), online at culturedigitally.org/2017/08/platform-metaphor/.
Gotham, K. F. (2006) "The Secondary Circuit of Capital Reconsidered: Globalization and the U.S. Real Estate Sector" *American Journal of Sociology* 112 (1): pp. 231–275.
Greenfield, A. (2013) *Against the Smart City* (New York: Do Projects).
Grove, K. (2018) *Resilience* (New York: Routledge).
Grove, K., Krivý, M., Rickards, L., Schliwa, G., Collier, S., Cox, S. and Gandy, M., "Interventions on Design and Political Geography" *Political Geography* 74 (2019), 102017.
Haila, A. (2016) *Urban Land Rent: Singapore as a Property State* (Oxford: Wiley-Blackwell).
Hall, S. (2003) "New Labour's Double-Shuffle" *Soundings* 24: pp. 10–24.
Halpern, O., Mitchell, R. and Geoghegan, B. D. (2017) "The Smartness Mandate: Notes Toward a Critique" *Grey Room* 68: pp. 106–129.
Harvey, D. (1982) *The Limits to Capital* (Oxford: Basil Blackwell).
Harvey, D. (1997) "The New Urbanism and the Communitarian Trap" *Harvard Design Magazine* 1: pp. 68–69.
Hörl, E. (2018) "The Environmentalitarian Situation: Reflections on the Becoming-Environmental of Thinking, Power, and Capital" *Cultural Politics* 14 (2): pp. 153–173.
Jones, R., Pykett, J. and Whitehead, M. (2014) "The Geographies of Policy Translation: How Nudge Became the Default Policy Option" *Environment and Planning C: Government and Policy* 32 (1): pp. 54–69.
Kaminer, T. (2011) *Architecture, Crisis and Resuscitation: The Reproduction of Post-Fordism in Late-Twentieth-Century Architecture* (London: Routledge).
Kloeckl, K. (2013) "Senseable Cities: City as a Platform," *The Platform Strategy Executive Symposium, MIT Media Lab*, presentation, 26 July, online at http://ebusiness.mit.edu/platform/agenda/slides/9%20Kristian-Kloeckl_City-as-a-platform.pdf.
Knorr-Cetina, K. (1999) *Epistemic Cultures: How the Sciences Make Knowledge* (Cambridge, MA: Harvard University Press).
Kontokosta, C. (2016) "The Quantified Community and Neighborhood Labs: A Framework for Computational Urban Planning and Civic Technology Innovation" *Journal of Urban Technology* 23 (4): pp. 67–84.
Krivý, M. and Ma, L. (2018) "The Limits of the Livable City: From Homo Sapiens to Homo Cappuccino" *Avery Review* 30, online at https://averyreview.com/issues/30/limits-of-the-livable-city.
Lees, L., Slater, T. and Wyly, E. (2008) *Gentrification* (Abingdon: Routledge).
Lefebvre, H. (1991 [1974]) *The Production of Space* (Oxford: Basil Blackwell).
Lefebvre, H. (2003 [1970]) *The Urban Revolution* (Minneapolis, MN: University of Minnesota Press).
Lefebvre, H. (2014 [1947, 1961, 1981]) *Critique of Everyday Life* [The One-Volume Edition] (London: Verso).
Liesegang, J. (2014) "The City as a Sphere for Action," *Osthang Project Summer School Symposium*, presentation, 24 July, online at http://raumlabor.net/city/.

Lobsinger, M. L. (2013) "Two Cambridges: Models, Methods, Systems, and Expertise" In Dutta, A. (Ed.) *A Second Modernism: MIT, Architecture, and the 'Techno-Social' Moment*, pp. 651–685 (Cambridge, MA: MIT Press).

Marshall, A. (2018) "Sidewalk Labs' Bid to Reinvent Toronto Starts with Shape-Shifting Streets," *Wired*, 17 August 2018, online at https://www.wired.com/story/google-sidewalk-labs-toronto-quayside-flexible-streets/.

Martin, R. (2010) *Utopia's Ghost. Architecture and Postmodernism, Again* (Minneapolis: University of Minnesota Press).

Mattern, S. (2017) "A City Is Not a Computer" *Places Journal*, online at https://placesjournal.org/article/a-city-is-not-a-computer/.

Morozov, E. (2013) *To Save Everything, Click Here: The Folly of Technological Solutionism* (New York: PublicAffairs).

Morozov, E. and Bria, F. (2018) *Rethinking the Smart City: Democratizing Urban Technology* (New York: Rosa Luxemburg Stiftung).

Mudge, S. L. (2018) *Leftism Reinvented: Western Parties from Socialism to Neoliberalism* (Cambridge, MA: Harvard University Press).

Peck, J. (2016) "Economic Rationality Meets Celebrity Urbanology: Exploring Edward Glaeser's City" *International Journal of Urban and Regional Research* 40 (1): pp. 1–30.

Peck, J. and Theodore, N. (2015) *Fast Policy: Experimental Statecraft at the Thresholds of Neoliberalism* (Minneapolis: University of Minnesota Press).

Peck, J. and Tickell, A. (2002) "Neoliberalizing Space" *Antipode* 34 (3): pp. 380–404.

Peck, J., Theodore, N. and Brenner, N. (2009) "Neoliberal Urbanism Redux?" *International Journal of Urban and Regional Research* 37 (3): pp. 1091–1099.

Pinder, D. (2015) "Lefebvre, Utopia and the Urban Question" *International Journal of Urban and Regional Research* 39 (1): pp. 28–45.

Plehwe, D. (2009) "Introduction" In Mirowski, P., Plehwe, D. (Eds.) *The Road from Mont Pèlerin. The Making of the Neoliberal Thought Collective*, pp. pp. 1–44 (Cambridge, MA: Harvard University Press).

Ratti, C. and Claudel, M. (2016) *The City of Tomorrow: Sensors, Networks, Hackers, and the Future of Urban Life* (New Haven: Yale University Press)

Ratti, C. and Townsend, A. (2011) "The Social Nexus" *Scientific American* 305 (3): pp. 42–49.

Robinson, J. (2016) "Comparative Urbanism: New Geographies and Cultures of Theorizing the Urban" *International Journal of Urban and Regional Research* 40 (1): pp. 187–199.

Rodgers, S. and Moore, S. (2018) "Platform Urbanism: An Introduction" *Mediapolis* 3 (4), online at http://www.mediapolisjournal.com/2018/10/platform-urbanism-an-introduction/

Rouvroy, A. (2012) "The End(s) of Critique: Data-Behaviourism vs. Due-Process" In Hildebrandt M. and DeVries E. (Eds.) *Privacy, Due Process and the Computational Turn*, pp. 143–168 (New York: Routledge).

Rouvroy, A. (2018) "Governing Without Norms: Algorithmic Governmentality" *Psychoanalytical Notebooks* 32: pp. 99–101.

Rouvroy, A. and Berns, T. (2013) "Gouvernementalité algorithmique et perspectives d'émancipation: Le disparate comme condition d'individuation par la relation?" *Réseaux* 31 (177): pp. 163–196.

Rouvroy, A. and Stiegler, B. (2015) "Le régime de vérité numérique. De la gouvernementalité algorithmique à un nouvel État de droit" *Socio. La nouvelle revue des sciences sociales* 4: pp. 105–132.

Schorske, C. E. (1981) *Fin-de-Siècle Vienna: Politics and Culture* (New York: Vintage Books).

Scott, F. (2016) *Outlaw Territories. Environments of Insecurity/Architectures of Counterinsurgency* (New York: Zone Books).
Sennett, R. (2012) "The Stupefying Smart Sity" *LSECities*, online at http://lsecities.net/media/ objects/articles/the-stupefying-smart-city/en-gb/
SenStadt Senatsverwaltung für Stadtentwicklung (Ed.) (2007) *Urban Pioneers. Temporary Use and Urban Development In Berlin* (Berlin: Jovis Verlag).
Speck, J. (2013). *City 2.0: The Habitat of the Future and How to Get There* (New York: TED Books, 2013).
Srnicek, N. (2017) *Platform Capitalism* (Cambridge, MA: Polity Press).
Stehlin, J. (2018) "Urban Platforms, Rent, and the Digital Built Environment" *Mediapolis* 3 (4), online at http://www.mediapolisjournal.com/2018/10/urban-platforms-rent-and-the-digital-built-environment/.
Sunstein, C. R. (2019a) "Trump Is Right to Warn Democrats About 'Socialism'," *Bloomberg Opinion*, 7 February, online at https://www.bloomberg.com/opinion/articles/2019-02-07/what-is-socialism-trump-s-closer-to-the-truth-than-democrats.
Sunstein, C. R. (2019b) "Algorithms, Correcting Biases" *Social Research* (online before print)
Swenarton, M., Avermaete, T. and Van Den Heuvel, D. (Eds.) (2015) *Architecture of the Welfare State* (New York: Routledge).
Tafuri, M. (1976) *Architecture and Utopia. Design and Capitalist Development* (Cambridge, MA: MIT Press).
Thaler, R. H. and Sunstein, C. R. (2008) *Nudge: Improving Decisions about Health, Wealth, and Happiness* (New Haven, CT: Yale University Press).
The Behavioral Insights Team [BIT] (2016) *Behavioral Insights for Cities* (New York: Behavioral Insights Team), online at https://www.bi.team/wp-content/uploads/2016/10/Behavioral-Insights-for-Cities-2.pdf.
Tonka, H. (2011 [1968]) "Critique of Urban Ideology" In Buckley, C. and Violeau, J.-L. (Eds.) *Utopie. Texts and Projects, 1967–1978*, pp. 155–177 (Los Angeles, CA: Semiotexte).
Townsend, A. M. (2013) *Smart Cities: Big Data, Civic Hackers, and the Quest for a New Utopia* (New York: W. W. Norton & Company).
Turner, F. (2006) *From Counterculture to Cyberculture: Stewart Brand, the Whole Earth Network, and the Rise of Digital Utopianism* (Chicago, IL: University of Chicago Press).
Utopie (2011 [1967]) "The Logic of Urbanism" In Buckley, C. and Violeau, J.-L. (Eds.) *Utopie. Texts and Projects, 1967–1978*, pp. 111–123 (Los Angeles, CA: Semiotexte).
Vidler, A. (1998) "The Third Typology" In Hays, K. M. (Ed.) *Architecture Theory since 1968*, pp. 288–294 (Cambridge, MA: MIT Press).
Wachsmuth, D. and Weisler, A. (2018) "Airbnb and the Rent Gap: Gentrification Through the Sharing Economy" *Environment and Planning A: Economy and Space* 50 (6): pp. 1147–1170.
West, G. (2017) *Scale: The Universal Laws of Growth, Innovation, Sustainability, and the Pace of Life in Organisms, Cities, Economies, and Companies* (London: Penguin Books).
Wirth, L. (1938) "Urbanism as a Way of Life" *American Journal of Sociology* 44 (1): pp. 1–24.
Yeung, K. (2017) "'Hypernudge': Big Data as a Mode of Regulation by Design" *Information, Communication & Society* 20 (1): pp. 118–136.
Yeung, K. (2018) "Algorithmic Regulation: A Critical Interrogation" *Regulation & Governance* 12 (4): pp. 505–523.
Zuboff, S. (2015) "Big Other: Surveillance Capitalism and the Prospects of an Information Civilization" *Journal of Information Technology* 30 (1): pp. 75–89.

12

WIKI-URBANISM

Curating a slum resettlement colony with open knowledge platforms

Padmini Ray Murray and Ayona Datta

Introduction: an "Analogue" editathon

In September 2018, we brought together a group of 12 young women living in a slum resettlement colony in Delhi's urban edge to participate in a Wikipedia editathon. The aim of the editathon was to produce a page on "Madanpur Khadar JJ Colony", where all the women had lived since childhood. This exercise in itself was an achievement, given the constraints on space and mobility imposed on these women by their families. We went to several homes of these women explaining the purpose of the editathon, assuring their families that this was a "legitimate" event, so that they gave "permission" to these women to travel for over an hour to the cultural and archival centre of the city – Indira Gandhi National Centre for the Arts (IGNCA) and participate in a much-anticipated day-long workshop. As part of the project team, we waited in anticipation to feedback for the first time, directly into open-knowledge platforms from the grassroots, based on the ordinary everyday experiences of women in the margins.

As soon as we started the editathon, we came upon our first challenge – digital capacity. While the women used Android phones, and most were adept at finding their way around the phone interface for familiar applications such as WhatsApp, YouTube, Google and other mobile applications using their Android phones, they did not know how to open an internet browser, search for and access Wikipedia. Thus, they could not make a distinction between search and browser – indeed, it seemed that many of them were unfamiliar with the notion of the browser as a portal to the Internet, and of the web itself. They perceived the icon for Google Chrome only as an application that facilitates search.

Consequently, we established that the browser-based potential of the World Wide Web (WWW) for most of these women was completely absent. Clicking on the icon for Chrome takes one to the default home page which is Google Search, which gave them the impression that the only way to find content was via the search engine. While they were capable of a rudimentary job search on Google, they were thus still ghettoised within the closed spaces of the marketplace of apps that they knew to download (via peer and other forms of onboarding) and were unable to navigate outwards into the vast spaces of the internet and open access knowledge platforms such as Wikipedia. This unfamiliarity of Indian users with Wikipedia is an established finding – a study conducted by the Wikimedia Foundation in 2016 found that only 33% of Indian internet users had heard of Wikipedia (Cruz, 2018).

The second challenge was infrastructural – network connectivity. The women only had access to mobile phones, but the network was slow and intermittent. This meant that even for those who finally managed to access Wikipedia, it was near impossible to register, log in and contribute to information online and in real-time as a group. As a way of moving forward, we switched to an analogue editathon, by using post-it notes to mimic co-authoring wiki entries. Based on the format of existing Wikipedia pages for more prominent neighbourhoods, the women focused on describing different aspects of the resettlement colony – history, local neighbourhood, accessibility, demographics, education and employability as well as physical and social infrastructure, in English or Hindi (Figure 12.1).

These notes could be written in English or Hindi, and once we created a basic structure for the Wikipedia page, we began to infill it with the rich information curated by the women. We aimed to publish a dual language page in English and Hindi – the former for a global audience, and the latter for a Hindi speaking audience, particularly for participants and residents of the colony who could update the information on the page regularly.

As soon as we uploaded the Wiki page, we faced our third challenge – authenticity. Wikipedia marked it with two notices at the top – one which requested copyediting for "*grammar, style, cohesion, tone or spelling*" and second more

FIGURE 12.1 Post-it notes for an analogue editathon. Photo: Ayona Datta.

FIGURE 12.2 Screenshot of older version of the Wikipedia page.

serious one regarding its citation trail (see Figure 12.2). The second notice concerned us as a team since Wikipedia threatened to remove the article if it did not cite "reliable" published sources, and in the context of an editathon where most of the page was written based on the first-hand experiences of the young women, we could only conclude that their knowledge was not, according to Wikipedia's editorial policy, deemed as "legitimate".

In this chapter we will take up each of these challenges to argue that (a) Wikipedia provides a way to curate the city as text, which is synchronous with how marginal citizens usually experience the city – through verbal directions and landmarks, rather than a 2-dimensional Google map and geolocated pins connected to the network in real-time; (b) just as a map-based platform (such as Google maps) the Wikipedia platform is not necessarily neutral but it enables a form of local nuance that is often lost in a visualisation of 2-dimensional relations in a map. Local place names can be elaborated, historicised and embodied through the textual and narrative capacities of Wikipedia in ways that are not enabled in map-based platforms; yet (c) despite being constructed as an open knowledge platform, Wikipedia's curatorial requirement of "authentic" knowledge as citational excludes oral and embodied knowledge and experience of those living in the urban margins. This goes against the emancipatory potential offered in textualisation of the city through open knowledge platforms.

"Madanpur Khadar JJ Colony": what's in a name?

Early on in our work we agreed to write the city along the terms of those in the margins of the digital and material spaces of the city. The women we worked with were young millennials who were second-generation migrants to the city – their parents had settled in slums and unauthorised colonies in the city since the 1960s, which were demolished during the slum eviction drives of the 2000s when Delhi prepared for hosting the Commonwealth Games (Datta 2012).

The women themselves had little recollection of the move, yet were marked by the stories and struggles of the resettlement colonies in the peripheries of the city where their parents had been relocated to by the Delhi Development Authority. Unlike the slums where they came from (which were usually depicted as blank spaces on the map), these resettlement colonies were visible on Google Maps with clear road networks and some landmarks. Yet they were still unmapped in most other ways – there was no information about these colonies in digital format, apart from the odd media article which would mention these are exceptional spaces of crime or poverty (The Hindu 2004) or scant scholarship that would frame them as spaces of "grey economy" (Gidwani and Kumar 2019). The editathon was intended to correct this (mis)representation of their colony through an open knowledge platform – Wikipedia.

As we began the editathon, we were immediately confronted with the politics of representation. What would the header page be titled? The women were insistent that this should be "*Madanpur Khadar JJ Colony*" as the signpost says at the start of the road that leads into the neighbourhood (see Figure 12.3). JJ standing for "*Jhuggi Jhopri*" is a particular Delhi policy rhetoric for all sorts of temporary squatter settlements (Datta 2012). The research team was concerned about this label which we felt incorrectly represented a resettlement colony where residents owned property titles, and therefore suggested that this should be titled as "*Madanpur Khadar*". But as it emerged this was also the name of the village whose agricultural land had been acquired to create the resettlement colony. The women felt that it was the JJ Colony whose history and lived experiences had so far remained invisible and therefore important to have the exact name of the colony on the header.

FIGURE 12.3 Entry to Madanpur Khadar JJ Colony. Photo: Rohit Madan.

Wikipedia as "Platform Urbanism"

While digital platforms have been around for a few decades, it is the increasing capacity of platforms to transform urban environments, which have directed recent research towards "platform urbanism". As Stehlin notes "platforms deterritorialise the built environment, reterritorialising it in a new physical-digital composition" (Stehlin 2018, online). They direct our navigational experience and thereby how we visualise the city through digital infrastructures. Platforms are now a popular technology of governance with city governments, used to efficiently manage a network of urban infrastructures and services to citizens. This approach has been critiqued for framing the "city as visualised facts" (Kitchin, Lauriault, and McArdle 2015, p. 6) whereby urban policy makers have a disembodied and panoptic engagement with the city. Indeed critiques of platforms arise from their support of technocratic governance and their co-optation for corporate interests while bypassing deeper questions of social inequalities (Bratton 2015; Anttiroiko 2016; Barns 2018; Janowski, Estevez, and Baguma 2018; Caprotti and Liu 2019; Leszczynski 2019). Thus as Krivy notes, platforms can actually produce gentrification, poverty and social exclusion through its digital environments (Krivy 2018).

Yet, despite increasing agreement over the exclusionary impacts of platforms, the social and material impacts of low-tech non-visual or textual platforms on the city are rarely studied. Recent scholarship has shown that the proliferation of locative media with "the ability to find oneself relative to everywhere else" (Bleecker and Nova 2006, p. 17), social media platforms (Wyche 2015), WhatsApp (O'Hara et al. 2014; Malka, Ariel, and Avidar 2015; Dixon 2018; Omanga 2018) and a range of other mobile apps enables those in the margins to access information and knowledge through relatively low-cost mobile phones (Datta 2020). It is also emerging that claims to citizenship are made from the margins increasingly through geotagging and countermapping techniques enacted through basic mobile phones (Levy 2018; Luque-Ayala and Neves Maia 2018), and that marginal citizens are increasingly using global corporate platforms such as Google, Facebook and Twitter to raise issues around absent infrastructure, curate and highlight their grievances with infrastructure (Omanga 2018).

This comes with its own challenges. There exists a deep digital divide in access to and use of technology across gender (Counterview 2017), generation (Rangaswamy 2013), caste (Kamath 2018), religion (Sarkar 2016), class and urban-rural divides (Gilbert 2010) that restricts open and free access to information and knowledge. And even if working-class women have access to mobile phones and digital infrastructures, the power of corporate technologies is threatening their open access to knowledge on the web. By using strategies such as funnelling search by an algorithm and discouraging the use of the web browser by making their offerings app first, and website later, platforms such as Twitter, Facebook, Uber and others have monopolised much of what we understand as knowledge on the web. Furthermore, most of the platforms offer possibilities to geolocate information on a map-based application. This offers tremendous potential to corporate giants such as Google to spatialise consumer behaviour and patterns to then push products and services to relevant people and places. And if we agree with current scholarship (Gurumurthy and Chami 2014) that phone ownership is not necessarily a pathway to empowerment, rather it is the capacity to reflect on available information and develop critical consciousness that has transformatory gender potential – we can certainly see the inherent politics of platforms in shaping knowledge in and of the city.

For women with access to limited resources such as education and infrastructure, these "dark patterns" of geolocation and web funnelling, posit a particular threat, allowing them no recourse to other forms of information outside the "walled garden" created by proprietary algorithms. In order to further tighten their grasp on new users, especially women, these dark patterns are re-inscribed by education programmes run by Google (Purnell 2016), ostensibly to help ease their transition to the digital, but in reality, indoctrinate them away from the values at the heart of the open web.

"Googlisation" of Indian platforms

The advent of Google Maps in India has radically transformed the lens through which Indians encounter their cities and understand space itself. While the mapping of India was a core motivation of the British in order to understand and chart their colonial dominion, maps that use co-ordinates for wayfinding is not culturally embedded mode of understanding the world. However, despite a chequered history of initial resistance by the Indian government (Rajan 2013; Choudhury 2016), Google Maps eventually established itself a lodestone essential to the growth of the digital economy in India. As Kumar writes, this was possible through the presentation of private interests in the monopolisation of data as a global good, bypassing "traditional controls on information flow" and leveraging "network power" (Kumar 2010, p. 157).

Google is fundamental to the proliferation of platforms in India and globally and is core to the workings of several map-based platforms such as Uber and Ola. The Indian government is also increasingly dependent on Google's infrastructure and mapping for realising landmark policies such as "Digital India" and "100 Smart Cities". The Google map is woven into a narrative of technologically driven development championed by the "Digital India" strategy that constantly sets it up in opposition to local, contextual understandings of the urban environment. This is appropriately captured in a report supported by Google (Dalberg Global Development Advisors 2015, p. 4).

> Maps help us search for places we are interested in, pinpoint their locations, optimize routes to get there, understand surrounding neighborhoods

> better, and communicate better with others. When maps cannot answer these questions, we rely on other sources that may be costly, time-intensive, or incomplete.

This "Googlisation" of geo-spatial data has overdetermined the way in which cities are increasingly read through and as only legible by technological informational monopolies in a digital age. This is the city understood solely by its potential for hyper-productivity and optimisation of capitalist endeavour, with very little regard for the lived realities of its inhabitants and their local knowledge, seen as "time-intensive". This lacuna is magnified further when viewed through the lens of gender, which remains invisible in the move towards universalising digital citizenship (Datta 2018, 2020). Inequality is fundamentally inscribed into the exclusionary nature of the smart city agenda in India, which fails to account for gendered experiences, and how it shapes one's relationship with the city.

At face value, Wikipedia seems to offer a model of platform urbanism through which the city can be known outside its googlisation. Wikipedia is a narrative medium. Even its option of a map-based addition does not offer geo-location. Through its narrative medium it was possible to evoke smells, sounds and other sensory experiences that are very different from how the Google car maps the city. We positioned Wikipedia as a platform of "urban" knowledge that could enable a descriptive narrative and push against what we imagine as a "map" based platform for understanding the city. As a textual platform it enabled an active archive, a palimpsest of historical and social narrative that could emerge from the ground. Yet as a digital platform it too enabled algorithmic funnelling of information in the way that it was structured and organised across categories.

Analogue pasts and digital "Urban" archives

The shift to an urban archive enabled by the Wikipedia page moves us away from the notion of the archive as animated upon the death of its author/subject to one that is constantly "live" – an affordance that we understand as enabling the democratisation of the archive. Ray Murray describes how the post-Web 2.0 archive complicates this notion of "death" as the starting point of the archive – where event, documentation and comment collapse the category of the "mortal" by inhabiting a state of always becoming (Ray Murray 2019). She observes that social media, and the digital archive then, are constant narrations of historical event, accelerating our relationship with a polyphonic archive. This shift, she argues, necessitates a reconsideration of the archive as an epistemological site, and its implications for feminist praxis in India (Ray Murray 2017).

This becomes particularly crucial in the face of the apathy displayed towards the active building and creation of archives for the poor and marginalised. Indeed, there is active resistance against archiving the struggles of slum dwellers in India. Records are destroyed, document trails are broken, paper is no longer considered valid. In fact, there is a rich scholarship on urban informality and marginality that suggest that the removal of archives particularly of the dispossessed is a state-driven strategy that seeks to deny them of their rights (Das and Poole 2004). It is particularly instructive, then, to understand that increasingly, our principal threshold to history is the world wide web, and thus for the purposes of the young women, Wikipedia was the only archive that allowed, with its relatively low barrier to entry, a site to inscribe the history of their struggles over two generations in the city.

Through the editathon, we spoke of *Madanpur Khadar JJ Colony* as a way to go beyond the imagination of commercial and the corporate platforms (such as Google) as active players in digital space, and as a gesture to the possibilities that Wikipedia and other ideologically similar movements, such as OpenStreetMap can have for platform urbanism. What Wikipedia does, in this case, is to assert through its mandate that anyone can write, curate and archive the city. This conception of Wikipedia as an example of an urban infrastructure finds credibility in Benkler and Nissenbaum's formulation of such platforms as "commons-based peer production" (Benkler and Nissenbaum 2006, p. 394).

In the editathon, this was evident in detailed descriptions of the chowks (squares) mostly named after different food items (such as Jalebi or Samosa) as significant places in the colony, which could only be represented through a narrative and accompanied by photos. The excerpt below from the final Wikipedia page (Wikipedia 2018) illustrates the nuance and embodied attributes that these places hold for the women.

> The first square on entering the colony is known as Samosa Chowk, followed by Nirman Chowk, after which is Jalebi Chowk and Sri Ram Chowk, which then leads on to the road to Kalindi Kunj. Shops around the Nirman Chowk sell building material. These are found between Jalebi Chowk and Samosa Chowk.
>
> While there are grocery and vegetable shops on each block, the main markets are the vegetable market near Jalebi Chowk and the Saturday market (Shani bazaar), which is set up near the Shani mandir (temple). The vegetable market opens daily. Jalebi Chowk and the adjacent market is the most popular destination for evening snacks and vegetables, which runs on a daily basis; it also has an SBI, Canara Bank and ICICI ATMs.

In addition to this, the introduction of Wikidata to the suite of Wikipedia-based projects, further consolidates the possibilities of a feasible alternative to corporate platforms. Wikidata is a machine-readable database that contains the data that is often "locked" away in the content of Wikipedia entries, for use in different applications, including maps. Of particular interest given the context of our project, was Wikidata's progressive approach to multilingual content. When we first introduced the notion of working with Wikipedia to the young women, we realised that even though they had some rudimentary understanding that Wikipedia existed (mostly because it often came up as a top result in a Google search) they were not aware that anyone could contribute to it, nor were they aware that Wikipedia existed in Indian languages, including in Hindi, their mother tongue. The publication of the "Madanpur Khadar JJ Colony" page simultaneously in Hindi (Wikipedia 2019) enabled plurality by design. This allowed conflicting data to co-exist and provided mechanisms to organise this plurality on one Wikidata site that hosts multilingual content.

The limits to "Open" urban archives

Despite the opportunities offered for peer production, when our Wiki page was slapped with the notice of citation and the threat of removal, we had to consider how archiving and mapping one's own space itself may be a struggle for rights of those marginalised in the city. We could not find any independent digital or analogue sources that mentioned Madanpur Khadar JJ Colony that could provide citational validity to the page. However, Datta's earlier work had charted the violence of law through which slums were evicted in the 2000s in Delhi (Datta 2012), and Bhan and Menon-Sen's documentation of the early struggles by those relocated to these resettlement colonies in Delhi's peripheries (Bhan and Menon-Sen 2008) provided context for the making of the settlement. We positioned these two key texts as the "authentic" citations on the Wiki page, drawing upon them to position Madanpur Khadar JJ Colony within the urban politics of eviction, demolition and settlement in the early 2000s. Not surprisingly, our page was accepted and the notice was taken off. Since then many volunteers have contributed to the page and updated the style and grammar. In a sense, our facilitation of the authoring of the Wikipedia article, and the use of our privilege as published authors with insight into academic conventions to ensure the page's continued existence on the site can be seen as what Vetter and Pettiway would classify as "direct praxis, which can be accomplished through both assimilationist intervention and/or speculative praxis" (Vetter and Pettiway 2017, online). They define this as a tactic that still colours inside the lines, as it were, by working within the existing frameworks of rules and conventions, but to tacitly subvert the site's ideological landscape.

Core to the production of an open urban archive then are Wikipedia's politics of citation which prioritises sources that exist only in print which means that in its current form, it is unsuitable for the recording and dissemination of indigenous and non-Western modes of knowing the world, given that these largely exist in oral form. This inadequacy, particularly in the context of the representation of place, has been highlighted by Graham who has critiqued the encyclopaedia on the grounds that "Wikipedia is characterized by uneven geographies, uneven directions, and uneven politics influencing the palimpsests of place". (Graham 2011, p. 271). The suitability of Wikipedia as a home for marginalised knowledges is actually further compromised by its radically open mandate. These knowledge systems are better served by initiatives that treat the community it serves as owners of, and the core audience of the archive, rather than capitulating to the "radically open" agenda of Wikipedia that deems everyone an expert rather than members of the community being written about (Van der Velden 2013).

Moving forward: disruptive platforms?

In a context where Google has begun to dominate much of the public life of Indian cities, Wikipedia offers ways to reimagine the relations of some of its most marginalised citizens with the city. The Wikipedia editathon prised open Google's monopoly on geographic data (through maps) and challenged the established digital norms of "seeing" the city. This was achieved through two modes: the first was to learn from the women of their local, embodied experience of the city and record that as part of the Wikipedia article, as this knowledge and perspective is far more embodied and experiential than Google's geolocated map data; the second was to use Wikipedia itself as an intervention to subvert simplistic non-nuanced representations of the city as seen through the Google's monoscopic vision.

However, the designs of knowledge infrastructures on our devices and of the Web are built to discriminate and perpetuate non-feminist approaches. As activist-scholars we should be thinking more deeply about the building and design of platform infrastructures that dominate knowledge and our experience of the Web itself. If one is not alert to the different ways in which knowledge is known and made in the Global South, Wikipedia, can reinforce "data-driven discrimination" unless approached with the sufficient nuance that addresses non-Western approaches to making knowledge. This nuance is required to ensure that Wikipedia as a platform reinforces data justice, rather than skew from it, which it is susceptible to do in its current form.

As Kitchin et al. suggest, we need to create contextual and relational grids of knowledge embedded through platforms that "openly recognize their contested and negotiated politics and praxis" (Kitchin, Maalsen, and McArdle 2016, p. 93). This is resonated in Vetter & Pettiway's (Vetter and Pettiway 2017) introduction of the idea of speculative media praxis as a possible intervention that can move us beyond assimilationist logic to more transformative action. The second phase of this project, located in Bengaluru, is informed by such an approach, which aims to conceptualise an urban platform informed by insights from grassroots and local communities. Through workshops and interviews that facilitate co-design activities, we aim to produce a platform architecture that prioritises "engagement, privacy and security" through "interactive, inclusive and continuous engagement methods to practice transparency" (Calzada 2017).

Inspired by Barcelona's governance platform Decidim which describes itself as an "abstract, functionally open, modular and configurable framework" (Barcelona Muncipality 2019, online) for participatory democracy, we seek to use design and algorithmic interventions (such as careful metadata tagging, linked open data or creating more content on sites like Wikipedia) in the next phase to push back against the epistemological violence being waged against women and the marginalised by technology corporations.

Acknowledgements

We wish to acknowledge funding from the Arts and Humanities Research Council (Ref: AH/R003866/1) that facilitated the editathon and other activities on a research network titled "*Gendering the Smart City: Curating alternative networks for addressing Gender based violence* (GBV)". This project would not have been possible without the participation and enthusiasm of the young women who were part of this research. We are grateful to Arya Thomas, our local research assistant for her hard work and support in the workshops. We are also grateful to Jagori and Safetipin for partnering on this project and providing support and access to the community, recruiting participants, organising the workshops and providing continuous feedback on our work. Parts of this chapter were presented in conferences in Oxford and Sri Lanka and we are grateful to audiences there for feedback and comments that have contributed to the arguments.

References

Anttiroiko, Ari-Veikko. 2016. 'City-as-a-platform: The rise of participatory innovation platforms in Finnish cities'. *Sustainability* 8 (9): p. 922. doi:10.3390/su8090922.

Barcelona Muncipality. 2019. 'Decidim: Free open-source participatory democracy for cities and organizations'. *Decidim*. 2019. https://decidim.org.

Barns, Sarah. 2018. 'Smart cities and urban data platforms: Designing interfaces for smart governance'. *City, Culture and Society*, Innovation and identity in next generation smart cities, 12 (March): pp. 5–12. doi:10.1016/j.ccs.2017.09.006.

Benkler, Yochai, and Helen Nissenbaum. 2006. 'Commons-based peer production and virtue'. *The Journal of Political Philosophy* 14 (4): pp. 394–419.

Bhan, Gautam, and Kalyani Menon-Sen. 2008. *Swept off the Map: Surviving Eviction and Resettlement in Delhi*. New Delhi: Yoda Press.

Bleecker, Julian, and Nicolas Nova. 2006. 'A synchronicity design fictions for asynchronous urban computing'. The Architectural League of New York.

Bratton, Benjamin H. 2015. 'Cloud megastructures and platform Utopias'. In *Entr'acte: Performing Publics, Pervasive Media, and Architecture*, edited by Jordan Geiger, pp. 35–51. Avant-Gardes in Performance. New York: Palgrave Macmillan US. doi:10.1057/9781137414182_2.

Calzada, Igor. 2017. 'The techno-politics of data and smart devolution in city-regions: Comparing Glasgow, Bristol, Barcelona, and Bilbao.' *Systems* 5 (1): pp. 1–18. doi:10.3390/systems5010018.

Caprotti, Federico, and Dong Liu. 2019. 'Emerging platform urbanism in China: Reconfigurations of data, citizenship and materialities'. *Technological Forecasting and Social Change*, July, 119690. doi:10.1016/j.techfore.2019.06.016.

Choudhury, A. 2016. 'Why Google's street view has been rejected by the Indian government'. 2016. https://yourstory.com/2016/06/google-street-view-banned.

Counterview. 2017. 'India's "Push" for digitized economy may widen gender gap, warns UNICEF: Just 29% internet users are females'. 2017. https://www.counterview.net/2017/12/indias-push-for-digitized-economy-may.html.

Dalberg Global Development Advisors. 2015. 'Smart maps for smart cities: Urban India's $8 billion opportunity'. 2015. https://static1.squarespace.com/static/543bf912e4b067de38e4b116/t/580f98a5c534a5c14402ca35/1477417131556/Google-Smart-Maps-Report.pdf.

Das, Veena, and Deborah Poole. 2004. *Anthropology in the Margins of the State*. Santa Fe: School of American Research Press.

Datta, Ayona. 2012. *The Illegal City: Space, Law and Gender in a Delhi Squatter Settlement*. Ashgate Publishing, Ltd.

———. 2018. 'The "Digital Turn" in postcolonial urbanism: Smart citizenship in the making of India's 100 smart cities'. *Transactions of the IBG* 43 (3): pp. 405–419.

———. 2020. 'The "Smart Safe City": Gendered time, speed and violence in the margins of Delhi's urban periphery'. *Annals of the American Association of Geographers*.

Dixon, Natalie. 2018. 'Stranger-ness and belonging in a neighbourhood WhatsApp group'. *Open Cultural Studies* 1 (1): pp. 493–503. doi:10.1515/culture-2017-0046.

Gidwani, Vinay, and Sunil Kumar. 2019. 'Time, space and the subaltern: The matter of labour in Delhi's grey economy'. In *Subaltern Geographies*, edited by Tariq Jazeel and Stephen Legg, pp. 142–166. Athens, GA: University of Georgia Press.

Gilbert, Melissa. 2010. 'Theorizing digital and urban inequalities'. *Information, Communication & Society* 13 (7): pp. 1000–1018. doi:10.1080/1369118X.2010.499954.

Graham, Mark. 2011. 'Wiki space: Palimpsests and the politics of exclusion'. In *Critical Point of View: A Wikipedia Reader*, edited by Geert Lovink and Nathaniel Tkacz, pp. 269–282. Amsterdam: Institute of Network Cultures.

Gurumurthy, Anita, and Nandini Chami. 2014. 'Digital technologies and gender justice in India - An analysis of key policy and programming concerns'. IT for Change. http://www.itforchange.net/sites/default/files/IT%20for%20Change%20-%20HLPC%20Submission%20-%2016%20April%202014-1.pdf.

Janowski, Tomasz, Elsa Estevez, and Rehema Baguma. 2018. 'Platform governance for sustainable development: Reshaping citizen-administration relationships in the digital age'. *Government Information Quarterly*, Platform governance for sustainable development, 35 (4, Supplement): pp. S1–S16. doi:10.1016/j.giq.2018.09.002.

Kamath, Anant. 2018. '"Untouchable" cellphones? Old caste exclusions and new digital divides in peri-urban Bangalore'. *Critical Asian Studies* 50 (3): pp. 375–394. doi:10.1080/14672715.2018.1479192.

Kitchin, Rob, Tracey P Lauriault, and Gavin McArdle. 2015. 'Knowing and governing cities through Urban Indicators, City Benchmarking and Real-Time Dashboards.' *Regional Studies, Regional Science* 2 (1): pp. 6–28.

Kitchin, Rob, Sophia Maalsen, and Gavin McArdle. 2016. 'The praxis and politics of building urban dashboards - ScienceDirect'. *Geoforum* 77: pp. 93–101. https://www.sciencedirect.com/science/article/pii/S0016718515300245.

Krivy, Marcos. 2018. 'We are all platform urbanists, but not all in the same way'. *Mediapolis: A Journal of Cities and Culture* 3 (4): np. https://www.mediapolisjournal.com/2018/11/platform-urbanists-not-in-the-same-way/.

Kumar, Sangeet. 2010. 'Google earth and the nation state: Sovereignty in the age of new media'. *Global Media and Communication* 6 (2): pp. 154–176.

Leszczynski, Agnieszka. 2019. 'Glitchy vignettes of platform urbanism'. *Environment and Planning D: Society and Space*, October, 0263775819878721. doi:10.1177/0263775819878721.

Levy, Helton. 2018. 'Online self-representation in Brazil's Favelas: Personalising the periphery'. *First Monday* 23 (8). doi:10.5210/fm.v23i8.8200.

Luque-Ayala, Andrés, and Flávia Neves Maia. 2018. 'Digital territories: Google maps as a political technique in the re-making of urban informality'. *Environment and Planning D: Society and Space*, April, 0263775818766069. https://doi.org/10.1177/0263775818766069.

Malka, Vered, Yaron Ariel, and Ruth Avidar. 2015. 'Fighting, worrying and sharing: Operation "Protective Edge" as the first WhatsApp war'. *Media, War & Conflict* 8 (3): pp. 329–344. https://doi.org/10.1177/1750635215611610.

O'Hara, Kenton P., Michael Massimi, Richard Harper, Simon Rubens, and Jessica Morris. 2014. 'Everyday dwelling with WhatsApp'. In *Proceedings of the 17th ACM Conference on Computer Supported Cooperative Work & Social Computing*, pp. 1131–1143. CSCW '14. New York, NY, USA: ACM. https://doi.org/10.1145/2531602.2531679.

Omanga, Duncan. 2018. 'WhatsApp as "Digital Publics": The Nakuru analysts and the evolution of participation in county governance in Kenya:' *Journal of Eastern African Studies* 13 (1). https://www.tandfonline.com/doi/abs/10.1080/17531055.2018.1548211.

Purnell, N. 2016. 'How Google's bicycle-riding internet tutors are getting rural Indian women online.' *The Wall Street Journal*, 2016. https://blogs.wsj.com/indiarealtime/2016/10/03/how-googles-bicycle-riding-internet-tutors-are-getting-rural-indian-women-online/.

Rajan, N. 2013. '12 Indian cities get street view, but it's not from Google.' *Business Today*, 2013. https://www.businesstoday.in/technology/internet/12-indian-cities-get-street-view-but-its-not-from-google/story/199665.html.

Rangaswamy, Nimmi. 2013. 'Digital life of the youth in India - What is the current scenario and where are we headed?' *Digital Equality* (blog). 6 December 2013. https://digitalequality.in/internet-magic-in-indian-slums/.

Ray Murray, Padmini. 2017. 'Writing new Sastras: Notes towards building an Indian feminist archive.' In *New Feminisms in South Asia: Disrupting the Discourse through Social Media, Film and Literature*, pp. 105–117. London: Routledge.

———. 2019. 'Envisioning inclusive alternatives to smart cities.' Blog. Gendering the Smart City. 2019. https://gendersmartcity.com/exhibitions-and-workshops/bangalore-workshop-gendering-the-smart-city/.

Sarkar, Sreela. 2016. 'Beyond the "Digital Divide": The "Computer Girls" of Seelampur'. *Feminist Media Studies* 16 (6): pp. 968–983. doi:10.1080/14680777.2016.1169207.

Stehlin, John. 2018. 'Urban platforms, rent and the digital built environment.' *Mediapolis: A Journal of Cities and Culture* 3 (4): np. https://www.mediapolisjournal.com/2018/10/urban-platforms-rent-and-the-digital-built-environment/.

The Hindu. 2004. 'Madanpur Khadar residents to get power connections', 2004. https://www.thehindu.com/2004/07/13/stories/2004071317280300.htm.

Van der Velden, Maja. 2013. 'Decentering design: Wikipedia and indigenous knowledge'. *International Journal of Human-Computer Interaction* 29 (4): pp. 308–316.

Vetter, Mathew. A, and Keon Pettiway. 2017. 'Hacking hetero/normative logics: Queer feminist media praxis in Wikipedia.' *Technoculture: An Online Journal of Technology in Society* 7.

Wikipedia. 2018. 'Madanpur Khadar JJ Colony'. Open Knowledge Platform. Wikipedia. 2018. https://en.wikipedia.org/wiki/Madanpur_Khadar_JJ_Colony.

———. 2019. 'मदनपुर खादर जेजे कॉलोनी'. Open Knowledge Platform. हिन्दी विकिपीडिया. 2019. https://hi.wikipedia.org/wiki/मदनपुर_खादर_जेजे_कॉलोनी.

Wyche, Susan. 2015. 'Exploring mobile phone and social media use in a Nairobi Slum: A case for alternative approaches to design in ICTD'. In *Proceedings of the Seventh International Conference on Information and Communication Technologies and Development*, 12:1–12:8. ICTD '15. New York, NY, USA: ACM. doi:10.1145/2737856.2738019.

13

FROM PANOPTICONS TO THE PARTIAL

Digital and blockchain mapping in platform urbanism

Clancy Wilmott

> FOAM provides the tools to enable a crowdsourced map and decentralized location services.
>
> (foam.space, 2019)

> Hyperion, a decentralized map platform, aims to achieve the "One Map" vision – to provide an unified view of global map data and service, and to make it universally accessible just like a public utility for 10B people.
>
> (hyn.space, 2019)

Introduction

Digital maps form a significant part of the apparatus of platform urbanisms. Their presence can be found across multiple scales, from city-wide to personal (Leszczynski, 2019). They are, for instance, crucial to the display and calculation of urban data in large informatics dashboards found in cities such as Sao Paolo (Mattern, 2015) while also forming a fundamental component of the mechanics of many platform economies, including ride-share, bike-share and delivery services at the person-to-person scale of the mobile phone application (Richardson, 2015). As the technologies of digital mapping transform, they influence the political and infrastructural impact of platforms on urban spaces and lives. This chapter considers the political consequences of the emerging application of blockchain technologies, as decentralised, write-only, peer-to-peer, cartographic *and* geolocative services to digital mapping in platform urbanisms. This includes the tension between top-down traditional cartographic structures and peer-to-peer exchange, the ethics of visibility and open location data as a mechanism for transparency, the dependence on universal systems of participation and the question of responsibility in decentralised systems that are both digital *and material.*

Rodgers and Moore (2018) argue that although conceptual understandings of platforms and infrastructures are increasingly intertwined in urban discourses, thinking of platforms as a kind of "doing", rather than simply "a thing" allows for understanding the platform as a situated action that is cultural, social and political – as well as economic. Platform urbanisms, characterised by location-specific and real-time digital production of urban spatial relationships (whether economic, social, cultural or political), depend on digital mapping, which provides a crucial bridge between the material processes of the city and the digital or online world. This occurs across multiple forms: the use of cartographical interfaces within mobile, wearable or desktop applications (Lammes and Wilmott, 2016); urban dashboards used for planning, governance and decision-making (Mattern, 2015); and back-end data cartographies such as geocoded place names (Zook and Graham, 2007). Digital mapping, as a mode of spatially datafied "doing", is key to constructing the real-time city (Kitchin, 2015) – whether this is classifying, measuring, counting, looking or moving in space and time (Wilmott, 2016) – which points to the entanglement of digital platforms and urban infrastructures. The relationship between data and urbanism has received significant attention from geographers, from the governmental role of informatics (Barns, 2016) to the conversion of the city into an operating system (Marvin and Luque-Ayala, 2017) and the use of anticipatory measures by urban governments for pre-empting urban crises through data (Merricks-White, 2016). The last is of particular interest, since blockchain paradigms are often opposed not just to urban governance through data, but to government altogether (Dallyn, 2017; Baldwin, 2018; Garrod, 2019).

Thus, platform urbanism poses two major challenges for digital developers and cartographers across both the production of platforms and the specific relationship between urban spaces and digital mapping technologies. In the first case, a large number of peer-to-peer platforms – from ride-hailing apps, to real estate and property or work – require at least a basic use of location services, which are increasingly dominated by a small number of mapping platforms. Traditional modes of digital mapping (such as Google Maps or Tom Tom) have been critiqued for the centralisation of mapping data into a limited few organisations and their limited ability to provide accurate readings in urban environments (see Hyperion, 2018). Even open-source mapping platforms such as Open Street Map have been critiqued for clunky cooperative decision-making mechanisms, and the reliance on a few dedicated mappers over a broader and more diverse group (Perkins, 2014). Secondly, the growing density of built urban spaces complicates the veracity of digital mapping systems. Since technologies such as global positioning systems (GPS) require clear lines of exchange between devices, radio towers and satellites, urban spaces present particular issues for geolocative mapping due to the density and height of the urban landscape. As a peer-to-peer or peer-to-object exchange, blockchain mapping platforms claim to offer more adaptability, efficiency and resolution for near real-time tracking of location and movement.

Deploying the transparent, consensus-based, decentralised and append[1]-only philosophy of blockchain systems, blockchain mapping start-ups such as FOAM and Hyperion make claims that blockchain could not only address these major issues in urban platforms, but could herald a new age of digital mapping altogether (FOAM, 2018; Hyperion, 2018; Mattern, 2019). In technological terms, blockchain claims to signal the end of so-called "big centralised platforms" in location data, cutting out the middleman, so to speak, in favour of decentralised networked location-sharing. In theoretical terms, technological decentralisation represents a transition from the cartographic as a panoptic mechanism, in Foucault's (1995) terms, to a partial set of knowledges, in Haraway's (1988) terms. At the same time, the liquidation of the visual panoptic gaze into a partial situated mapping is undermined by the total visibility of the blockchain and sensitive information, and the ambiguity of the responsibility and the role that Haraway (1997) calls "the modest witness". As such, I argue that while the move from centralised commercial and governmental digital platforms towards peer-to-peer/object location sharing and tracking sees a critical dispersal of responsibility and accountability in terms of data, privacy and control, the role of location in blockchain mapping creates specific material and ethical concerns and cannot be extricated from the political economy of blockchains in general.

Using two case studies, FOAM and Hyperion, this chapter considers how the consequences of emerging shifts from top-down traditional forms of geo-locative mapping towards blockchain mapping within platform urbanisms become entangled with the politics of knowledge, visibility and responsibility by unpacking the social and ethical implications of their implicitly market-based claims of the democratisation of cartographic knowledge. First, I consider the current position of digital mapping applications and how these are applied to urban platforms. Then, I analyse some of the propositions made by both Foam and Hyperion about the way in which blockchain mapping would provide "proof of location" (POL) and cartographic data. Finally, returning to platform urbanisms, I investigate how decentralised blockchain mapping might work in urban platforms and everyday digital mobilities, along with some of the possibilities and pitfalls of the blockchain approach.

Digital mapping, urban platforms and blockchain

Digital mapping, broadly defined as the use of digital devices for cartographical purposes, has long been a crucial aspect to the provision of forms of platform urbanism. Digital maps have specific geo-locative capabilities beyond analogue maps, due to the algorithmic structures of Global Positioning Systems (GPS), and the calculative abilities of Geographic Information Systems (GIS), in that they are able to track and trace urban processes in near real-time. Near real-time, or quasi-real-time are critical terms, here. The transfer of information through GPS is not instantaneous, and therefore, not real-time, even if it is only a few milliseconds off. Following on from Moore and Rodgers's description of the platform, digital maps used within platform urbanism can be understood as discursive-material (Barad, 2003), in that they are both software and hardware, and form crucial links between the visualisation of presence on the cartographic interface and the situated presence of a body in (urban) space (Elwood and Leszczynski, 2013). Yet, because these links between data and bodies are lived and near-real-time, rather than simply hypothetical, they are bound up in *both* the surveillance mechanisms of watching, classifying and calculating, and the biopolitical mechanism of disciplinary control through ordering bodies in space. With the same technology, it is equally plausible and possible to locate a suspected person via their mobile and dispatch police to arrest them as it is to track a parcel delivery service to a house and hear the doorbell. Thus, digital maps are a key junction between digital and material space in platform urbanism, creating concrete political implications for the production of digital urbanisms (and urban digitalities).

Since geographic data is expensive to both gather and maintain to an adequate accuracy for widespread use, most cases of platform urbanism tend towards overlaying platform-specific data on top of a limited number of commercially or freely available cartographic backgrounds, tiles or rasters. Digital mapping is generally composed of three levels: the base layer which provides the *carto-graphic interface* colouring roads, waterways, buildings etc...; a *database* of place names, points of interest, businesses etc..., usually part of a geographic information system (GIS), which can be searched; and finally, real-time *location data* gathered usually via global positioning systems (GPS), or Global Navigation Satellite Systems (GNSS) more broadly, used to track deliveries, arrivals, positions and so on. While there is a flourishing community of mapping application programming interfaces (APIs) which make it possible for developers to record, collate, analyse and visualise location data, the ownership of cartographic infrastructure – satellites, maps, towers, GPS devices – remains relatively concentrated. For those making apps in a platform urbanism environment, there are really only four "big" cartographic options: Google Maps, OpenStreetMap, Here (run by Nokia) and TomTom (see Table 13.1). Uber, for instance, overlays the specific information of the location of nearby ride shares, the users' particular ride, and proposed route using the Google Maps API: Uber's data on top of Google Maps' cartographic infrastructure. Although users have no choice in which platform they use, Uber does allow drivers to choose between Google Maps and Waze – which is also owned by Google Maps (Gekker and Hind, 2019). This base-level oligopoly, blockchain advocates argue, stifles the democratic and participatory potential of location data.

In comparison, blockchain mapping companies such as FOAM and Hyperion are heralding themselves as the new generation of "decentralized" mapping technologies designed to both disrupt traditional systems of cartographic data

TABLE 13.1 Platform maps

Platform	*Maps API*	*Basemap*
Uber	GoogleMaps	Google Maps
Shopify	Mapbox	OpenStreetMap
Facebook	-	TomTom
AirBNB	GoogleMaps	Google Maps
Delib	Google Maps	Google Maps
Foursquare	Mapbox	OpenStreetMap
WeWork	Mapbox	OpenStreetMap
Craigslist		OpenStreetMap

APIs and basemaps used by urban platforms (Author, 2019).

tracking, storage and representation as well as the political economy of digital mapping itself. FOAM, is a Brooklyn-based US startup established in 2017 and run by architects, entrepreneurs and urban designers. In their initial blog post, they introduce FOAM Protocol as an open-access standard for proof of location, adding economic incentives to provide proof of location within classic cartographical structures:

> Backwards compatible with OSM and connected to their API, we add a monetization layer to open source mapping to enable new geospatial markets of exchange. This will allow people to build applications that are connected to verifiable physical addresses, and allow any coordinate to be turned into a blockchain wallet that can hold a balance and be tagged with crowdsourced data.
>
> *(foam.space, 2017)*

Hyperion, also established in 2017, is a Hong-Kong based startup run in English and Mandarin by mathematicians, entrepreneurs and computer scientists. They advocate the "Hyperion Digital Location Right" (HDLR) (https://www.hyn.space) model of blockchain mapping, based around three key "Trinity"[2] principles: (1) crowd-building, allowing anyone to partake in mapping; (2) crowd-sharing incentives and rewards for contributors; and (3) crowd-governing, self-government ecosystem based on community structures. A white paper released by Hyperion in 2018 articulates a "mapping 3.0" in which decentralised users are incentivised to participate in large-scale proof of location services peer-to-peer, rather than via digital mapping corporations like Google or third-party user apps like Uber.

Both companies are mostly interested in proof-of-location (PoL), which focuses largely on the database and locative data layers of digital maps. Foam suggests that while their PoL is currently based around confirming (and staking

TABLE 13.2 Evolution of digital mapping platforms

1.0	Digital mapping	Interactive, searchable, personalisable. Control and data centralised from community contribution, concerns over privacy, transparency and openness.	Google, Apple, Baidu
2.0	Open mapping	Map generated by and for users through volunteer contributions, changeable Some voting mechanisms to change map but still fairly traditional use of GPS tracks and satellite. Open data, centralised to community.	Mapbox, Open Street Map
3.0	Decentralised mapping	Based on exchange, peer-to-peer technology, appendable but immutable. "self-governing", incentives, decentralised control.	Blockchain mapping – Foam etc.

Adapted from Hyperion's (2018) white paper.

money on) landmarks and other fixed structures, in the future, it is envisioned that a network of sensors will enable GPS-replacement services for use in ride-share and delivery apps, autonomous vehicles, shipping and logistics management, verification of home address etc..., with each contract exchange (or verification) costing a small amount of their bespoke currency. Foam sells FOAM tokens or currency, Hyperion sells HYN. Users must purchase tokens or currency in order to participate. Both start-ups rely on a combination of investment by large investors and venture capital and individuals who can buy HYN or FOAM currency using the Ethereum crypto-currency system upon which both FOAM and Hyperion piggyback. Hyperion tends towards larger investors, for instance, Hayak, Float Capital and Biaozhun Capital. Comparatively, in 2018, Foam undertook a USD $16.5 million FOAM token sale that largely attracted small investors, but was followed by a large investment from public crypto-currency investor KR1.

Blockchain mapping is built from the principles of blockchain computing as a peer-to-peer information transfer system. Rather than centralising data exchange through a single regulatory institution (say, a government bank, corporate exchange or military satellite system), whose data and databases are usually hidden, blockchain computing offers open, market-driven protocols for users to use as exchange agreements, all of which are written into a public, or user-accessible ledger. Blockchain computing can be found across a variety of fields including cryptocurrency (e.g. BitCoin), file sharing (torrenting), contract law (smart contracts) and even video games. There are three kinds of blockchain: public, private and consortiums. Blockchain works on the argument that centralised control of exchange of money, services or goods is both opaque and inequitable, allowing unfair advantage to central organisations (such as national banks, legislature, central markets) to set prices and rates, to be the sole record-keepers of exchanges, and to be able to go back and alter those records. In real terms for mapping, this means that companies like Google both produce and maintain the datasets (or logs) of GPS coordinates and timestamps of individual users' comings and goings. To access this data real-time, it is necessary for the user to go through Google or a third-party app, and it is near impossible (without location spoofing software) to change that data. Once logged, furthermore, Google maintains management of both access and amendment to those GPS tracks, not the user, meaning that it is possible that the location history of the user might be altered by a company either through accident or hostility.

Theoretically, blockchain sidesteps the requirement for centralisation by offering an alternate model. A chain is simply a ledger (in economic terms) or a log (in cartographic terms). The blocks are individual transactions between users which are added to the chain and once added, cannot be altered (append-only). Any investor in the blockchain (whether it is bitcoin or Foam) can view all the updates in the chain in near-real-time (Figure 13.1).

Finally, both FOAM and Hyperion argue that blockchain mapping offers a situated alternative to GPS as a spatial verification system. Since the signals are not being sent upwards to satellites in order to be triangulated and located on the coordinate plane, there is less chance for signal interference via buildings and other tall urban infrastructure (Figure 13.2). Instead, devices connect with each other without the need for satellite, triangulating across the surface of the earth, rather than between the ground and the satellite.

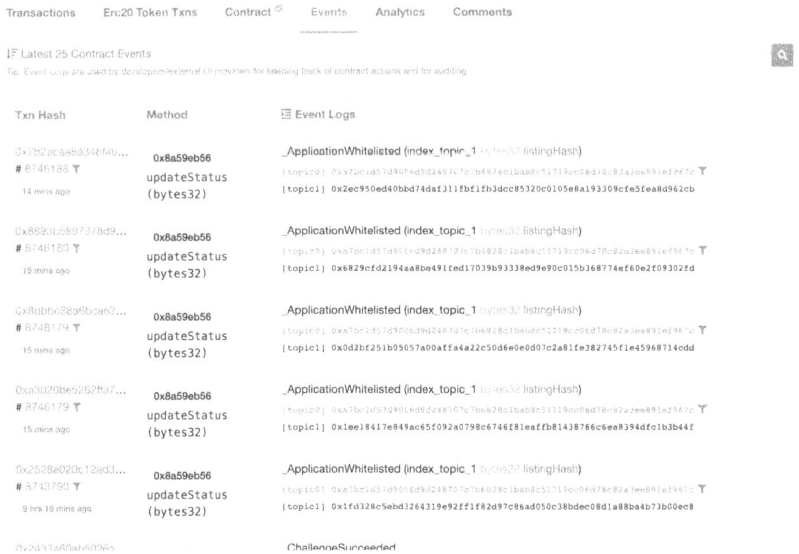

FIGURE 13.1 *Chain.* List of contract events from FOAM (2019).

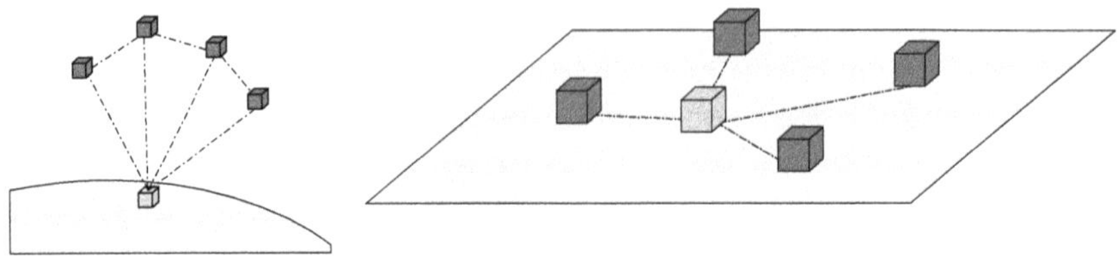

FIGURE 13.2 Traditional top-down digital mapping using satellites (left) and peer-to-peer blockchain mapping (right). Image credit: Author.

In practical terms, blockchain mapping technologies harness sensors that work within the landscape (like Bluetooth, RFID or LiDAR) to "ping" a user's phone in order to track movement through space. Instead of the traditional method of sending signals between devices via satellites or radio towers, pinging involves sending out signals *between* devices, which are then either received and sent back, or reflected back to the original device; the time taken to return is a good measurement of distance, much like sonar and radar. It is still unclear exactly how and where these sensors would work: they might be other people's phones or wearable devices which would ping constantly looking for devices. They may also be embedded into the urban infrastructure, either specific for purpose (like standalone poles) or piggy-backing on other structures (like lamp posts or CCTV cameras). Furthermore, these sensors do not necessarily require centralised infrastructure like transmission towers, internet servers or mobile data like 4G to work: in theory, a peer-to-peer signal (like bluetooth) would suffice.

Panopticons, god-tricks and universality

There is significant attention within critical geography about the power of surveillance provisioned through geolocative technologies and Foucauldian conceptions of power (Crampton, 2003). As Foucault argues:

> Disciplinary power … is exercised through its invisibility; at the same time it imposes on those whom it subjects a principle of compulsory visibility. In discipline, it is the subjects who have to be seen. Their visibility assures the hold of the power that is exercised over them.
>
> *(Foucault, 1977: p. 187)*

Importantly, in Foucault's work, disciplinary power enacted through the act of watching, or surveillance. Various comparisons have been made between the assertive presence of the central column of the panopticon and the encroaching presence of CCTV cameras on our streets (Koskela, 2003; Dobson and Fisher, 2007) as well as to the general use of digital and visual surveillance in space (Ganesh, 2016). To a degree, this is reiterated in the kinds of anxieties that emerge through the use of mobile phones (Leszczynski, 2015), and ambient data gathering practices, including geolocation, by technology firms (Thatcher, 2016). A thread of research on digital mapping underscores the panoptic potential of information technologies (Elmer, 2003) locative media (Zeffiro, 2006), platforms like Google Maps (Fitzpatrick and Reynolds, 2009; Mittman, 2012), and cartography itself (Reynolds and Fitzpatrick, 1999). Whether the phone is turned on or off, and whether location-tracking is enabled or not, there is always the potential to be watched, and so people modify both their behaviour, and their settings as a result (Lezczynski, 2015). Added to this is the possibility of triangulating location via other sensors (CCTV cameras equipped with facial recognition, or RFID tags for instance) which are able to give approximations of the current location, or exact locations in the near-past. As such, they can be understood to operate in multiple ways:

1. As quasi real-time trackers which hybridise urban, material and digital spaces
2. As territorial machines to capture, calculate and contextualise data
3. As surveillance mechanisms through which urban processes are watched and controlled

The distinction here is not one between disciplinary power as a subsection of biopolitics, but rather the relationship between the visual power of surveillance (as in the clinic or the panopticon), and the diffuse disciplinary structuring of lived experience. As Crampton (2007) writes, cartographic power was central to how geosurveillence became an embedded mechanism of biopolitics. Cartography, as a technique of power "that questions, monitors, watches, spies, searches out, palpates, brings to light" (Foucault, 1978: p. 45) generates data, which is transferred upwards through ordered hierarchies, and ultimately, dialogues the political will of elites who use that data to frame populations as "risky" or "at risk", defining languages, producing normativities and setting up examinations or checks to manage them and maintain control.

A key difference between the structure of the panopticon and the digital qualities of platform urbanism is that in digital maps, the line of view between the eye of the guard and the body of the prisoner is better understood as a line of transmission, mediated via a complex apparatus of public/private infrastructure, technology and organisations. Here, the question of governance in geospatial data as it is deployed in platform urbanism becomes further complicated. There is a distinct blurring of responsibility between commercial and governmental bodies in determining who is accountable for the collection, management and fair use of geo-locative data in line with social and political expectations. Furthermore, how that data is then used to influence and control populations – as Dalton et. al. (2019) have argued elsewhere – is yet another consideration. Indeed, the vast majority of geolocative data is collected and leveraged by commercial organisations and big tech firms such as Google, Apple, Baidu and others, which is then sold to local authorities in order to produce what Kitchin (2014) has called "technocratic urban governance" – in short, planning, governing and disciplining through data via the expectation, if not the actualisation, of the real-time city.

In a related vein, Haraway (1988) has described the absolute top-down objectivity assumed by scientific discourses as a "god-trick". Similar to the panopticon, this god-trick is predicated upon kind of looking through a universalist, and often imagined gaze. Like Foucault's analysis of the clinic and the panopticon, of the myriad of technological extensions to the human eye – satellites, video, microscopes, telescopes – Haraway writes:

> Vision in this technological feast becomes unregulated gluttony; all seems not just mythically about the god trick of seeing everything from nowhere, but to have to put the myth into ordinary practice.
>
> *(Haraway, 1988: p. 581)*

Here, the shift of the scientific paradigm – of cartographers and geographers, making their maps in studios – reaches the everyday mobilities of ordinary people, through their phones. Furthermore, Haraway argues that vision, however, does not simply involve seeing from nowhere: it is seeing from nowhere *through* someone, something or some time. The technologies and governance of who looks *through what* (and who owns it) is what is at stake in blockchain mapping: the "instruments of vision". Against god-tricks, she argues "for politics and epistemologies of locating, positioning and situation where partiality and not universality are the condition of being heard to make rational knowledge claims" (Haraway, 1988: p. 589).

There are elements of both the panopticon and the god-trick in the current articulations of digital mapping within platform urbanisms across political, economic, discursive and symbolic forms. For instance, the use of the map in Uber's ride-share app allows the user to survey the location of the driver, their estimated arrival, and their route through the cartographic interface. This deploys the gaze of the customer as one mechanism to discipline the driver into efficiency in navigation through data sharing (and reinforced by the rating system at the end of trip), but also decontextualises the situated experience of the driver into a cartographic pseudo-objectivity. This pseudo-objectivity, which "second-guesses" any action of the driver that is not determined by the map, is highly fallible, and does not take into account the changing nature of urban landscapes including traffic lights, road works and detours, traffic incidents or other reasons why a driver might choose a different path. It also outweighs the expertise of the driver-as-navigator, a local and embodied form of knowledge and a crucial skill in the provision of traditional taxi services, into the so-called objectivity of cartographic data and calculation into location tracking and navigational algorithms – in this case, Google/Alphabet. This duality of god-trick abstraction via cartographic and locational data and panoptic discipline via real-time location tracking or monitoring is typical of digital urban mobilities, whether it is everyday navigation (Duggan, 2018), mobile labour such as deliveries (Richardson,

2018), vehicular mobilities, autonomous or otherwise (Gekker and Hind, 2019) or playful cartographic apps (Lammes and Wilmott, 2016).

However, the combination of the god-trick/panoptic duality with the agglomeration of cartographic knowledge, data and infrastructure into a few globalised services, renders a model of digital mapping in platform urbanism that is very centralised and deeply unequal. This has the ongoing effect of creating and then reinforcing dependence on digital mapping services like Google by unsettling local and embodied knowledges (Wilmott, 2016), thereby creating both a supply *and a demand* for their own cartographic services. Furthermore, while companies like Google are the most egregious examples of this dynamic, attempts to move against the centralisation of cartographic data – like OpenStreetMap – are still bound by a discursive foundation in absolute and universal systems of cartographic representation (Wilmott, 2019), as well as the centralisation of the production of cartographic data in other ways (Perkins, 2014).

Partiality is why the claims made by blockchain mapping start-ups are so disruptive. Start-ups like FOAM and Hyperion claim that the way in which blockchain works is more secure and stable: data entered cannot conflict with other data already in the blockchain; and data within the chain is immutable and cannot be changed by individual segments of the chain. The result for governance is that blockchain systems are both decentralised and also consensus based: nothing can be changed in the block-chain without the agreement of *all* parties. This is evidenced in the increasing role of blockchain for cybersecurity. Since no single device contains all the information, blockchain networks are much harder to hack and the risk of failure is distributed across all parties; while devices might ping one another, information (including packets, encryption keys, etc) about where this user is in space is not stored centrally. Thus, the transfer of geolocative data between devices is always contingent and partial, as the GPS track is distributed between hundreds of devices, rather than being a single line of connection between the phone and the satellite.

This is not only a technical capacity rendered by blockchain, but part of a broader ideological push towards different models of cartographic and digital knowledge/power. However, from the perspective of HPN, decentralisation is not an objective itself, but a feature for other system design goals: censorship resistance, open participation and fault-tolerance.

> However, from the perspective of HPN [Hyperion], decentralization is not an objective itself, but a feature for other system design goals: censorship resistance, open participation and fault-tolerance.
>
> *(Hyperion, 2018: p. 15)*

Thus, while in theory, the aspect of decentralisation might indicate some dispersion of traditional cartographic panoptic or god-trick power, in practice, this is secondary to other, competing – and I argue, contradictory – goals. An entire section of both white papers focus on "interoperability", that is, translating information across different digital systems and languages, from the practicalities of optimising mobile phone operating systems (like iOS) to use the tokens, to the idealogical push to embed FOAM into crypto-currency systems by producing, what I might term here, "carto-currencies".

Thus, in practice, there is some viability in the claim that blockchain mapping systems herald a major shift in mapping technology in dismantling the panoptic gaze of governmental and corporate organisations by offering partial views of emergent spatial processes, created by and for users. There is also a crucial distinction between the use of urban data *for* governance and the governance *of* urban data. Blockchain challenges both, by offering both economic alternatives to the use and collection of geospatial data (whether physical attributes like buildings or streets or semantic attributes like geolocation or points of interest) and a self-governing, quasi-autonomous collective political alternative to top-down collection and control of this data. Blockchain mapping platforms enter into a space where it is already extremely tricky to moderate how urban data is used, and for what, with little transparency or oversight about the underlying functions and verification of big tech platforms like Google or Baidu.

In economic terms, blockchain also loosens the grip of third-party hardware and software firms who mediate between data producers and data analysis (or in mapping, the users and the digital map infrastructure itself), making blockchain more cost-effective. The protocol layer – or, the codes used to exchange data – are based on blockchain. Here, arguably, cartographic agency is distributed between multiple actors, all of whom only get a partial view of the digital map rather than being funneled into companies like Apple and Google, where users only have limited means to opt in or out of such tracking (and even then, often barely).

Blockchains have shown some potential to evade state governance of economic systems, as been shown in some Indigenous blockchain currencies like MazaCoin. At the same time, Tekobbe and McKnight (2016) argue that the need for investment in blockchains to maintain their value still drives a paternalistic, colonial and affective capitalism in how these chains are discursively sold - decentralisation is not the same as decolonisation. The same issue punctuates blockchain mapping: a deeper understanding of cartographic knowledge/power in a situated or decolonising context is not at all recognisable in either FOAM or Hyperion. Hegemonic universality reappears in various ways within formations of blockchain mapping. For instance of the major criticisms of blockchain mapping is how likely users are to participate. Like Perkins (2013), FOAM (2018) argues that the lack of economic incentives for platforms such as OpenStreetMap – or even What3Words – does little to encourage anyone but a marginal few to use the platforms, much less contribute or maintain the cartographic infrastructure. In response, they offer an incentive for participation in blockchain mapping structured around economic paradigms inherent in crypto-currency discourses:

> In other words, just as the growth of Bitcoin, Ethereum, and many other blockchains were assisted by crypto-economic incentives, so too is the

> FOAM protocol assisted by incentives to build out the hardware to provide a decentralized alternative to GPS. Similar to other blockchain mining, Zone operators on the FOAM protocol are in essence providing comparable work to Bitcoin miners.
>
> *(FOAM, 2018: p. 18)*

At the same time, Zone operators who invest, claim and maintain infrastructure This includes charging money for exchanging contracts while offering reductions or even income for providing key infrastructure such as towers. Of course, this requires everyone to either use the same currency, or for equivalences to be set at the time of exchange, governed by either the platform, the users or the market (or all three).

Furthermore, the transparency of the ledger and the necessity of being open to all devices on the network (and exchange accordingly) operate with a different form of panopticism, or god-trick through absolute visuality. All transactions between devices on the blockchain are visible – either within the network or to the general public. While in theory, the transparency of transactional data may be useful for financial, currency or other exchange domains, the sensitivity of locational data in terms of security, autonomy and privacy raises different concerns. Since it costs real money to map and challenge claims made on spaces and places (i.e. cryptocurrency Ethereum exchanged into FOAM tokens), already there is evidence of domination of the mapping platform by fin-tech, business education, industrial and other economically and technologically privileged sectors over people impacted by this mapping. A screenshot (Figure 13.3) of London

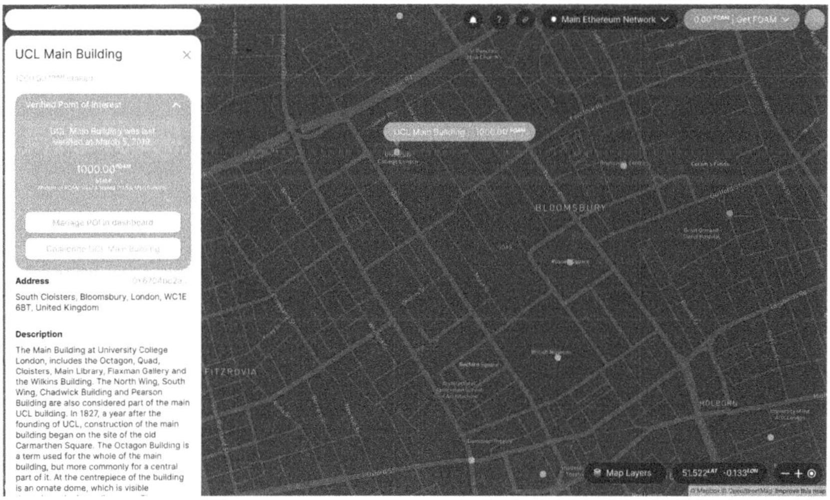

FIGURE 13.3 *Staked.* Screenshot of FOAM point of interest (POI) dashboard for UCL Main Building. The map layer for the POIs is overlaid onto Open Street Map.

in FOAM shows a range of green, verified points of interest, including the main building of University College, London.

In order to challenge this claim, a certain number of FOAM must be spent, and there is no guarantee that it will be successful. Although the centralisation of data has shifted from the totalising grasp of Google into a multidimensional voter space like Open Street Map, the mechanism for place-claims is monetary rather than participatory. The critique of Open Street Map that the majority of mapping activity is engaged by a limited number of users can be mirrored in the critique of FOAM that already the majority of mapping activity will be limited to a small number of richer purchasers. As an individual or small collective in such a space, it will be extremely difficult to compete.

As noted above, location data plays a crucial role in linking the material world of bodily presence and action with the digital world of data and algorithms (Elwood and Leczszynski, 2013). Furthermore, the ledgers like that shown in Figure 13.1. use alphanumeric identifiers for each transaction, and its participants which largely render this data incomprehensible to the non-expert eye, but can be collated and calculated usefully with that already have this capacity. The politics and ethics of monitoring become further complicated when understood in light of systemic oppression (race, gender, class, sexuality) and biopolitics. Not all people operate in space the same way – and visibility to the eye of not only the users within chain, but possibly also any individual, state or corporation may pose specific risks for those who are more likely to be subjugated by those processes (c.f. Wilmott, 2016).

Like many platform models, blockchain mapping operates on dispersed responsibility. This has follow on implications for how location (rather than cartography) is verified: it is still unclear what happens when two competing claims about the location of an object (say a person or a car) are made, and if one is incorrect and, for instance, leads to a road accident – who is responsible? The determination of location involves at least two actors (peers, sensors), plus the moving object itself: is it the object, or one of the actors who verified the location? Or, is it the platform which provides the contract exchange, such as FOAM or Hyperion? Yet, shying away from responsibility is a key phenomenon emerging from platforms. Facebook claims that they are not a publisher, but a platform, while Uber distances themselves from taking responsibility for the actions of their drivers citing that they are contractors, not employees. Within platform *urbanisms* specifically, the ambiguity over responsibility has real, material, bodily and social implications. Furthermore, since in both whitepapers, FOAM and Hyperion are touting integration with driverless vehicles, recent incidents such as the Uber self-driving car accident in Arizona which killed Elaine Herzberg, indicate how these implications could result in violence or death – as mapping, and digital mapping, often already have.

Thus, while blockchain mapping may appear to represent Haraway's notion of the "modest witness" (1997) as a "response-able" actor in and of the world, its distributed nature evacuates responsibility altogether rather than establishing a sense of social responsibility. Accountability is at once dispersed throughout the mapping community, while at the same time, participation is limited to mappers who can afford to participate. Mapping here is no longer akin to territorialising – as it has been traditionally understood by cartographic scholarship – but more about a libertarian vision of mutual contracts and market exchange. Thus, in this case, partiality is not used as a way of humbling the vision of totalising scientific knowledge, but rather a system that erases corporate responsibility through partiality and partial participation, at the same time maintaining the same universal structures of the god trick and demanding total visibility of the panoptic while only making it legible to the most powerful: the worst of all worlds.

In summary, blockchain mappings do have potential to disrupt the stranglehold of big digital mapping companies in platform urbanism. The promise of decentralised, collectivised and transparent networks offer a different, more horizontal way of thinking about digital mapping, without the pitfalls of some of the more major open-source cartographic platforms like Open Street Map. Yet, on the level of the politics of representation, the flattening requirement for a universal language remains remarkably familiar, and the dream of total transparency does not necessarily equate to more accountability or responsibility. In fact, in the case of both FOAM and Hyperion, the question of accountability is ambiguous, obscured under claims about technological superiority and ideological arguments towards dispersion and decentralisation. Thus, partiality is not bound up in the articulations of responsibility offered by Haraway, but instead is dispersed through purchase and diminished by contract exchange.

Notes

1 In 'append-only' systems, information once entered cannot be changed or deleted, like a ledger rather than a document.
2 I will leave a discursive unpicking of the language used by Hyperion for another time.

References

Baldwin, J. (2018) In digital we trust: Bitcoin discourse, digital currencies, and decentralized network fetishism. *Palgrave Commun* 4: p. 14.

Barad K. (2003) Posthumanist performativity: Toward an understanding of how matter comes to matter. *Signs: Journal of Women in Culture and Society* 28: pp. 801–831.

Barns, S. (2016) Mine your data: open data, digital strategies and entrepreneurial governance by code, *Urban Geography* 37:4, pp. 554–571.

Crampton J. (2003) Cartographic rationality and the politics of geosurveillance and security. *Cartography and Geographic Information Science* 30: pp. 135–148.

Crampton J. (2007) The biopolitical justification for geosurveillance. *The Geographical Review* 97(3): pp. 389–403.

Dallyn, S. 2017. Cryptocurrencies as market singularities: The strange case of Bitcoin. *Journal of Cultural Economy* 10(5): pp. 462–473.

Duggan, M., 2018. Navigational mapping practices: Contexts, politics, data. *Westminster Papers in Communication and Culture* 13(2): pp. 31–45.

Elwood, S. and Leszczynski, A. (2013) New spatial media, new knowledge politics. *Transactions of the Institute of British Geographers* 38: pp. 544–559.

FOAM (2018) *FOAM Whitepaper: FOAM – The Consensus Driven Map of the World.* https://foam.space/publicAssets/FOAM_Whitepaper.pdf [Accessed 19 December 2019].

Foucault, M. (1978) *History of Sexuality: Volume 1*, New York: Random House.

Foucault, M. (1995) *Discipline and Punish: The Birth of the Prison*, New York: Vintage Books.

Garrod, J. (2019) On the property of blockchains: Comments on an emerging literature. *Economy and Society* 48:4, pp. 602–623.

Gekker, A., & Hind, S. (2019). Infrastructural surveillance. *New Media & Society.* doi:10.1177/1461444819879426.

Haraway, D. (1988) Situated Knowledges: The Science Question in Feminism and the Privilege of Partial Perspective. *Feminist Studies* 14: pp. 575–599.

Haraway, D.J. (1997) *Modest−Witness@Second−Millennium.FemaleMan−Meets−OncoMouse: Feminism and Technoscience*, New York: Psychology Press.

Hyperion (2018) *Hyperion Whitepaper v.1.15.* https://whitepaper.io/document/417/hyperion-whitepaper [Accessed 19 December 2019]

Kitchin, R. (2015) The real-time city? Big data and smart urbanism. *GeoJournal* 79(1): pp. 1–14.

Lammes, S., & Wilmott, C. (2018). The map as playground: Location-based games as cartographical practices. *Convergence 24*(6): pp. 648–665.

Leszczynski, A. (2015) Spatial big data and anxieties of control. *Environment and Planning D* 33: pp. 965–984.

Lezcyznski, A. (2019) Platform affects of geolocation. *Geoforum* 107: pp. 207–215.

Marvin, S. and Luque-Ayala, A. (2017) Urban operating systems: Diagramming the city. *International Journal Urban Regional Research* 41: pp. 84–103.

Mattern, S. (2015) Mission control: A history of the urban dashboard. *Places Journal.* March 2015. https://doi.org/10.22269/150309 Accessed 19 December 2019.

Mattern, S. (2019) A map that tracks everything blockchain-based mapping hopes to replace GPS. Can it be trusted? *The Atlantic.* https://www.theatlantic.com/technology/archive/2018/11/can-blockchain-maps-replace-gps/576985/ [Accessed 19 December].

Merricks White, J. (2016) Anticipatory logics of the smart city's global imaginary. *Urban Geography* 37(4): pp. 572–589.

Perkins, C. (2014) Plotting practices and politics: (im)mutable narratives in OpenStreetMap. *Transactions of the Institute of British Geographers* 39: pp. 304–331.

Richardson, L. (2015) Performing the sharing economy. *Geoforum* 67: pp. 121–129.

Rodgers, S and Moore, S. (2018) Platform urbanism: An introduction. *Mediapolis* 3. https://www.mediapolisjournal.com/2018/10/platform-urbanism-an-introduction/ Accessed 19 Dec 2019.

Tekobbe, C and McKnight, J. (2016) Indigenous cryptocurrency: Affective capitalism and rhetorics of sovereignty. *First Monday* 21(10). https://firstmonday.org/ojs/index.php/fm/article/view/6955/5632 Accessed 19 Dec 2019.

Wilmott, C. (2016) Small moments in spatial big data. *Big Data and Society* 3(2): pp. 1–16.

Zook, M. A., & Graham, M. (2007). Mapping DigiPlace: Geocoded internet data and the representation of place. *Environment and Planning B: Planning and Design* 34(3): pp. 466–482.

SECTION 4

How are platforms re-shaping everyday urban experiences?

14

PLATFORM PHENOMENOLOGIES

Social media as experiential infrastructures of urban public life

Scott Rodgers and Susan Moore

Introduction

Platforms have emerged as some of the most disruptive, yet also increasingly ordinary, dynamics of contemporary urban life. As Barns (2019) notes, even as platforms might be understood as proprietary technological ecologies, which thrive on user surveillance and extracting value from user data, they are also reshaping everyday socio-spatial experience with indefinite consequences. For Barns, negotiating this "pivot" between the technical, commercial and embodied implications of an emergent platform urbanism will require diverse epistemologies (see also contributions to Rodgers and Moore, 2018). In this chapter, we argue for a phenomenological perspective on social media, approaching them as experiential infrastructures of everyday urban communication. While our argument is primarily conceptual, we will also draw on recent research we have conducted into the role social media has played in mediating public exchanges about a controversial cycling program in Walthamstow, East London, UK. We will pay particular attention to how such exchanges emerge through real-time-like experiences of locality, mediated by both the technical features of social media platforms and their everyday practical dynamics. These temporalities of social media amount to relatively novel forms of urban public life, the stakes and consequences of which are ambiguous.

Early accounts of social media and public life often tended towards more celebratory narratives of a new participatory media culture. But a prominent counter-narrative has developed, in both academic scholarship and popular discourse, in which the power of social media platforms is progressively being challenged. Like other platforms, social media are increasingly understood as interoperable "walled gardens" built up by large, monopolistic companies devoted to extracting economic value from user contributions and metadata (e.g. Van Dijck, 2013). For van Dijck et al. (2018), the unprecedented size and power of social media, as companies and technological ecosystems, demands we think about how these and other platforms might be encouraged or even forced to better serve public values and the common good. Concerns are increasingly being directed, in other words, to "platform governance". As Gorwa (2019) points out, platform governance entails thinking about how the algorithmic architectures of platforms govern the behaviour of their users, and in turn how platform companies might be pressured to self-govern, or otherwise be governed through local legal mechanisms.

This chapter contributes to these critical literatures on social media as emergent infrastructures of governance beyond the state. Yet in making our argument we will also sound a cautionary note around how we should approach the power of social media platforms, not only as companies, but in particular as algorithmically-mediated infrastructures. As Langlois (2013: pp. 102–103) points out, the "participatory" claim around social media is in part true: in principle, anyone can establish an account and express views within minimal censorious intervention. However, such platforms become mediums of governance not just by enabling new voices, but by establishing new technical and cultural environments with a relatively internalised logic to manage the flow and visibility of meaning and information (ibid). The norms and values that inform these internalised logics in social media are – like in other data-driven, cloud-based digital platforms – black-boxed into their functioning algorithms (see Lee and Larsen, 2019). The proliferation of algorithms, as a form of "secondary agency" (encoded to make subsequent decisions without human use or authorisation, cf. Kitchin and Dodge, 2011; Mackenzie, 2006), has become one of the most prominent and even mythical problems of contemporary scholarship. Algorithms are a problem not only because they appear to be so powerful in their effects, but also because they are opaque in their inner workings (Ziewitz, 2016; Bucher, 2018: pp. 41–65). This problem of algorithmic opacity often leads to an interpretation of social media (and other digital media) as "subtle" forms of urban power, in which communication and everyday experience is silently structured by pervasive, computational mediation (cf. Rodgers, Barnett and Cochrane, 2014: p. 1061).

There are however conceptual and empirical limits to analysing the politics of algorithms and software in these ways. As Ananny and Crawford (2018) argue, even if we could pin down a given platform's technical architectures or data, or make an algorithm's workings transparent, that would not sufficiently address questions of political or public accountability. The algorithmic architectures of social media platforms clearly matter, but they do not work in isolation: technical designs or autonomous decision-making capacities only become political as part of situated practices and material spaces (cf. Crawford, 2016; Willems, 2019). Our response is to put forward a phenomenological perspective on social media as infrastructures of urban public life. This helps us to negotiate an apparent tension between two ways of thinking about social media as "infrastructures" of urban communication: technical and experienced. On the one hand, social media are translocal, standardised technical systems that structure and stylise social interaction in particular ways (cf. Alaimo and Kallinikos, 2019). Yet on the other hand, the translocal standardisation of social media is experienced and enacted through localised, embodied activities at particular moments. As we will show, a phenomenological approach does not choose between these two ways of thinking about social media infrastructures but rather approaches the technical through the experiential.

Focusing on the experience of social media entails an emphasis on the contingencies and even ambiguities surrounding their political implications for urban life. In turn, such an approach also raises questions about what form our scholarly critiques should take. We argue in particular against what Sedgwick (2003: pp. 123–151) might term a "paranoid" reading of social media platforms – one oriented from the outset to anticipating, unveiling, theorising and critiquing platforms as oppressive or violent. Instead, the phenomenological approach we outline provides the basis for "reparative" (Sedgwick, 2003) or "postcritical" (Felski, 2015) readings[1] of the more indefinite political implications that social media platforms might have in practice for urban communication and public life.

Planning, participation and social media platforms

Our conceptual argument is informed by recent collaborative research[2] we have undertaken on the relationships of urban planning, participation and social media platforms in Walthamstow, East London, UK. We have focused specifically on the local implementation of the "Mini-Holland" transportation scheme of the London Borough of Waltham Forest. This scheme, officially named "Enjoy Waltham Forest" by the Council, was one of three funded by Transport for London to the tune of £30 million, and entailed making a series of significant changes to the borough's road infrastructure, in order to enhance the environment for cyclists and pedestrians. Combining qualitative and data analytics approaches, we have examined how the scheme and related issues have been promoted, explained, mobilised, consulted on, ridiculed, ephemerally mentioned and perhaps above all antagonistically debated via different social media platforms. The "antagonistic" dimension has been particularly important since social media helped make possible some animated and often theatrical protests against the scheme, surprising the Council and attracting national media attention (e.g. see Hill, 2015).

We focused in particular on three prominent digital platforms through which different kinds of publics convened around the scheme: Twitter, Facebook and a more bespoke platform called Commonplace. Our examination of Twitter, quantitatively our largest dataset, clearly showed that the platform was dominated by cycling campaigners and key politicians. While we found many instances of acrimonious exchanges on the platform, Twitter appeared to be a locus for campaigners and politicians to consolidate their support for the scheme or expand their network through reciprocal mentions and retweets. Facebook, by contrast, hosted a wider range of perspectives. In part because it is a platform environment more embedded into and interwoven with its users' broader everyday experience and identity maintenance, and in part, because the platform permits and indeed encourages more extensive and recursive contributions. On Facebook, exchanges often centred on disputes about the value of cycling and its possible ties with encroaching gentrification. These disputes exemplified increasingly familiar forms of contemporary political division, for example, between middle class/working class, elite/ordinary, young/old and facts/emotions. Although apparently antagonistic, such Facebook exchanges could also be interpreted as agonistic, as proposed by Mouffe (2013): defined by political pluralism rather than consensus. Meanwhile, the Council managed much of their interaction with local publics using Commonplace, a platform produced by London-based developers that is specifically designed for urban regeneration consultations. Commonplace is based around a map interface, where participants are presented with a geographically delineated area – a "commonplace" – and invited to pin comments and emotional metrics to locations. What is notable about Commonplace is that its avowed mission involves disrupting conventional approaches to public consultation. While its paying clients are able to present their project with information of their choosing, access a real-time project dashboard and easily generate infographics, they also must accept the Commonplace terms and conditions, in which clients agree to cede control over the resulting, publicly-displayed map of user comments.

Through this research we have found that social media appear to support new kinds of "participation" in urban public life. From the perspective of local politicians, bureaucrats and activists, these new forms of participation were unpredictable. They could embody a useful source of public consensus, aligning with local council proposals. But just as easily, social media could be hotbeds of divisive, antagonistic exchange; places in which so-called "post-truth" claims flourish within digitally-mediated "echo chambers" or "filter bubbles". They could also be an emotional and organisational catalyst for organised protest (cf. Gerbaudo, 2012; Papacharissi, 2015). On the basis of social media exchanges, hundreds of residents opposing the scheme turned up at two notable protests: one at the Town Hall vote on the scheme; and a second months later at its official opening, on a normally quiet local street. But regardless of whether they supported consensus, antagonism or protest, social media seem to challenge the status, authority and relatively predictable logic of more formal consultative exercises (see Moore et al., forthcoming).

Platforms such as Facebook, Twitter or Commonplace are not just, however, venues where like-minded people might cloister (e.g. to oppose or support a local issue), or organisational and emotional catalysts for occasional protests. They also afford a new *pace* to local public affairs, built around "tiny acts of political participation" (Margetts et al., 2017: p. 34): innumerable, publicly expressed – and often archived – claims, commentaries, anxieties, or bits of content relating to an urban neighbourhood and its transformation. These tiny acts, particularly in conjunction with mobile technologies, subsist especially well when users react, comment and share with others in apparent "real-time". Elsewhere, we have termed this involvement as "ambient participation" (Rodgers et al., forthcoming). Social media provide a venue for participation that is not usually defined by discrete objectives, nor clearly directed towards institutions, but instead is dispersed, incidental and – importantly – most often orientated to the moment. Our argument, in other words, is that social media platforms are not only helping to bring about new types and spaces of urban publicness, but also new temporalities.

Publicness, social media and temporality

Paradigmatic city spaces such as the agora, salon or coffeehouse have often been advanced as models of the public sphere. As (Sharma, 2014: p. 13) argues, however, such spatially-inclined embodiments of publicness were also temporal. They required their participants to have free time, for one, but they also organised public discourses temporally. As Warner (2002) points out, most uses of the word "public" fall back on either a narrowly *social* idea of the public as a body of people (notably the social totality of "the public"), or a narrowly *spatial* idea of a bounded public (for example a speaker and audience in a delimited physical space). Warner proposes an additional, *temporal* sense of publics, which often works in conjunction with public bodies and spaces. He does so through a discussion of "textual publics", which are defined not so much by social groups or spaces, nor by texts per se, but by practices of addressing a public via texts (which for our purpose can include digitally-mediated communication). Importantly, practices of public address involve a "chicken-and-egg circularity". The public will only survive insofar as it continues to be addressed into the future. Yet to address that public, one must also take for granted its existence in the past. The "dailiness" of broadcasting (Scannell, 1996) and traditional print news are archetypal examples of how this temporal circularity of publicness is mediated. However, this kind of temporally punctuated publicness, along relatively clear and reflexively understood intervals, is comparatively absent in social media.

While the extensive literature on social media and publicness has not ignored questions of time and temporality, it has more often tended to emphasise sociality and spatiality. Social media are seen to make possible, for example, alternative types or categories of publics that augment or challenge the mass public sphere, such as personal, calculated, hyperlocal, ad hoc or ephemeral publics (Schmidt, 2013; Bruns and Burgess, 2015; Jenkins, Itō and boyd, 2015; Highfield, 2016). Social media are also seen to blur the boundaries of private and public spheres, providing a means through which private matters increasingly become public concerns, and public matters more frequently enter into private life (Papacharissi, 2010). In as far as the public sphere finds its material "shape" through the specificities of different mediums (Carpignano, 1999), social media have been seen to reshape publicness through their novel structural affordances. boyd's (2010) account of "networked publics" argues that social media are distinct from analogue forms of publicness in how they automatically archive user contributions, which can then be easily retrieved, replicated, modified, searched and also potentially (and unpredictably) scaled up to very large audiences.[3]

Often more implicit in the literature outlined above is that practices of addressing publics through social media involve unprecedented speed. John Tomlinson (2007, cited in Kaun, 2015) observes that, while industrial-era technologies were characterised by speed, it was a form of speed which was mechanical and effortful. One notable example of this is the railway post office, seen in the UK and elsewhere. The railway post office located mail sorting practices within the moving train carriage, while also deploying a bag-and-hook apparatus that allowed pick up and drop off without stopping. This example epitomises what Harvey (1989) labelled "time-space compression", in which the problem of time lag in communication across distance is minimised, if through considerable effort. For Tomlinson, post-industrial technologies are by comparison characterised by an experience of relative immediacy, and (seemingly) effortless speed. Immediacy and speed are common themes in discussions around so-called smart cities, the proponents of which often claim we are on the brink of, or might have even realised, a "real-time city" (Kitchin, 2014). Weltevrede et al. (2014) have argued that totalising and often messianic claims that social media offer a "real-time" experience (of urban life, or other sociospatial phenomena) have important limitations. Asserting that events on social media happen in real-time assumes an actual correspondence between events enfolding through platform environments and objective clock time. Weltevrede et al. (2014: p. 143) suggest that instead, we should conceptualise "realtimeness" as an experiential condition, a practical understanding of time fabricated through "and immanent to platforms, engines and their cultures".

Realtimeness as an experiential condition becomes clearer by studying practices of public address within nested threads on Facebook. Threads related to the Mini-Holland transportation programme, on place-based Facebook groups such as Walthamstow Residents News, frequently began with a particular image or video alongside commentary. For example, an image of attractive new cobblestone street pavements and planted trees, paired with a positive affirmation of the cycling scheme. Or a picture showing a long line of car lights, used to bemoan the traffic congestion the scheme was often alleged to produce. These occasionally provocative, but often relatively mundane, initial posts might produce over 100 comments within nested threads across 2 or 3 days, from initial post to the last comment. The temporal structure of such Facebook threads, if measured according to clock time, is clearly asynchronous. In other words, such threads comprise a succession of "nows" rather than a single "now". Yet our analysis indicates that users affectively experience, and sometimes explicitly understand themselves to be partaking in, a real-time-like environment. As Kaun and Stiernstedt (2014: p. 116) argue, despite the constant archiving of public contributions, Facebook users primarily have a "temporal experience … of immediacy, ephemerality, "liveness," and flow … [they are] immersed in an atmosphere and an interface of rapid change and forgetfulness, rather than remembrance and preservation". Indeed, antagonistic Facebook exchanges around the Mini-Holland scheme most often centred on the content (or alleged motives) of other users' contributions, apparently made in that moment (i.e. at a proximate point in the thread), rather than addressing arguments built-up cumulatively.

Coleman (2018: p. 68) describes social media as affording users "a 'temporal present' where bodies, technologies and the socio-cultural matrix are intertwined and experienced in terms of 'aliveness', and 'always-on-ness'". Of course, it could quite plausibly be argued that such a temporal presence of aliveness and always-on-ness could be associated with other mediated settings that existed long before, and continue alongside, social media. For example, face-to-face conversations amongst neighbours about all manner of public matters. Yet as the above example of Facebook threads illustrate, sharing practices through social media, experienced as real-time-like, are not the same as corporeal involvement at particular moments with others (Zingale, 2013). Participants encounter one another, for example, through reading, typing, liking and sharing rather than hearing, speaking and gesturing. To better specify social media as a distinct form for experiencing and addressing urban public life, we will now elaborate on the conceptual features, and political implications, of a phenomenological perspective on social media as infrastructures of everyday communication.

Approaching social media infrastructures phenomenologically

Social media are increasingly seen as complex infrastructures, as are platforms more generally, evidenced by many of the contributors to this book. One of the most direct ways of approaching media infrastructures is to better examine how the media we experience as content, texts or interfaces depend on often-hidden or usually-ignored material and technical artefacts, such as satellites, signal towers, printing presses and data centres (Parks and Starosielski, 2015). But media infrastructures are not always identifiable artefacts. For Larkin (2013: p. 329), infrastructures are anything which "create the grounds on which other objects operate" including for example financial instruments, biological conditions or social conventions. Similarly, Easterling (2016) describes "infrastructure space" as extending from myriad physical (e.g. roads, canals, water pipes) and wireless networks (e.g. microwaves) to phenomena such as shared industrial standards. Infrastructures, in other words, "can be lightweight and portable as well as heavy and fixed" (Peters, 2015: p. 32), and perhaps because of this, "infrastructure" has become one of the more elastic concepts in the social sciences and humanities (see Mattern, 2017: xxv–xxvii).

Plantin et al. (2018) argue that, while "platform" and "infrastructure" may seem like two alternative lenses for understanding digital objects, they are actually complementary, since platforms and infrastructures are increasingly swapping traits. As platforms such as Facebook have evolved from websites into interoperable, technical ecosystems, they have enrolled a wide range of infrastructures external to the platform itself (Helmond, 2015). And in this process of enrolling infrastructures, platforms have themselves managed to become new kinds of coordinating infrastructures. Plantin et al. (2018) usefully outline two main ways scholars have approached infrastructure, which might inform the study of platforms. The first (and the one they explicitly emphasise) is to examine the historical development and evolution of infrastructures as large technical systems. The second is to examine the phenomenology of infrastructures, which means thinking about infrastructures more so in terms of what they *do* across a series of practical milieus, rather than what they are. Two political frames that potentially correspond with these approaches are well captured by Dourish and Bell (2011: pp. 96–98). The first is "sociopolitical", which sees infrastructures as "crystallisations of institutional relations", and accordingly entails a focus on access to and control of infrastructures as such. The second focuses on experience, and draws attention to the politics of our everyday practical uses of and dependencies on infrastructures.

To further flesh out these approaches and political frames, let us momentarily consider a seemingly unexciting example: the wooden pallet. In February 2019, it emerged that, should the United Kingdom leave the European Union without an interim deal, it would not have enough pallets of the correct standard in order to move the commodities it needs (Neate, 2019). The first approach outlined by Plantin et al. (2018) might think about wooden pallets as a technical infrastructure. They might be seen for their sheer materiality, as "stuff you can kick" (Parks, 2015). Having such wooden pallets meet the correct technical specification might be what Callon (1984) calls an "obligatory passage point": the narrow point of a funnel, around which a much larger network must converge. The second, the more phenomenological approach might study how it was that these pallets remained so hidden in the first place. How they became what Susan Leigh-Star (1999: p. 377) calls "boring" things: the often-invisible entities that support certain practical and institutional arrangements, here relating to logistics systems and goods distribution.

Both of these approaches are valid, and our argument for the importance of the second approach entails combining elements from the first. Emphasising the frequent invisibility of media infrastructures in practical settings does not preclude their study as technical artefacts, a point made in phenomenological perspectives on media (e.g. see (Markham and Rodgers, 2017a). Phenomonelogical perspectives on media have been exemplified by the "practice turn" in media studies (e.g. Couldry, 2004; Bräuchler and Postill, 2010) and related arguments for "non-media-centric" approaches (Morley, 2009; Moores, 2018). These approaches emphasise the activities and settings through which phenomena emerge as media, rather than seeing these activities and settings as effects generated by media (e.g. media texts, technologies or producers). Such approaches thus shift attention from what media do to people, to what people do with media; or what Martìn-Barbero (1993) conceptualised as "mediation" over media (cf. Kember and Zylinska, 2012). But this priority given to mediation does not advocate a naïve user- or human-centrism against the technical agency of media infrastructures. Instead, it avoids characterising media and in particular digital technologies as encroaching on "humans", who are often seen in a position of "resistance" (cf. Kember and Zylinska, 2012; Rose, 2017). Approaching social media as experiential infrastructures entails seeing them as a technological "enframing" for everyday activities (Heidegger, 1977); as the often-hidden supports making everyday practices endure and recur over time (Sterne, 2003).

The phenomenological perspective we have outlined on social media also has implications in terms of political temperament. It focuses less on the "hidden causes, determining conditions, and noxious motives" of platforms, and more on what urban communication via social media infrastructures "unfurls, calls forth, makes possible" (Felski, 2015: p. 12). Since Husserl, phenomenological approaches have been associated with "bracketing out": the suspension of judgment, in order to focus on the study of experience. In studying platforms and other media, bracketing out means focusing on the dynamics of situated mediated experience and, if only temporarily, disregarding forces outside of that experience as explanations (Markham and Rodgers, 2017b). This approach is illustrated by our brief analysis of social media temporalities, as novel forms of real-time-like participation in urban public life (for a more detailed account, see (Rodgers et al., forthcoming). We refrained from offering, for example, platform capitalism, Facebook's corporate motives, or algorithmic calculations as explanations. Our analysis does not preclude or deny the relevance of such analyses. Instead, it defers the form of critique they imply, while making possible another: one focused on the political epistemology of everyday knowledge and communication, mediated by social media platforms as experiential infrastructures of urban life.

Conclusion

This chapter has put forward a phenomenological perspective for examining the emergence of social media platforms as increasingly ordinary, taken-for-granted infrastructures of urban communication. Through this lens, we explored how social media afford a real-time-like experience of locality, illustrating this with reference to recent research we conducted on the mediation of often-contentious public exchanges around a cycling program in Walthamstow, East London. This temporally-focused analysis of social media practices helps make visible relatively novel ways of experiencing and addressing matters of urban public life. While we have emphasised the experience of social media infrastructures, our approach also retains an inherent concern for background technical qualities. The functional architectures of platforms such as Facebook both enframe and depend on user practice. As a result, our interest has been less in how social media make new urban phenomena or agents public, and more in how they constitute new environments reshaping how urban public life is mediated as a spatiotemporal process (cf. Carpignano, 1999).

The approach we have taken represents just one way of critiquing the urban politics of social media. Without doubt, it is a form of critique with blind spots, notably towards the politics we might associate with the practices and ideals of platform companies. Popular attention, for example, is increasingly being devoted to the intentionally "addictive" design of social media interfaces, apparently substantiated by reports that executives are discouraging family members from having accounts on the very platforms they help create (Price, 2018). And despite the obvious role social media play in mediating politics today, for good or ill, platform companies continue to treat their "participating" users as consumers rather than citizens of the state (Butt et al., 2016). We recognise and affirm the value of these and other critical issues and approaches relating to social media, and platforms more generally. But we would argue that scholarly analyses of a "platform urbanism" would do well to reject a "deus ex machina" (see Markham and Rodgers, 2017b: p. 11). By this, we mean resorting to a vague underlying driver, such as algorithmic agency or neoliberal logic, for observed practices, events and phenomena related to platforms. When carefully articulated, these and other deep explanations can have a valid place, but too often they act as a convenient plot device, abruptly closing the analysis and seemingly precluding other interpretations. Platforms and digital technologies are not seamless means of instrumental power, and a growing chorus of critical scholarship has begun to point to their "glitches" (Leszczynski, 2020), "quirks" (Beer, 2019) and "lossiness" (Payne, 2018). If publicness is transforming with social media, it will not be enough to challenge the power of platforms. Instead, we need to think about ways of negotiating practices of public address that are intrinsically, and also problematically, reliant on platform technics. This might mean thinking through the political and ethical responses that might be appropriate and viable, in and through these new urban environments of everyday knowledge and communication.

Notes

1 While writers such as Eve Kosofsky Sedgwick and Rita Felski come from a literary tradition, their arguments are also clearly relevant for developing alternative approaches to the politics of digital technologies (e.g. Ash, 2018; Velkova and Kaun, 2019). We are grateful to Niels van Doorn for encouraging us, at the University of Manchester workshop leading to this book, to consider the possible links that queer theory and reparative/postcritical approaches might have with our own emphasis on the phenomenology of public address, which relies in significant part on the work of Michael Warner (2002).

2 This research project was funded by an EPSRC pilot grant. Besides the authors of this paper, the other main project member was Andrea Ballatore at Birkbeck, University of London.

3 It should be noted that boyd's analysis was written before the 'ephemeral turn' of social media (Haber, 2019). Newer, non-archiving social media such as Snapchat and TikTok entail different structural affordances, and arguably different forms of publicness.

References

Alaimo, C. and Kallinikos, J. (2019) 'Social media and the infrastructuring of sociality', in Kornberger, M. et al. (eds) *Thinking infrastructures*. (Research in the Sociology of Organizations), pp. 289–306.

Ananny, M. and Crawford, K. (2018) 'Seeing without knowing: Limitations of the transparency ideal and its application to algorithmic accountability', *New Media & Society*, 20(3), pp. 973–989. doi: 10.1177/1461444816676645.

Ash, J. (2018) *Phase Media: Space, Time and the Politics of Smart Objects*. New York: Bloomsbury.

Barns, S. (2019) 'Negotiating the platform pivot: From participatory digital ecosystems to infrastructures of everyday life', *Geography Compass*, 13(9). doi: 10.1111/gec3.12464.

Beer, D. (2019) *The Quirks of Digital Culture*. Bingley: Emerald Publishing.

boyd, danah (2010) 'Social network sites as networked publics: Affordances, dynamics, and implications', in Papacharissi, Z. (ed.) *A Networked Self: Identity, Community, and Culture on Social Network Sites*. New York: Routledge, pp. 47–66.

Bräuchler, B. and Postill, J. (eds) (2010) *Theorising Media and Practice*. Oxford: Berghahn Books.

Bruns, A. and Burgess, J. (2015) 'Twitter hashtags from ad hoc to calculated publics', in Rambukkana, N. (ed.) *Hashtag Publics: The Power and Politics of Discursive Networks*. New York: Peter Lang, pp. 13–28.

Bucher, T. (2018) *If...Then: Algorithmic Power and Politics*. Oxford: Oxford University Press.

Butt, D., McQuire, S. and Papastergiadis, N. (2016) 'Platforms and public participation', *Continuum*, 30(6), pp. 734–743. doi: 10.1080/10304312.2016.1231777.

Callon, M. (1984) 'Some elements of a sociology of translation: Domestication of the scallops and the Fishermen of St Brieuc Bay', *The Sociological Review*, 32(1_suppl), pp. 196–233. doi: 10.1111/j.1467–954X.1984.tb00113.x.

Carpignano, P. (1999) 'The shape of the sphere: the public sphere and the materiality of communication', *Constellations*, 6(2), pp. 177–189.

Coleman, R. (2018) 'Social media and the materialisation of the affective present', in Sampson, T. D., Maddison, S., and Ellis, D. (eds) *Affect and Social Media: Emotion, Mediation, Anxiety and Contagion*. London; New York: Rowman & Littlefield International, pp. 67–75.

Couldry, N. (2004) 'Theorising media as practice', *Social Semiotics*, 14(2), pp. 115–132.

Crawford, K. (2016) 'Can an algorithm be agonistic? Ten scenes from life in calculated publics', *Science, Technology, & Human Values*, 41(1), pp. 77–92. doi: 10.1177/0162243915589635.

Van Dijck, J. (2013) *The Culture of Connectivity: A Critical History of Social Media*. Oxford: Oxford University Press.

Van Dijck, J., Poell, T. and de Waal, M. (2018) *The Platform Society*. New York: Oxford University Press.

Dourish, P. and Bell, G. (2011) *Divining a Digital Future: Mess and Mythology in Ubiquitous Computing*. Cambridge, MA: MIT Press.

Easterling, K. (2016) *Extrastatecraft: The Power of Infrastructure Space*. London; New York: Verso.

Felski, R. (2015) *The Limits of Critique*. Chicago, IL: The University of Chicago Press.

Gerbaudo, P. (2012) *Tweets and the Streets: Social Media and Contemporary Activism*. London: Pluto Press. Available at: http://www.oapen.org/search?identifier=642730 (Accessed: 30 August 2018).

Gorwa, R. (2019) 'What is platform governance?', *Information, Communication & Society*, pp. 1–18. doi: 10.1080/1369118X.2019.1573914.

Haber, B. (2019) 'The digital ephemeral turn: Queer theory, privacy, and the temporality of risk', *Media, Culture & Society*, 41(8), pp. 1069–1087. doi: 10.1177/0163443719831600.

Harvey, D. (1989) *The Condition of Postmodernity: An Enquiry into the Conditions of Cultural Change*. Oxford: Wiley-Blackwell.

Heidegger, M. (1977) *The Question Concerning Technology and Other Essays*. Translated by W. Lovitt. New York: HarperCollins Publishers (Harper Perennial modern thought).

Helmond, A. (2015) 'The platformization of the web: Making web data platform ready', *Social Media + Society*, 1(2), p. 205630511560308. doi: 10.1177/2056305115603080.

Highfield, T. (2016) *Social Media and Everyday Politics*. Cambridge, UK; Malden, MA: Polity.

Hill, D. (2015) 'Waltham Forest "mini-Holland" row: Politics, protests and house prices', *The Guardian*. Available at: https://www.theguardian.com/uk-news/davehillblog/2015/nov/07/waltham-forest-mini-holland-row-politics-protests-and-house-prices.

Jenkins, H., Itō, M. and boyd, danah (2015) *Participatory Culture in a Networked Era: A Conversation on Youth, Learning, Commerce, and Politics*. Cambridge: Polity Press.

Kaun, A. (2015) 'Regimes of time: Media practices of the dispossessed', *Time & Society*, 24(2), pp. 221–243. doi: 10.1177/0961463X15577276.

Kaun, A. and Stiernstedt, F. (2014) 'Facebook time: Technological and institutional affordances for media memories', *New Media & Society*, 16(7), pp. 1154–1168. doi: 10.1177/1461444814544001.

Kember, S. and Zylinska, J. (2012) *Life After New Media: Mediation as a Vital Process*. Cambridge, MA: MIT Press.

Kitchin, R. (2014) 'The real-time city? Big data and smart urbanism', *GeoJournal*, 79(1), pp. 1–14. doi: 10.1007/s10708-013-9516–8.

Kitchin, R. and Dodge, M. (2011) *Code/space: software and everyday life*. Cambridge, MA: MIT Press.

Langlois, G. (2013) 'Participatory culture and the new governance of communication: The paradox of participatory media', *Television & New Media*, 14(2), pp. 91–105. doi: 10.1177/1527476411433519.

Larkin, B. (2013) 'The politics and poetics of infrastructure', *Annual Review of Anthropology*, 42(1), pp. 327–343. doi: 10.1146/annurev-anthro-092412–155522.

Lee, F. and Björklund Larsen, L. (2019) 'How should we theorize algorithms? Five ideal types in analyzing algorithmic normativities', *Big Data & Society*, 6(2), p. 205395171986734. doi: 10.1177/2053951719867349.

"Leszczynski, A. (2020) 'Glitchy vignettes of platform urbanism', Environment and Planning D: Society and Space, 38(2), pp. 189–208. doi: 10.1177/0263775819878721.

Mackenzie, A. (2006) *Cutting Code: Software and Sociality*. New York: Peter Lang.

Margetts, H. et al. (2017) *Political Turbulence: How Social Media Shape Collective Action*. Princeton: Princeton University Press.

Markham, T. and Rodgers, S. (eds) (2017a) *Conditions of Mediation: Phenomenological Perspectives on Media*. New York: Peter Lang.

Markham, T. and Rodgers, S. (2017b) 'Theorizing media phenomenologically', in Markham, T. and Rodgers, S. (eds) *Conditions of Mediation: Phenomenological Perspectives on Media*. New York: Peter Lang, pp. 1–18.

Martìn-Barbero, J. (1993) *Communication, Culture and Hegemony: From Media to Mediations*. London: Sage.

Mattern, S. (2017) *Code + Clay... Data + Dirt: Five Thousand Years of Urban Media*. Minneapolis: University of Minnesota Press.

Moores, S. (2018) *Digital Orientations: Non-Media-Centric Media Studies and Non-Representational Theories of Practice*. New York: Peter Lang (Digital formations).
Moore, S., Rodgers, S. and Ballatore, A. (no date) 'Competing logics? Social media platforms as fields of public engagement in urban change', *In preparation*.
Morley, D. (2009) 'For a materialist, non-media-centric media studies', *Television & New Media*, 10(1), pp. 114–116.
Mouffe, C., Wagner, E. and Mouffe, C. (2013) *Agonistics: Thinking the World Politically*. London; New York: Verso.
Neate, R. (2019) 'Unpalatable news? UK faces pallet crisis if there is no-deal Brexit', *The Guardian*. Available at: https://www.theguardian.com/politics/2019/feb/26/uk-pallet-crisis-no-deal-brexit.
Papacharissi, Z. (2010) *A Private Sphere: Democracy in a Digital Age*. Cambridge: Polity.
Papacharissi, Z. (2015) *Affective Publics: Sentiment, Technology, and Politics*. Oxford: Oxford University Press.
Parks, L. (2015) '"Stuff you can kick": Towards a theory of media infrastructures', in Svensson, P. and Goldberg, D. T. (eds) *Between Humanities and the Digital*. Cambridge, MA: The MIT Press, pp. 355–373.
Parks, L. and Starosielski, N. (eds) (2015) *Signal Traffic: Critical Studies of Media Infrastructures*. Urbana: University of Illinois Press.
Payne, R. (2018) 'Lossy media: Queer encounters with infrastructure', *Open Cultural Studies*, 2(1), pp. 528–539. doi: 10.1515/culture-2018–0048.
Peters, J. D. (2015) *The Marvelous Clouds: Toward a Philosophy of Elemental Media*. Chicago, IL; London: the University of Chicago Press.
Plantin, J.-C. et al. (2018) 'Infrastructure studies meet platform studies in the age of Google and Facebook', *New Media & Society*, 20(1), pp. 293–310. doi: 10.1177/1461444816661553.
Price, C. (2018) 'Trapped – the secret ways social media is built to be addictive (and what you can do to fight back)', *Science Focus*. Available at: https://www.sciencefocus.com/future-technology/trapped-the-secret-ways-social-media-is-built-to-be-addictive-and-what-you-can-do-to-fight-back/.
Rodgers, S., Barnett, C. and Cochrane, A. (2014) 'Media practices and urban politics: Conceptualizing the powers of the media-urban nexus', *Environment and Planning D: Society and Space*, 32(6), pp. 1054–1070. doi: 10.1068/d13157p.
Rodgers, S. and Moore, S. (2018) 'Platform urbanism: An introduction', *Mediapolis: A Journal of Cities and Culture*, 4(3). Available at: http://www.mediapolisjournal.com/201.
Rodgers, S., Moore, S. and Ballatore, A. (no date) 'Ambient participation: Time-spaces of neighbourhood politics through social media platforms', *In preparation*.
Rose, G. (2017) 'Posthuman agency in the digitally mediated city: Exteriorization, individuation, reinvention', *Annals of the American Association of Geographers*, 107(4), pp. 779–793. doi: 10.1080/24694452.2016.1270195.
Scannell, P. (1996) *Radio, Television and Modern Life: A Phenomenological Approach*. Oxford: Blackwell.
Schmidt, J.-H. (2013) 'Twitter and the rise of personal publics', in Weller, K. (ed.) *Twitter and Society*. New York: Peter Lang (Digital formations), pp. 3–14.
Sedgwick, E. K. (2003) *Touching Feeling: Affect, Pedagogy, Performativity*. Durham: Duke University Press.
Sharma, S. (2014) *In the Meantime: Temporality and Cultural Politics*. Durham: Duke University Press.
Star, S. L. (1999) 'The ethnography of infrastructure', *American Behavioral Scientist*, 43(3), pp. 377–391.

Sterne, J. (2003) 'Bourdieu, technique and technology', *Cultural Studies*, 17(3–4), pp. 367–389. doi: 10.1080/0950238032000083863a.

Tomlinson, J. (2007) *The Culture of Speed: The Coming of Immediacy*. Los Angeles; London: SAGE (Theory, culture & society).

Velkova, J. and Kaun, A. (2019) 'Algorithmic resistance: Media practices and the politics of repair', *Information, Communication & Society*, pp. 1–18. doi: 10.1080/1369118X.2019.1657162.

Warner, M. (2002) *Publics and Counterpublics*. New York: Zone Books.

Weltevrede, E., Helmond, A. and Gerlitz, C. (2014) 'The politics of real-time: A device perspective on social media platforms and search engines', *Theory, Culture & Society*, 31(6), pp. 125–150. doi: 10.1177/0263276414537318.

Willems, W. (2019) '"The politics of things": Digital media, urban space, and the materiality of publics', *Media, Culture & Society*, 41(8), pp. 1192–1209. doi: 10.1177/0163443719831594.

Ziewitz, M. (2016) 'Governing algorithms: Myth, mess, and methods', *Science, Technology, & Human Values*, 41(1), pp. 3–16. doi: 10.1177/0162243915608948.

Zingale, N. C. (2013) 'The phenomenology of sharing: Social media networking, asserting, and telling', *Journal of Public Affairs*, 13(3), pp. 288–297. doi: 10.1002/pa.1468.

15

URBAN CONSUMPTION, MARKETS AND PLATFORMS AS FLEXIBLE SPATIAL ARRANGEMENTS

Lizzie Richardson

Introduction

The city of Newcastle-upon-Tyne, like other English cities, has recently seen an expansion in takeaway prepared food choice thanks to meal delivery through companies such as Deliveroo and UberEats. These companies have what they call a "technology platform" that enables customers in certain urban areas to order and purchase a meal which is then delivered to their location, usually within 30 minutes. This results in two spatial tendencies in the urban geography of consumption. The first is one of *dispersal* in the consumption of prepared food, as customers eat meals at a distance from the restaurant. The second is one of *concentration* of new physical sites for the collective preparation of meals, such as dark kitchens where menus from different restaurants are offered solely for delivery. These tendencies in the urban geographies of consumption require an examination of the relationship between platforms and markets. Deliveroo is involved in the construction of a market for takeaway food delivery in the cities where it operates. Following studies of marketisation (Berndt and Boeckler, 2009; Boeckler and Berndt, 2013), such market making constitutes processes of arrangement that illustrate how platforms manifest through the temporary coordination of a variety of actors. The chapter firstly seeks to show how interrogating such market making in cities results in a different approach to platforms, one that shifts away from the interface – the on-screen representation of the perfect meeting of supply and demand through the platform – and instead focuses on the material practices through which goods are constituted via contingent arrangements of buyer and seller.

Examining these material practices of marketisation through the platform requires attending to the algorithm's role in building contingency by putting specific demands on space and time, which paradoxically allow for and indeed enable flexibility. Secondly then, the chapter shows how such market making opens an understanding of platforms as *flexible arrangements* of urban consumption space. Food service flexibly connects dispersed restaurants, customers and riders every time a meal is delivered. However, reliable flexible food service requires certain standard processes that support the temporary coordination of different actors. To develop this, the chapter firstly establishes how Deliveroo is altering the geographies of urban consumption of prepared food in England with a particular focus on Newcastle, the location for the collection of the empirics informing the chapter. It then shows how these changes are tied to processes of market making for the delivered meal, arguing that any understanding of such markets must go beyond the on-screen interface to examine the coordination of networked actors. The final section sets out the delivered meal as a flexible arrangement of urban consumption space that requires standardisation of its different composite actors to enable their concentrated coordination, followed by a brief conclusion.

Eating out: reconfiguring urban consumption spaces

The change to urban consumption practices enabled by food service platforms such as Deliveroo alters an existing geography of "leisure spaces" in Newcastle-upon-Tyne. Consumption spaces have been vital to the city centre regeneration of Newcastle since the late 1990s. Similar to other cities in the north of England, a combination of industrial decline and the construction of an out-of-town shopping mall, the Metro centre, saw the departure of much economic activity from the city centre. One important attempt to revitalise the city centre has been policies that support the growth of pubs, bars and restaurants in what is termed the "night-time economy" (Shaw, 2015). Newcastle now hosts a variety of "casual dining" restaurants that have contributed to a rise in the frequency of "eating out" in Britain; from roughly 80% eating out less than once a month or not at all in 1989, to nearly 70% eating out once a month or more by 2015. However, from 2017 onwards, a "crunch" on such casual dining restaurants has occurred, with closures of sites by several well-known restaurant brands (Naylor, 2018). These closures are the result of a combination of factors including the overly ambitious expansion of restaurant chains, falling "consumer confidence", rising labour costs after the introduction of a "national living wage" and the requirements for multi-channel access including delivery.

The possibilities for food delivery enabled through the Deliveroo app are altering these existing post-industrial urban geographies of consumption, with two spatial tendencies emerging. The first tendency, and currently most evident, is one of *dispersal*, constituted by the increasing numbers of UK restaurants that are now producing food for consumption off-site. Prior to the launch of Deliveroo in the UK in 2013, aggregator platforms such as Just Eat and Hungry House provided a portal for online selection, ordering and payment from existing take-away food establishments, with the delivery of the order to the place of consumption to be carried out either by the restaurant or by the customer. Deliveroo similarly aggregated restaurants in one virtual place, but with two key differences that feed into the changing spaces of urban consumption. One was that the interface for the display of meal choices began life as a mobile app rather than a website, and the other was that it partnered (for a commissioning fee per order) with higher quality restaurants that had previously not offered delivery. Deliveroo was providing options that diverged from the pizza, Indian and Chinese (and of course "fish and chips") meals that had previously dominated UK takeaway food, and were making it easy for customers to choose of food almost anywhere in the city through the tap of a screen. The opportunity for customers to order online was pitched by Deliveroo to restaurants as a means of "virtually" expanding their site. To enable this dispersal of restaurant space into the city the company not only provided the technology for meal selection, ordering and payment but also for delivery.

The second, more nascent, spatial tendency is one of *concentration* in which new physical sites have been constituted for the production of prepared food. In 2016 Deliveroo established "dark kitchens" that could supply areas in cities that were not serviced by existing restaurants networks. Deliveroo provided restaurant trained staff with a kitchen from which to furnish orders from their usual menu. Concurrently, the company encouraged existing restaurants to develop "virtual brands". These brands would appear under a new "Deliveroo Editions" section on the customer app, providing menus separate from the existing (physical) restaurant and therefore offering a further revenue stream for a given restaurant. The preparation of meals for these brands could either take place in the restaurant's existing kitchen (but only supply to customers ordering through Deliveroo) or in one of the separate dark kitchens. Most recently, Deliveroo launched a "virtual food market" in 2018, allowing customers to place an order with multiple restaurants in a single transaction. Then in Hong Kong in November 2018, Deliveroo opened up a bricks-and-mortar location called Deliveroo Food Market premised on the same idea; hosting five restaurant groups offering 15 dining "concepts" that serve a variety of "global dishes" (CNBC, 2018). The Deliveroo Food Market in Hong Kong is one early indication of a physical concentration for collective consumption, albeit with a different configuration and concentration of ownership. In order to understand the emergence of these dual processes of dispersal and concentration constituting the urban geography of platforms, we must turn to questions of markets.

Making a market for the delivered meal

The purchase of delivered prepared food through Deliveroo can be understood as a process of marketisation (Berndt and Boeckler, 2009; Çalışkan and Callon, 2010; Boeckler and Berndt, 2013). Deliveroo has made – or at least has significantly extended – a market for the delivery of prepared meals in the city of Newcastle. This marketisation is significant for understanding the relationship between platforms and the urban geography of consumption because market making processes are necessarily spatial, even if not tied to a physical market place. They involve the coordination of different dispersed elements, most notably the buyer, the seller and the goods. The terms of such coordination require processes of calculation to enable the buyer and the seller to meet (although not necessarily physically) to resolve their opposing interests, generally through the agreement of a price. Thus, markets can but do not have to be territorialised in a physical market place. To ensure exchange though, they nonetheless require elements of *spatial arrangement*: the coordination of the goods, the buyer and the seller in "virtual" or "physical" space.

Thus, to understand the tendencies of dispersal and concentration constituting the urban geography of the platform, it is necessary to consider exactly what is meant by a market in the context of food delivery. The arrangement of actors associated with the Deliveroo platform combine, following Callon (2016), two broad and competing views of markets and goods, set out below. In the first, labelled "interface-markets" by Callon[1] and drawing on neoclassical economics, the market operates to determine the meeting of supply and demand so that the constitution of goods (i.e. their value, how they are made) is black-boxed. In the other, "market-agencement" from economic sociology, the market is continually being remade through the contingent coordination of actors that also exposes the negotiation of goods. This perspective starts from the ground to consider how actors involved in markets are changeable rather than fixed (e.g. as buyer or seller), resulting from their position in wider networks (Granovetter, 1985). In this latter type of market organisation, for the networked agents involved, their:

> identities, interests and objectives, in short, everything which might stabilise their description and their being, are variable outcomes which fluctuate with the form and dynamics of relations between these agents.
>
> *(Callon, 1998: p. 8)*

It is this second view that is vital to making sense of the platform's urban geography as a flexible spatial arrangement.

Interface-markets: articulating supply and demand

The first view of markets is at the level of the *interface*, or the screen-based representation involved in the coordination of different actors. This representation pre-determines the roles of actors as buyers and sellers. Such a view of markets from the interface is distinctly different from an understanding of markets in which there is an adaptability to the actors involved. When represented via the interface, the number of buyers and sellers active at any one time might fluctuate (i.e. the volume of supply and demand), but the identity of these actors as buyers or sellers is fixed before any interaction, and so, therefore, is their intended coordination and the price for exchange. What is the market role of Deliveroo's platform then when viewed from the interface, if it is neither clearly buyer nor seller nor necessary for the resolution of contingencies if representation means the coordination of these particular actors is already decided? The answer to this lies in an examination of the goods being exchanged. If the buyer ("eater") and seller ("restaurant") are relatively easy to identify, then the goods for purchase is more complex. The buyer is purchasing the goods of a "delivered meal", and the totality of that purchase lies not in the restaurant that produced the meal, but in the platform that combines the meal with its delivery.

Therefore, it is possible to understand the platform as a good, or what Callon (2016) terms a "platform-good". These platform-goods are both a link between blocs of supply and demand, and a service supplier. This is evident in the Deliveroo example where the platform-good of the "delivered meal" is the link between the supplier (restaurant) and the demander (consumer), but also a service of food delivery. Such a good does not, of course, have to be limited to the "digital platform"; Callon offers the daily free press as an example of a platform-good. These newspapers articulate three different groups to one another: the readers of the newspaper; the firm that produces and distributes these papers; and the advertising agencies that buy the space. Thus the main function of the platform-good is to articulate supply and demand outside of it such that:

> the platform-good maintains the separation between blocs [of supply and demand], while keeping them linked. Conversely, when the blocs of supply and demand are independent of one another, they require platform-goods for a market to exist. This combination of blocs and platforms constitutes the structure of interface-markets.
>
> *(Callon, 2016: p. 21)*

So the "interface-market" is the particular arrangement constituted through the good of the platform and is thus the view of the market visible from the interface. In this type of market, supply and demand can be interfaced – made to meet physically or virtually – but remain autonomous blocs, and crucially there is little or no contingency in terms of the valuation or definition of the good transacted. The market, and thus the calculation of goods, is separate from processes of market-making, or marketisation.

The "delivered meal" as a platform-good is therefore defined and valued outside of the interaction it creates between supply and demand, outside of the market *per se*. The result is that:

> it is not essential to know how the good is maintained as a platform, through superficial or profound changes, and how as it changes it creates attachments, that is, how it constantly reproduces its function as a platform articulating supply and demand.
>
> *(Callon, 2016: p. 23)*

Any modification of the goods is therefore aimed at redefining the supply and demand that it articulates, which in the case of Deliveroo involves accruing more customers and restaurants. This view of markets from the interface facilitates the closure of questions concerning the goods themselves, including the labour constituting them. The volume of supply and demand that is articulated is of primary importance, rather than how the "platform-goods" are qualified as such to carry this out. For Deliveroo, such a prioritisation of the quantity of exchanges of the delivered meal, therefore, enables the labour (e.g. of delivery, of making the food) through which these exchanges occur to be bracketed off, or black-boxed. The platform-good, by definition, functions to connect supply and demand such that this can be understood as an automatic process. In the case of the newspaper, this platform-good can be conceived as a relatively stable physical object that comprises the functions of link and service provider. However, the delivered meal, when examined beyond the interface, is less intuitively an object but rather a flexible spatio-temporal arrangement that occurs through the calculated coordination of the different actors, opening up the other view of markets.

Market-agencements: calculating the delivered meal

The second view of markets decentres the interface and widens the lens to take in the different entities that are collectively arranged so that the transactions of a good can take place. The urban market for food delivery is a fluid configuration of restaurants, riders, digital devices, vehicles and so on that acts to coordinate the goods of the delivered meal in their transaction from restaurant to customer. This configuration is simultaneously an action and a formatting of action that Callon (2016) terms a *market-agencement*. When viewed from this perspective, the discrete blocs of supply and demand, and their articulated goods of the delivered meal, appear substantially more messy and contingent than their representation through the interface. This is perhaps most apparent when focusing on the goods of the delivered meal themselves, that far from being determined in advance, as in the imaginary of the interface-market, are part of continuous processes of calculation and qualification necessary for the broader constitution of the market as such. For meal delivery to occur, the journey from the restaurant to the customer must be made calculable in order to guarantee the articulation of supply and demand through the platform. The solution to this problem of delivery involves not simply mapping the route but ensuring that there is an available delivery rider for each order placed. Given the perishability of a meal, this allocation of a rider to an order must occur as quickly as possible, which requires utilising a delivery rider who is close to the restaurant. This, in turn, means ensuring that there are sufficient numbers of riders on the road to meet the demand for meals.

Calculation to resolve these problems of delivery occurs through an algorithm designed by Deliveroo called "Frank". Frank has two roles, one of which is selecting the appropriately located rider to pick up a meal, and the other is:

> Matching the number of orders … [Deliveroo] expect customers will make (order volume) with the number of riders needed to deliver them … As well as order volume, the tool takes a number of others factors into account, such as seasonal trends and holidays, e.g. Valentines Day to help [us] determine the number of sessions to make available in each zone, for every hour, in all [our] cities and countries around the world!
>
> *(Deliveroo Tech Round-Up, June 2018, from author's field notes)*

Frank is, therefore, a calculative agent that formulates a problem and its solution (what is to be done), and combines this with the strategy and instructions for carrying it out (how it should be done) (Kitchin, 2017: p. 16). Such calculative agency is greatest when it can establish a long but finite list of entities, among which rich and varied relations are allowed, and through which formal procedures for the classification of these entities are set out (Callon and Muniesa, 2005: p. 1238). This means that, in order to identify and classify these entities, the calculative agency is distributed across different human and non-human actors and their equipment. Thus, to calculate the delivered meal requires the collation and manipulation of data, which concerns a variety of different actors and concerning which the actors have different degrees of agency in producing.

For example, in the allocation of orders to different riders, this data concerns:

> The specific dish that is being prepared; the location of the restaurant; the time of day and the day of the week; the number of riders on the road; how many live customer orders there are; the distance from the restaurant to the customer.
>
> *(Deliveroo Tech Round-Up, May 2018)*

During the course of an order there are particular spatial and temporal points marking key moments to coordinate the meal, the rider and the customer. These are distributed switches – or exchanges – when two (or more) parties (including people and things) interact, such as the rider recording via a swipe on their phone that they have collected the meal from the restaurant (Ruppert, Law and Savage, 2013: p. 35). More broadly, to ensure that there are sufficient numbers of riders on the road, data concerning other factors are accounted for such as the weather, events like large football games and so on. The possible implications of some of these elements (e.g. for volume of orders or availability of riders) can be predicted further in advance than others. So the closer to the present that different factors can be accounted for (i.e. the more immediate the production and communication of data), the more accurate is the calculation for each individual order and the matching of riders with order volume. This processes of "taking factors into account" is central to understanding the contingency of the calculation of the delivered meal, and to market-agencement more broadly.

For the market to be structured through ongoing collective action, it requires a series of framings that take particular factors into account and therefore enable specific formats for action, such as the movement of goods, or the roles of buyers and sellers at a given time (Callon, 2016). In combining the competing perspectives of the market as an interface (from neoclassical economics) and as an agencement (from economic sociology), the platform could be understood as an attempt to control the contingencies of these processes of framing: the representation via the interface seeks to control the smooth articulation of the delivered meal from restaurant to customer. Whilst this is in part true – insofar as the realisation of the delivered meal is the intended outcome – for the delivered meal to occur, Deliveroo nonetheless relies on degrees of autonomy of the rider, restaurant and customer to ensure that the goods are a flexible arrangement. Building on Helmond's (2015) definition, platforms manifest as a dispersal of features through external networks – such as restaurants or delivery riders – and simultaneously a concentration of content from these networks according to the platform's own formatting – such as online menu choices, or delivery slots. Through the decentralised generation of content then, the platform reflexively creates relationships between these differently networked participants in a recentralised projection based on their real-time changing activities. This balance of decentralised networks of restaurants, riders and customers and their recentralised projection means that there is always a possibility that actors might escape the framings of the platform. Thus to realise the delivered meal as such as flexible arrangement, the rider, restaurant and customer must perform together in concert. When successful, this is an orchestration that is achieved through calculations that require these different actors to adhere to particular standards, discussed below.

The delivered meal as a flexible arrangement of urban consumption space

The meal, like the platform, does not have a fixed, territorialised structure, but rather is flexibly constituted through the spatial tendencies of dispersal and concentration. This flexibility enables restaurants to switch on or off the app to increase or decrease orders; customers to order when and where they want through their mobile device, and riders to decide when to work and whether to accept an order. Perhaps the most visible in urban public space of these actors constituting the flexible arrangement of the delivered meal is the delivery rider. The requirement that prepared food is collected almost as soon as the order is placed means that a rider must be mobilised immediately to go to the restaurant to collect the food. Such a system relies on riders being "free" (i.e. from other orders through the platform) to pick up a meal, but also, therefore, creates a situation where if riders are "free" they could potentially be engaged in other activities (e.g. resting, delivering orders through other platforms). There is thus no guarantee that an order will be accepted by the rider to which it is allocated by the algorithm. However, despite this apparent freedom of riders (Shapiro, 2017), the flexible arrangement of the delivered meal only occurs through processes of standardisation.

Standardisation is the creation of parameters for each type of actor to enable its interaction with other types of elements in the arrangement. These standards are necessary to inform the view of the market from the interface that matches supply and demand by controlling the status as buyers and sellers, and their method of articulation (i.e. riders in the case of Deliveroo). Standards are different for different actors, depending on their role in the arrangement. Riders are standardised through their submission into a complex system of shift allocation, based on certain data collected about their performance reliability, that produces layers of contingency in their capacity to work that enables flexibility elsewhere in the arrangement. Riders with Deliveroo are firstly only contingently able to make themselves available to work, and secondly when they are available they are not guaranteed the possibility to earn anything, and thirdly when they are offered earnings these are variable rather than fixed. Taking each of these in turn, the first is a foundational contingency concerning the possibility for an individual rider to be available to work. Riders have to be able to go "online" on the Deliveroo rider app so that they can receive orders (i.e. be able to work). However, riders can only go online when there is space available in an individual hourly slot on any given day. The purpose of the slots is to ensure both that there is the right number of riders on the road to meet the demand for orders to be delivered, and that the riders that are on the road will receive orders to deliver, to avoid an oversupply of delivery capacity.

The least contingent method for riders to ensure that they will be able to go online is to book a slot in advance when there are spaces available. This can theoretically be done at any time, but for most of the time slots remain full. Therefore, the time to book slots in advance is when the following week's slots are first made available, which is the Monday of the week prior. The opportunity to book slots in advance on Mondays though is not equally distributed across riders. Some riders are given "priority" booking based on their "rider statistics" collated from the previous weeks. These statistics are three measures calculated on a weekly basis: the percentage of attended sessions; the percentage of late cancellations; and the percentage of super-peak sessions[2] attended. Riders who have high overall and super-peak attendance, together with a low percentage of late cancellations, relative to other riders in the delivery zone, are given priority[3] booking of slots. Therefore, apparently contrary to the rhetoric of flexibility, the platform rewards riders who are more reliable by giving them the opportunity for more choice over when they want to work. For the riders, the implication is that those who have priority booking tend to book all slots that they think that they will (or even might) work, with the possibility of cancelling later if necessary.

The riders who are not able to book slots in advance when they are released can book slots when they become available, which is usually on the day (or at best the day before) of the slot. A notification can be set up on the app for each slot that will let riders know when spaces in the slot become available. In addition, it is also possible to go online without booking if there are spaces available in a slot at a given time. The opportunity for certain "priority" riders[4] to choose when to be available to work therefore demands greater flexibility from those who are unable to book as they scrabble to gain access to slots. The second area of contingency arises when riders are able to go online. Once online, there is no guarantee that they will earn anything during any given hour slot. Riders are paid per "drop" (i.e. per order delivered). However, Deliveroo offers no guarantee that riders will be assigned an order when they are online. Although one of the reasons for the pre-booking of slots is to try to ensure that there will be sufficient orders for riders (as well as vice versa), riders still have no certainty about the number of orders that they may be assigned.

This is compounded by the third dimension to rider flexibility which is the recent addition of distance-based fees in July 2018 which has changed the way the rider is paid. The distance-based fee is the total pay per order made up of a fixed pick-up and fixed drop-off fee, together with a variable fee for the distance travelled from the rider's location at the point of accepting the order to delivery at the customer. Previously there was a fixed fee per order, regardless of customer location. The implication of the new distance-based fee is that the rider does not know in advance of being allocated an order how much that they will receive for that order. Given the autonomy of each rider, the algorithm incentivises reliability to reduce contingency by ensuring that the riders statistically most likely to carry out orders are on the road. At the same time, the algorithm makes it possible for less reliable riders (i.e. those with inadequate or insufficient statistics) to meet short term spikes in demand. Thus, the three-fold possibility for change in the rider's conditions of work is standardised in their compliance with a system of shift allocation based on their performance reliability and is a necessary precondition to the flexible arrangement of the delivered meal.

Both restaurants and customers are also subject to processes of standardisation that aim to fix their status as sellers and buyers through the interface. For restaurants, standardisation occurs through their physical and virtual integration with the Deliveroo app. In part this means ensuring that orders through Deliveroo can be furnished by their existing restaurant service, including both the labour of meeting the orders but also the physical infrastructure that enables these orders to be received by the kitchen. Integration also occurs through the display of menus on the app that exists according to Deliveroo's standards of visual design. Whilst this does not normally result in a change to the content of a menu, it does mean that restaurants must accept their brand will not necessarily be foregrounded on the app. Food is often grouped to appear according to the type of cuisine provided as opposed to the name (or physical location) of the restaurant, with the aim of ensuring that the app retains customers through ease of use. The customer themselves – the autonomous choice-making agent – is not free from standardisation in their engagement in the flexible arrangement of the delivered meal. Most importantly, customers must adhere to a fixed location, that they have given through the app, at the designated delivery time. Although this is undoubtedly a minor form of standardisation, it indicates more broadly that in actualising each delivered meal, the possibility for flexibility for one actor depends upon certain standards shaping the response of another actor.

Conclusion

The emerging geographies of urban consumption exemplified by the market for the delivered meal indicate the necessity to examine how platforms operate as socio-spatial arrangements in cities beyond the capacities of the platform as a company, apparent through its on-screen interface and algorithm. Making a market for meal delivery involves arranging an array of dispersed elements, of which the platform as the company is part, in order to actualise the delivered meal through their temporary concentration. The arrangement constituting the platform is, therefore, a flexible structure for action that does not have a fixed territory – market*place* – but rather draws on other territorialised networks to actualise in urban form. Therefore, the platform, as such a flexible spatial arrangement, has a distinctive capacity to act; it carries out operations such as meal delivery through its ability to articulate together different more or less territorialised urban elements. These elements are temporarily enrolled in the arrangement of the platform without complete removal from their existing urban hierarchies and concentrations of agency.

The platform as a flexible spatial arrangement implies a reorganisation of urban operations, not necessarily through new physical infrastructures, but instead through novel technologies of coordination of existing urban elements. At present, as the example of the company of Deliveroo illustrates, these technologies of coordination are organised through an economy tailored to individualising rather than collective needs, involving forms of predominately private value generated from the coordination of differently networked elements and generally disguise forms of labour involved. However, as the chapter has illustrated by directing the focus away from the platform as a company, the flexible arrangement of urban operations that constitutes the platform does not have to sit in this type of economy premised on the accumulation of private value, it could equally exist in another form of economic organisation.

Notes

1 In putting forward this type of market, Callon (2016) is broadly caricaturing (and critiquing) understandings of markets and the role of market competition in strands of neoclassical economics. His own notion of "market-agencement" is a variant of economic sociology that seeks to conceptualise and understand the social construction of markets.

2 The super-peak slots can vary by location, but they tend to include 7–9 pm on Friday, Saturday and Sunday night. These not only regularly have high volumes of orders, but also as "weekends" are less desirable slots for riders to work. The inclusion of a measure of attendance during these hours within the currency of the "rider statistics" is therefore an incentive for riders to work, although the pay remains the same.

3 Priority means the rider is given access to booking earlier on the Monday, such as 11 am rather than 3 pm or 5 pm.

4 This priority status is shifting, a cumulative measure over a past few weeks, meaning that if a rider's performance declines, they will eventually lose access to early booking.

References

Berndt, C. and Boeckler, M. (2009) 'Geographies of circulation and exchange: Constructions of markets', *Progress in Human Geography*, 33(4), pp. 535–551. doi: 10.1177/0309132509104805.

Boeckler, M. and Berndt, C. (2013) 'Geographies of circulation and exchange III: The great crisis and marketization "after markets"', *Progress in Human Geography*, 37(3), pp. 424–432. doi: 10.1177/0309132512453515.

Çalışkan, K. and Callon, M. (2010) 'Economization, part 2: A research programme for the study of markets', *Economy and Society*, 39(1), pp. 1–32. doi: 10.1080/03085140903424519.

Callon, M. (1998) *The Laws of the Markets, Sociological Review Monograph Series*. Oxford; Malden, MA: Blackwell Publishers/Sociological Review.

Callon, M. (2016) 'Revisiting marketization: From interface-markets to market-agencements', *Consumption Markets & Culture*, 19(1), pp. 17–37. doi: 10.1080/10253866.2015.1067002.

Callon, M. and Muniesa, F. (2005) 'Peripheral vision: Economic markets as calculative collective devices', *Organization Studies*, 26(8), pp. 1229–1250. doi: 10.1177/0170840605056393.

CNBC (2018) 'Deliveroo is opening its first bricks-and-mortar location'. Available at: https://www.cnbc.com/2018/12/11/deliveroo-is-opening-its-first-restaurant-in-hong-kong.html.

Granovetter, M. (1985) 'Economic action and social structure: The problem of embeddedness', *American Journal of Sociology*, 91(3), pp. 481–510.

Helmond, A. (2015) 'The platformization of the web: Making web data platform ready', *Social Media + Society*, 1(2), p. 205630511560308. doi: 10.1177/2056305115603080.

Kitchin, R. (2017) 'Thinking critically about and researching algorithms', *Information, Communication & Society*, 20(1), pp. 14–29. doi: 10.1080/1369118X.2016.1154087.

Naylor, T. (2018) 'The casual dining crunch: Why are Jamie's Italian, Strada, Byron (and the rest) all struggling?', *The Guardian*, 22 February. Available at: https://www.theguardian.com/lifeandstyle/2018/feb/22/casual-dining-crunch-jamies-italian-strada-byron-struggling (last accessed 25 June 2018).

Ruppert, E., Law, J. and Savage, M. (2013) 'Reassembling social science methods: The challenge of digital devices', *Theory, Culture & Society*, 30(4), pp. 22–46. doi: 10.1177/0263276413484941.

Shapiro, A. (2017) 'Between autonomy and control: Strategies of arbitrage in the "on-demand" economy', *New Media & Society*, p. 146144481773823. doi: 10.1177/1461444817738236.

Shaw, R. (2015) '"Alive after five": Constructing the neoliberal night in Newcastle upon Tyne', *Urban Studies*, 52(3), pp. 456–470. doi: 10.1177/0042098013504008.

16

BETWEEN ALGORITHMS AND THE STREETS

The everyday politics of ride-hailing taxis in India

Anurag Mazumdar

Introduction

Ride-hailing platforms like Uber and Ola[1] were considered to be disruptors of traditional Indian transportation options in 2013 (Muralidhar, 2016; Kapoor, 2018). They have since become ubiquitous, with almost 100 Indian cities having ride-hailing platforms for taxis, three-wheeler autorickshaws, electric rickshaws and/or motorbikes by 2019. Platforms like Uber and Ola are supported and financed by multi-billion dollar venture-capital backed companies, while others like Namma TYGR are supported by political leaders and labour cooperatives.[2]

Such platforms have been shown to exacerbate existing economic inequalities and information asymmetries as well as add new dimensions of inequality. In India, research on ride-hailing platforms has largely focussed on the routine flexibilisation, feminisation and exploitation of labour and the continuities with earlier regimes of informal piecemeal work prevalent in India (Aneja, 2016; Surie and Koduganti, 2016). The studies have found that platform workers engage in complex negotiations that are not well captured by a binary between exploitation and entrepreneurship (Raheja and Deutsch, 2016; Surie, 2018). However, these analyses have not examined the contradictions and dissonances within the everyday practices of platforms nor how those contradictions shape and are shaped by the platforms themselves. Studying these contradictions helps to situate and contextualise platforms within India as it embraces techno-masculinist solutions (e.g. smart cities or digital governance) to long-standing social and economic problems (Datta, 2015; Gurumurthy, Chami and Thomas, 2016).

A structuralist or top-down perspective suggests that platforms are imposed on people through a set of material and discursive practices that manipulate and restructure labour relations through digital algorithms (Morozov, 2013;

Harding, Kandlikar and Gulati, 2016; Rayle et al., 2016) with little space for platform workers to express considered consent, agency or resistance (Huws, 2014; Rosenblat and Stark, 2016; Zwick, 2018). While platforms exercise subtle coercions there is also an element of self-initiated compliance; research conducted in the United States suggests that drivers have internalised the risk, freedom and celebratory rhetoric associated with digital labour while acknowledging their precarity (Neff, 2012; Malin and Chandler, 2017). Workers have accepted the new conditions associated with digital labour, both positive and negative, and also acknowledged its potential for resistance (Chen, 2018).

A contrasting perspective drawn from research on the affordances of social media platforms and their users sees workers, regulators and the state as platform architects (boyd, 2010; Nagy and Neff, 2015; Bucher and Helmond, 2018) with the capacity to co-create the platform in many ways. For example, "imagined affordances" (Nagy and Neff, 2015: p. 2) are new possibilities for a platform created when its technical and social functions are extended beyond their intended uses. Imagined affordances are not exceptional "slippages" (Wilson, 2018) and instead are the mechanisms that drive "the emergent, irreducible, co-generative dynamics between platforms and the urban" (Rodgers and Moore, 2018). They are the means by which platforms can be selectively contested, rejected, endorsed or built.

Examining the practices, strategies and policies of those involved in ride-hailing platforms as imagined affordances reveal how India's platform urbanism is co-constituted. For example, the negotiations of ride-hailing drivers with multiple techno-social actors and arrangements are engendered by both enabling and disabling neoliberalism. Similarly, the contingent and calculated strategies of ride-hailing drivers, state apparatuses and other actors reveal how the various participants interact to mutually govern platforms and themselves. Further, focusing on the selective endorsement and rejection of communitarian logic can help clarify the nature of ride-hailing platforms in India.

This analysis demonstrates that ride-hailing platforms are contributing to the disintegration of older motoring communities but are also creating newer forms of motoring communities by giving space to hitherto marginalised groups. In this way, corporate ride-hailing platforms are going beyond their intended role into an imagined affordance of social organising. The analysis also demonstrates that state-platform relationships are not exclusively confrontational nor merely as a matter of regulatory arbitrage. These relationships are co-constituted performances in which participants build each other while simultaneously subverting expectations through contingent and messy "imagined affordances" (Nagy and Neff, 2015) as platform realities become embedded into everyday urban life.

This research is part of a larger project studying the socio-spatial implications of the platform economy on labour and urban social relations in Indian cities. This chapter explores the contestations and affordances of the ride-hailing economy, the practices and discourses of performing "community" and the governance of and by platforms. The research presents original discourse and textual analysis of journalistic accounts of Uber and Ola drivers' narratives, advertisements, YouTube videos and corporate blogs published between 2013 and 2019. The analysis focuses on how ride-hailing platforms were framed for the consumption of driver-partner and passengers in India, with special attention paid to how Indian languages (such as Hindi) and English were used. Finally, media coverage and the Ministry of Skill Development and Entrepreneurship reports were also analysed to clarify the Indian government's strategies on governing ride-hailing platforms.

Breaking down "Community"

Traditional transport options, including *autorickshaws*, battery-operated rickshaws and *kaali-peeli* (black-and-yellow) taxis, are examples of long-standing collaborative consumption and ownership in Indian cities (Bhattacharya, 2014). An ethnographic study of Mumbai's taxi industry shows few taxi drivers own their vehicles, instead preferring to rent from members of a close hereditary network of their own kin, community or caste (Bedi, 2016a, 2016b). This allows economically weaker individuals to drive a rented vehicle and earn a livelihood, and so can be seen as egalitarian. At the same time, internal gatekeeping often restricts taxi driving opportunities to particular castes. For example, older taxi stands in Delhi restrict membership to relatively higher placed Yadavs and Singhs (caste names as well as last names).

The owners, operators and financiers pay salaries and incentives to drivers as well as provide support should a crisis arise in the drivers' lives. It is normal for owners to rent one car to multiple drivers on the same day, forming two or three shifts. Traditional Indian taxi industries are thus composed of drivers, middle-persons, permitting agents, garage owners, repair personnel, taxi operators and more and whose relationships illuminate the "material, communal, obligatory and sensorial circuits" (Bedi, 2016b: pp. 389–390). By collaboratively managing the physical and social structures of the taxi industry, the many and diverse actors create "structures of motoring labour" (Bedi, 2016a: p. 1013) with fewer "individualized ambits of identification" (Bedi, 2016b: p. 390) compared to taxi industries in cities of the Global North.

Disagreement around the term "sharing" by platform proponents, workers and consumers creates a definitional dilemma (Hern, 2015)[3] that allows ride-hailing platforms around the world to use communitarian language. In so doing, they frame their operations as part of the sharing ethic (Bostman and Rogers, 2010; Price and Belk, 2016) and as a "difference from" rather than a "difference within" a hyper-consumerist capitalist culture (Gibson-Graham, 2008; Richardson, 2015). Exactly what gets shared and who decides on the elements of sharing is not pre-decided and are constantly in flux. Despite the language of "sharing"

(Richardson, 2015), ride-hailing platforms can work to dismantle the existing "shared" networks and structures that they see as incompatible.

Platforms are typically designed around independent "driver partners" whose activities, travel behaviour, trips and payments can be closely monitored. To this end, Uber and Ola make it clear that they want drivers to be asset-owners rather than asset-renters (Mukherjee, 2016; Singh, 2016a, 2016b) and encourage ownership by offering drivers soft loans and leasing programs.

The fuzzy and complex networks of operation and ownership, that platforms discourage, are a source of middle-class anxieties (Anjaria, 2006); such urban informal transactions resist categorisation and are popularly described as dangerous and inefficient. Ride-hailing platforms, perceived as a Western and value-neutral solution to social problems, benefit from this popular discourse (Shahane, 2015; Kurmanath, 2018). Messaging, both overt and subtle, from Uber and Ola emphasises how they are reducing firm-level economic inefficiencies without mentioning how they may be introducing new economic and logistical inefficiencies (e.g. idle cars or unevenly distributed shifts). Narratives in print and digital media show that the sharing narratives and societal benefits promoted by ride-hailing platforms have been appropriated by drivers and are being incorporated into operating structures.

The niche corporate platform actors (Morozov, 2013) have "formalised" and undermined some of the traditional collaborative motoring practices, but have maintained others that are conducive to platform growth (Bhattacharya, 2014). For example, Uber's referral program, UberDost (translated as "Friends of Uber" from Hindi), draws from the same kin- and caste-networks that were utilised earlier to recruit new drivers. UberDost representatives return to hometowns and villages to recruit new drivers with "quick on-boarding and background checks" (Prabhat, Nanavati and Rangaswamy, 2019). The established drivers also teach their new recruits how to use smartphones and navigation apps as well as the "soft skills" of dealing with diverse passengers (Pillai, 2016).

Clearly, the rise of ride-hailing platforms has gradually dissolved some of the practices that sustained traditional collaborative taxi industries while also supporting other practices associated with those collaborative structures. The shifts to motoring practices have coincided with new patterns of economic inequality, concentrations of ownership and polarisation (Kashyap and Bhatia, 2018). These changes, including those that enable or constrain communitarian practices and ethos, often enjoy widespread support from drivers, even though they do not affect drivers homogeneously.

Gig labour is experienced differently by different actors, contingent on their social position. For example, Dalits[4] and other lower-caste communities are entering taxi driving in increasing numbers. These drivers welcome the ride-hailing platforms as providing opportunities through "faceless", "neutral" or non-discriminatory (Prabhat, Nanavati and Rangaswamy, 2019) recruitment process as opposed to the previous system of strict kin mediation (Kapoor, 2015). For instance, Mudassar, a 30-year-old driver, says,

> ...we are not from better off economic sections and we live in community ghettos..so many in my group have become Uber drivers..sometimes the community helps us with co laterals to buy a car...
>
> *(Prabhat, Nanavati and Rangaswamy, 2019)*

In this regard, London is not so different. The racially, ethnically and economically diverse Uber drivers in London have not responded to legislative changes in the same way as the relatively uniform black cab drivers; cabbies see ride-hailing platforms as a threat to their livelihoods while marginalised groups see them as opportunities for work, even if gig work is precarious (Bennhold, 2017). Racialised and gendered exploitation of labour is at the heart of the digital economy (van Doorn, 2017), so the on-demand labour features associated with urban platforms cuts across geographies.

The advent of ride-hailing platforms has led to a decline of existing collaborative taxi ownership and operation structures, but the implications for motoring "community" are complex. Uber and Ola discourage the traditional shared networks of taxi driver labour, finance, repair and maintenance as they are incompatible with the typical platform model of drivers as owners. This has contributed to the breakdown of communal ownership and an increase in newer concentrations of ownership. At the same time, these platforms also benefit from and encourage the (exploitative) communal ties between drivers and the drivers' own communities and networks. Their "facelessness" recruitment processes prevent the discrimination of traditional taxi communities and allow hitherto underrepresented and marginalised groups to access motoring labour opportunities, even as they acknowledge the precarity of gig labour. All of these new subjectivities, solidarities and social contestations can be understood as productive possibilities.

Performing platform governance

The Indian government has been captured by the idea of smart cities where digitised services are delivered, smart technologies are normalised and government performance is evaluated at multiple scales as users, space, services and networks are unbundled and re-bundled (Luque-Ayala and Marvin, 2015). The media in India has focused on the complex and often fractious relationship between the territorial sovereignty of the state and functional sovereignty of digital capitalism (Pasquale, 2017). At the same time, the growth of and reliance on platforms are often interpreted as apolitical and an objective sign of development. Thus, the "infrastructuralisation of platforms" and the "platformisation of infrastructures" (Plantin et al., 2018: pp. 295–307) have received relatively little critical attention, even though "the rise of ubiquitous, networked computing and changing political sentiment has created an environment in which platforms can achieve enormous scales, co-exist with infrastructures, *and in some cases compete with or even supplant them.*" [Emphasis original authors'] (Plantin et al., 2018: p. 301). This admixture of infrastructures and platforms allows for the creation of enclosures, innocuously at first (as when drivers' ratings determine insurance cover) but that can ultimately restrict access to previously universal services while still appearing to uphold public values (Chen and Qiu, 2019). In this context, platform governance means two things: *governance of platforms* and *governance by platforms*. Both aspects of platform governance will become increasingly important as platforms become more commonplace and how they "control, interact, and accumulate" influence becomes more pronounced, even as their media role also becomes stronger (Schwarz, 2017: p. 1).

In India, social protection and latent entrepreneurialism coexist and create an environment in which a growing distrust of the state's ability to provide public services sits alongside cautious state support for platforms and digital labour. As a result, the implications of platform governance and legitimacy are exemplified by, but go beyond, the ambit of ride-hailing platforms services. Such platforms do acquire legitimacy through recognised forms of "regulatory arbitrage" (Calo and Rosenblat, 2017: p. 1627), but also through less familiar processes of complex, recurring, multi-directional and diffuse negotiations between platform actors, online payment software, data gathering and mining organisations, geolocation software and other connected algorithms (Plantin et al., 2018). These negotiations are "performances" of sharing in which relevant stakeholders, including states, drivers, platforms and more, repeatedly articulate and interpret the sharing economy within a single framework (Richardson, 2015: p. 122). As their performances complement and contradict each other, they illuminate the complex web of relations known as the "sharing economy", provisionally resolving the confusions and conflicts of the sharing economy (Belk, 2014; Nadeem et al., 2015) and building legitimacy. Data asymmetries and imbalances of power are present in both regulatory arbitrage and "performance" negotiations, so legitimacy always creates tensions and contradictions.

The governance of both traditional taxis and ride-hailing platform is not entirely clear. The federal-level *Motor Vehicles Act 1988* in India lays out rules of national road transport and regulates privately owned and commercial vehicle permits, including those used by ride-hailing drivers. At the same time, practical road transport governance was placed in the State List of the Seventh Schedule of the Indian Constitution, making it mainly the responsibility of individual state governments. Many state governments have designed "newer" laws to govern traditional radio taxis, although they do not completely address the operations and requirements of ride-hailing taxi companies (actually registered as mobile application-based technology companies rather than taxi companies). Additionally, and despite public proclamations from the federal government to the contrary, some senior ministers at state levels argue that ride-hailing taxis are illegal and a threat to traditional black-and-yellow taxis (HT Correspondent, 2019).

The regulatory gaps are revealed in a conflict over rules instituted by Maharashtra and Karnataka state governments in 2017 meant to address concerns around corporate ownership, concentrations of power and loss of workers' rights. Specifically seeking to level the playing field between ride-hailing and conventional taxis, the *Maharashtra New City Taxi Rules 2017* require Uber and Ola drivers to apply for city-taxi permits to conduct business within designated limits. These rules also stipulate that Uber and Ola follow the price-fixing mechanisms for traditional taxis and avoid demand-driven algorithmic surge pricing. In protest, Uber started a campaign calling these regulations, "burdensome", "regressive" and "practically impossible to comply with" (FP Staff, 2016) and encouraged Mumbai residents to sign petitions against the Maharashtra government. Platform drivers filed a petition with the Bombay High Court, claiming that the regulations would hit platform drivers, rather than aggregators or platforms themselves, putting them at a disadvantage and marginalising their livelihoods. They argued that new drivers would face unfair costs because city-permits are charged according to engine capacity and private cars typically have bigger engines than traditional taxis. The Maharashtra state government responded by claiming that ride-hailing companies routinely flout local transportation rules (Express News Service, 2017) and that the "predatory, monopolizing and exploitative tactics" of ride-hailing platforms "cannot be permitted by a welfare state like Maharashtra" (Express News Service, 2017).

Such conflicts demonstrate the institutional vacuum created when digital platform behaviour changes more rapidly than states can regulate (Bapna and Saxena, 2017; Dudley, Banister and Schwanen, 2017; Flores and Rayle, 2017). In this gap, ride-hailing platforms may act above the law, question regulatory legitimacy or critique the poor enforcement of "weak" regulations. However, they do not directly question the importance of the state or other public institutions in supporting public values—defined as the sum total of organisations' contribution to achieve common good (Moore, 1995). Instead, platforms seek a relationship with the state as they facilitate platform-centred interactions through "platform-play" to "create public value for coming generations" (Barns, 2019) via a "dynamically evolving societal arrangement", with each party struggling to create and enforce a "negotiable social contract" (Van Dijck, Poell and Waal, 2018: p. 26).

By engaging and cooperating with the state, platforms may be contributing to the common good but are definitely furthering their own economic interests. They simultaneously call for both "reduction" and "enforcement" of the state by "assisting the self-organization of people online" and by immersing themselves in traditional domains of governments and communities (Van Dijck, Poell and Waal, 2018: p. 22–26). Notably, urban transport in India has historically been overburdened and social protection and welfare policies for marginalised populations are virtually non-existent. When present, the implementation of these "essential" functions and services remains a challenge. Platforms are deriving legitimacy by stepping into these gaps and providing these services, rather than simply advocating the retreat of the state.

For instance, the Union Ministry of Labour and Employment partnered with Uber to sign-on and train ride-hailing platform drivers through the National Career Service online portal (Ministry of Skill Development and Entrepreneurship 2016). This initiative titled UberSHAAN was launched in 2016 to support the Skills India Mission to "empower" a million driver-entrepreneurs by 2018 (Uber India, 2016) by encouraging micro-entrepreneurship. Similarly, the southern Indian state of Telangana subsidises 60% of the price of vehicles when Uber imparts the necessary training driving and navigation training. This scheme involves the Backward Classes Welfare Corporation and Tribal Welfare Corporation of Telangana, which oversee state efforts to empower marginalised citizens of Telangana. Schemes like this show that close to 70% of applicants are people from lower castes[5] and religious minorities, a feature of platform labour highlighted previously.

Uber has launched Uber Care, a driver support programme offering life and health insurance policies for drivers and their family as well as reduced interest rate loans, mainly for children's education. These material benefits show how platforms are actively managing "risks" of the platform economy through new "formal" webs of operation. The semi-formalisation and "contractualisation" of previous structures of labour have raised expectations and, in turn, generated some protests against ride-hailing platforms centred on the raised expectations resulting from these partial formalisations (Surie, 2017). These activities suggest that welfare policies are no longer the responsibility of the state alone as platforms will increasingly determine what constitutes welfare and who provides it.

The non-core activities of ride-hailing platforms, such as insurance, banking, investment advice and more, are not coercive and instead are tacitly accepted by drivers' groups. The Sarvodaya Drivers Association of Delhi (SDAD), a vocal ride-hailing driver union, has argued for stronger links between the state government and ride-hailing platforms; they demand that Uber and Ola undergo the government-run Standardisation Testing and Quality Certification (STQC) which certifies both public and private hardware and software. Such demands show that driver groups are not calling on the government to unravel the opaque nature of algorithms or achieve old-school labour regulations, suggesting the practices and discourses underlying platform governance are not top-down. By demanding the government safeguard platform capital and investments and "maintain the rationality of their investments, and hold larger market players to their promises", drivers groups are working "to sustain the logic and rationality of drivers' investments" as aligned with platform companies' rationalities (Surie, 2018).

Conclusions

The micro-politics, contestations and dissonances of ride-hailing taxi platforms in India reveal how motoring structures and infrastructures are organised as well as how practices and discourses illuminate their operation. One important conclusion from this study is that the strategies and practices to enable and disable "sharing" do not map onto individual actors, leaving the "sharing" ethic as fundamentally contingent and messy. Thus, platforms are part of the demise of shared community-managed taxi networks as well as the repurposing of existing informal labour community ties. Platforms are also linked to the rise of digital labour for marginalised populations, along with the precarity and productive possibilities such digital labour entails. Another important conclusion is that the governance of and by platforms involves contradictory demands for a reduced role of platforms and simultaneously greater inclusion within public utilities that operate on data and information. These paradoxical positions can be explained as strategic alliances between diverse actors and institutions that strive for co-existence of infrastructures and platforms. The scalable sociability of platforms provides these actors and institutions multiple ways to expand the affordances of platforms or otherwise act as platform architects. Both conclusions, highlight how the social, economic and political structures that undergird platform urbanism are simultaneously and selectively invoked and negated. This suggests that a transition to a "platform society" is neither neat nor straightforward.

Notes

1 Ola is a digital platform organisation focusing on ride-hailing cars, public transportation and food delivery that was launched in 2010 and is headquartered in Bengaluru, India. Ola is Uber's main competitor in India and has more drivers in many Indian cities.

2 Namma TYGR is an app-based ride-hailing taxi service launched in September 2017 as a joint venture between the Bengaluru Drivers' Union and the Indian subsidiary of Singapore-based technology start-up Savetur Digital. Namma TYGR claimed to be different from other ride-hailing platforms because it was supported by an established workers' union and promised to take only 12% of a drivers' earning per trip. However, Namma TYGR discontinued operations in mid-2018, purportedly due to non-compliance with government regulations and lack of interest from drivers. For more, see Anupam (2017) and Philip (2018).

3 Many writers have taken issue with how Uber, Airbnb and other platforms use the word 'sharing'. Hern (2015) argues that tenuous labour contracts shield platform companies by offloading risks to workers without sharing the benefits, while Sutzl (2014) argues that sharing cannot be applied to economic activities as sharing is fundamentally anti-economic. However, John (2013) argues that sharing can include communication and distribution, therefore allowing 'sharing' to bring ethical connotations and positive associations even in the context of for-profit corporations.

4 Dalits are considered to be the lowest caste of India's deeply entrenched four-tiered hierarchy in the Hindu religious and social order. Caste-based discrimination is illegal in India, but systemic inequalities mean that Dalits are only negligibly represented in professional positions and are often excluded from religious and social festivals. Thus, a section of Dalits feel that ride-hailing jobs offer them social mobility. For more, see Chugh's (2016) report on how Dalit women received a training scheme for ride-hailing drivers in Delhi.

5 Large numbers of applications from lower caste people to a government-sponsored training scheme for ride-hailing drivers are significant because opportunities to join professional roles in formal sectors are rare and subject to discrimination (Thorat and Attewell, 2007). This leaves people from the lower castes to work in informal sectors, often in dehumanising conditions and still subject to varying degrees of exclusion in informal jobs (Swaminathan, 2010).

References

Aneja, U. (2016) 'Social protection in the age of uber', *The Wire*, 2 December. Available at: https://thewire.in/labour/social-protection-uber-employment.

Anjaria, J. S. (2006) 'Street hawkers and public space in Mumbai', *Economic & Political Weekly*. 41(21), pp. 2140–2146.

Anupam, S. (2017) 'Namma TYGR comes to Bengaluru to take on Ola and Uber', *Inc42*, 30 November. Available at: https://inc42.com/buzz/cab-hailing-namma-tygr-bengaluru-ola-uber/ (Last accessed 3 February, 2020).

Bapna, N. and Saxena, A. (2017) 'Getting ola and uber drivers out of the regulatory grey zone', *The Wire*, 21 March. Available at: https://thewire.in/business/uber-ola-drivers-classified-consumers (Last accessed 3 February, 2020).

Barns, S. (2019) 'Negotiating the platform pivot: From participatory digital ecosystems to infrastructures of everyday life', *Geography Compass*, 13(9). doi: 10.1111/gec3.12464.

Bedi, T. (2016a) 'Mimicry, friction and trans-urban imaginaries: Mumbai taxis/Singapore-style', *Environment and Planning A: Economy and Space*. SAGE Publications Sage UK: London, England, 48(6), pp. 1012–1029.

Bedi, T. (2016b) 'Taxi drivers, infrastructures, and urban change in globalizing Mumbai', *City & Society*. Wiley Online Library, 28(3), pp. 387–410.

Belk, R. (2014) 'You are what you can access: Sharing and collaborative consumption online', *Journal of Business Research*. Elsevier, 67(8), pp. 1595–1600.

Bennhold, K. (2017) 'London's uber ban raises questions on race and immigration', *The New York Times*, 2 October. Available at: https://www.nytimes.com/2017/10/02/world/europe/uber-london-cab.html (Last accessed 3 February, 2020).

Bhattacharya, R. (2014) 'The informal sector the role of urban commons', *Yojana*, 58, pp. 46–48.

Bostman, R. and Rogers, R. (2010) *What's Mine is Yours - How Collaborative Consumption is Changing the Way We Live*. London: Collins.

boyd, danah (2010) 'Social network sites as networked publics: Affordances, dynamics, and implications', in Papacharissi, Z. (ed.) *A Networked Self: Identity, Community and Culture on Social Network Sites*. New York: Routledge, pp. 47–66.

Bucher, T. and Helmond, A. (2018) 'The affordances of social media platforms', in J Burgess, T. P. and Marwick, A. (eds) *SAGE Handbook of Social Media*. London, UK: Sage publications, pp. 233–253.

Calo, R. and Rosenblat, A. (2017) 'The taking economy: Uber, information, and power', *Colum. L. Rev*. HeinOnline. 117(6), pp. 1623–1690.

Chen, J. Y. (2018) 'Thrown under the bus and outrunning it! The logic of Didi and taxi drivers' labour and activism in the on-demand economy', *New Media & Society*. SAGE Publications Sage UK: London, England, 20(8), pp. 2691–2711.

Chen, J. Y. and Qiu, J. L. (2019) 'Digital utility: Datafication, regulation, labor, and DiDi's platformization of urban transport in China', *Chinese Journal of Communication*. Taylor & Francis, 12(3), pp. 274–289.

Chugh, N. (2016) '"How can a girl become a taxi driver?": Defying India's caste and gender taboos', *The Guardian*, 8 July. Available at: https://www.theguardian.com/global-development/2016/jul/08/india-dalit-how-can-a-girl-become-a-taxi-driver-defying-caste-gender-taboos. (Last accessed 3 February, 2020).

Datta, A. (2015) 'New urban utopias of postcolonial India: "Entrepreneurial urbanization" in Dholera smart city, Gujarat', *Dialogues in Human Geography*. SAGE Publications Sage UK: London, England, 5(1), pp. 3–22.

Van Dijck, J., Poell, T. and Waal, M. de (2018) *The Platform Society*. New York: Oxford University Press.

van Doorn, N. (2017) 'Platform labor: On the gendered and racialized exploitation of low-income service work in the "on-demand" economy', *Information, Communication & Society*. Taylor & Francis, 20(6), pp. 898–914.

Dudley, G., Banister, D. and Schwanen, T. (2017) 'The rise of Uber and regulating the disruptive innovator', *The political quarterly*. Wiley Online Library, 88(3), pp. 492–499.

Ministry of Skill Development and Entrepreneurship (2016) *Skill India, Uber & Maruti Join Hands for Driver Training in the Country; Promote Entrepreneurship*. New Delhi: Ministry of Skill Development and Entrepreneurship. Available at: https://www.msde.gov.in/assets/images/latest news/NSDC_Uber_Maruti Press Release.pdf.

Express News Service (2017) 'App-based cabs: Maharashtra govt defends new rules in high court', *The Indian Express*, 11 October. Available at: https://indianexpress.com/article/cities/mumbai/app-based-cabs-maharashtra-govt-defends-new-rules-in-high-court-ola-uber-4884355/ (Last accessed 3 February, 2020).

Flores, O. and Rayle, L. (2017) 'How cities use regulation for innovation: The case of Uber, Lyft and Sidecar in San Francisco', *Transportation Research Procedia*. Elsevier, 25, pp. 3756–3768.

FP Staff (2016) 'Uber starts campaign against Maharashtra draft rules, urges Mumbaikars to sign petition', *Firstpost.com*, 28 October. Available at: https://www.firstpost.com/business/uber-starts-campaign-against-maharashtra-draft-rules-urges-mumbaikars-to-sign-petition-3074332.html.

Gibson-Graham, J. K. (2008) 'Diverse economies: Performative practices for other worlds', *Progress in Human Geography*. SAGE Publications Sage UK: London, England, 32(5), pp. 613–632.

Gurumurthy, A., Chami, N. and Thomas, S. (2016) 'Unpacking digital India: A feminist commentary on policy agendas in the digital moment', *Journal of Information Policy*. JSTOR, 6(1), pp. 371–402.

Harding, S., Kandlikar, M. and Gulati, S. (2016) 'Taxi apps, regulation, and the market for taxi journeys', *Transportation Research Part A: Policy and Practice*. Elsevier, 88, pp. 15–25.

Hern, A. (2015) 'Why the term "sharing economy" needs to die', *The Guardian*, 5 October. Available at: https://www.theguardian.com/technology/2015/oct/05/why-the-term-sharing-economy-needs-to-die (Last accessed 3 February, 2020).

HT Correspondent (2019) 'App-based cabs like Ola-Uber illegal, says Maharashtra transport minister', *Hindustan Times*, 5 February. Available at: https://www.hindustantimes.com/mumbai-news/app-based-cabs-like-ola-uber-illegal-says-maharashtra-transport-minister/story-CBUEiy1NrnLoO5994AconJ.html (Last accessed 3 February, 2020).

Huws, U. (2014) *Labor in the Global Digital Economy: The Cybertariat Comes of Age*. New York: NYU Press.

John, N. A. (2013) 'The social logics of sharing', *The Communication Review*. Taylor & Francis, 16(3), pp. 113–131.

Kapoor, A. (2015) 'Driving past caste: End of the kaali peeli taxi and the stand', *Hard News*, 25 October. Available at: http://www.hardnewsmedia.com/2015/10/driving-past-caste (Last accessed 3 February, 2020).

Kapoor, M. (2018) 'The Lure of Ola-Uber economy is waning in jobs-starved India', *Bloomberg Quint*, 28 March. Available at: https://www.bloombergquint.com/global-economics/the-lure-of-ola-uber-economy-is-waning-in-jobs-starved-india#gs.SM85aC1S (Last accessed 3 February, 2020).

Kashyap, R. and Bhatia, A. (2018) 'Taxi drivers and taxidars: A case study of Uber and Ola in Delhi', *Journal of Developing Societies*. SAGE Publications Sage India: New Delhi, India, 34(2), pp. 169–194.

Kurmanath, K. (2018) 'Start-up to give Mumbai's kaali-peeli taxis a tech facelift', *Hindu Business Line*, 20 January. Available at: https://www.thehindubusinessline.com/info-tech/startup-to-give-mumbais-kaalipeeli-taxis-a-tech-facelift/article8258136.ece (Last accessed 3 February, 2020).

Luque-Ayala, A. and Marvin, S. (2015) 'Developing a critical understanding of smart urbanism?', *Urban Studies*. Urban Studies Journal Limited, 52(12), pp. 2105–2116.

Malin, B. J. and Chandler, C. (2017) 'Free to work anxiously: Splintering precarity among drivers for Uber and Lyft', *Communication, Culture & Critique*. Oxford University Press, 10(2), pp. 382–400.

Moore, M. H. (1995) *Creating Public Value: Strategic Management in Government*. Cambridge, MA: Harvard University Press.

Morozov, E. (2013) 'The "sharing economy" undermines workers' rights', *Financial Times*, 14 October. Available at: https://www.ft.com/content/92c3021c-34c2-11e3-8148-00144feab7de (Last accessed 3 February, 2020).

Mukherjee, S. (2016) 'Uber hails 200,000 cabs with eye to expand India drive', *The Economic Times*, 30 September. Available at: https://economictimes.indiatimes.com/small-biz/startups/uber-hails-200000-cabs-with-eye-to-expand-india-drive/articleshow/54594178.cms?from=mdr (Last accessed 3 February, 2020).

Muralidhar, S. H. (2016) 'How Ola disrupted taxi services in India?', *Review of Management*. Management Development Research Foundation, 6(3/4), pp. 5–17.

Nadeem, S. et al. (2015) 'On the sharing economy', *Contexts*, 14(1), pp. 12–19. doi: 10.1177/1536504214567860.

Nagy, P. and Neff, G. (2015) 'Imagined affordance: Reconstructing a keyword for communication theory', *Social Media+ Society*. SAGE Publications Sage UK: London, England, 1(2), p. 2056305115603385.

Neff, G. (2012) *Venture Labor: Work and the Burden of Risk in Innovative Industries*. Cambridge, MA: MIT press.

Pasquale, F. (2017) *From Territorial to Functional Sovereignty: The Case of Amazon, Law and Political Economy Blog*. Available at: https://lpeblog.org/2017/12/06/from-territorial-to-functional-sovereignty-the-case-of-amazon/ (Accessed: 3 February 2020).

Philip, C. Mathew (2018) 'A year on, cabbies who pinned hopes on Namma TYGR left in the lurch', *The Times of India*. 7 October Available at: https://timesofindia.indiatimes.com/city/bengaluru/a-year-on-cabbies-who-pinned-hopes-on-namma-tygr-left-in-the-lurch/articleshow/66111692.cms (Last accessed 3 February, 2020).

Pillai, S. (2016) 'Uber's new "friends": Recruiters who earn lakhs in referrals', *The Times of India*, 6 March. Available at: https://timesofindia.indiatimes.com/business/india-business/Ubers-new-friends-Recruiters-who-earn-lakhs-in-referrals/articleshow/52613104.cms (Last accessed 3 February, 2020).

Plantin, J.-C. et al. (2018) 'Infrastructure studies meet platform studies in the age of Google and Facebook', *New Media & Society*, 20(1), pp. 293–310. doi: 10.1177/1461444816661553.

Prabhat, S., Nanavati, S. and Rangaswamy, N. (2019) 'India's "Uberwallah" profiling Uber drivers in the gig economy', in *Proceedings of the Tenth International Conference on Information and Communication Technologies and Development*, pp. 1–5.

Price, L. L. and Belk, R. W. (2016) 'Consumer ownership and sharing: Introduction to the issue', *Journal of the Association for Consumer Research*. University of Chicago Press Chicago, IL, 1(2), pp. 193–197.

Raheja, N. and Deutsch, C. (2016) *India's Uber Dilemma: Entrepreneurship or Exploitation?*, *Sapiens Blog*. Available at: https://www.sapiens.org/technology/ride-sharing-gig-economy-india/ (Last accessed 3 February, 2020).

Rayle, L. et al. (2016) 'Just a better taxi? A survey-based comparison of taxis, transit, and ridesourcing services in San Francisco', *Transport Policy*. Elsevier, 45, pp. 168–178.

Richardson, L. (2015) 'Performing the sharing economy', *Geoforum*. Elsevier, 67, pp. 121–129.

Rodgers, S. and Moore, S. (2018) 'Platform urbanism: An introduction', *Mediapolis: A Journal of Cities and Culture*, 4(3), Available at: http://www.mediapolisjournal.com/201.

Rosenblat, A. and Stark, L. (2016) 'Algorithmic labor and information asymmetries: A case study of Uber's drivers', *International Journal of Communication*, 10, pp. 3758–3784.

Schwarz, J. A. (2017) 'Platform logic: An interdisciplinary approach to the platform-based economy', *Policy & Internet*. Wiley Online Library, 9(4), pp. 374–394.

Shahane, G. (2015) 'Why I gladly abandoned Bombay's kaali peeli cabs for Uber and Ola', *Scroll.in*, 17 June. Available at: https://scroll.in/article/734855/why-i-gladly-abandoned-bombays-kaali-peeli-cabs-for-uber-and-ola (Last Aaccessed 3 February, 2020).

Singh, M. (2016a) 'India's Ola partners with car maker Mahindra in its race against Uber', *Mashable*, 8 September. Available at: https://mashable.com/2016/09/08/ola-mahindra-india/ (Last accessed 3 February, 2020).

Singh, M. (2016b) 'Uber to make it easier for drivers to own a car in India', *Mashable*, 30 September. Available at: https://mashable.com/2016/09/30/uber-ola-drivers-leasing-cars-india/ (Last accessed 3 February, 2020).

Surie, A. (2017) 'Working on the app-based model', *The Hindu*. Available at: https://www.thehindu.com/opinion/op-ed/working-on-the-app-based-model/article19458973.ece (Last accessed 3 February, 2020).

Surie, A. (2018) 'Are Ola and Uber drivers entrepreneurs or exploited workers', *Economic & Political Weekly*, 53(24). Available at: https://www.epw.in/engage/article/are-ola-and-uber-drivers-entrepreneurs-exploited-workers.

Surie, A. and Koduganti, J. (2016) 'The emerging nature of work in platform economy companies in Bengaluru, India: The case of Uber and Ola Cab drivers', *E-Journal of International and Comparative Labour Studies*, 5(3), pp. 1–30.

Sutzl, W. (2014) 'The anti-economy of sharing', *Else Art Journal*, 1(0), pp. 122–140.

Swaminathan, M. (2010) 'Caste and the labour market', *The Hindu*. Available at: https://www.thehindu.com/books/Caste-amp-the-labour-market/article16550188.ece (Last accessed 3 February, 2020).

Thorat, S. and Attewell, P. (2007) 'The legacy of social exclusion: A correspondence study of job discrimination in India', *Economic & Political Weekly*, 42(41), JSTOR, pp. 4141–4145.

Uber India (2016) *#EmpowerAMillion Indians with UberSHAAN.*, *Uber India Blog*. Available at: https://www.uber.com/en-IN/blog/ubershaan/ (Accessed: 3 February 2020).

Wilson, M. W. (2018) 'On the slippages in platform urbanisms', *Mediapolis*, 3(4). Available at: https://www.mediapolisjournal.com/2018/11/slippages-in-platform-urbanisms/ (Last accessed 3 February, 2020).

Zwick, A. (2018) 'Welcome to the Gig Economy: Neoliberal industrial relations and the case of Uber', *GeoJournal*. Springer, 83(4), pp. 679–691.

17

PLATFORMS IN THE MAKING

Hacking the urban environment in Brazilian cities

Andrés Luque-Ayala, Rodrigo José Firmino, Tharsila Maynardes Dallabona Fariniuk, Gilberto Vieira and Juliana Marques

Introduction

This chapter unpacks a particular aspect of the emerging urban digital platform: the coming together of digital tools, digital/data activism and political asymmetries in both re-imagining and re-making the city's environment. Worldwide, the work of civic hackers, hackathons and other forms of data activism has eventually led to a range of city platforms—from municipal open data platforms to commercial and non-commercial applications (Luque-Ayala and Marvin, 2020). These platforms intervene in diverse urban issues, such as transport, public safety, citizen engagement and waste management. We argue that data activism and digital interventions prefigure urban platforms both materially and in terms of their political orientation. While not all digital platforms have roots in civic tech, we suggest that within practices of digital and data activism in cities there are always a range of *platforms in the making.*

We are particularly interested in forms of digital activism that seek to intervene in the broadly defined "urban environment" and the city's ecological flows (e.g. transport, energy, water, sanitation, waste and other infrastructures), as well as in the city's potential capacity for environmental stewardship and socio-ecological sustainability. These include hackathons, maker spaces, hacking collectives and other experimental efforts that, in the context of our study in large Brazilian cities, generated over five years more than 60 small-scale digital platforms and interventions aimed at urban ecological flows and socio-environmental issues. Our analysis suggests two critical findings. First, within the spaces of digital activism, both data and digital processes define how the city is to be understood and engaged with. Second, with few exceptions, there is a level of disconnect between the issues addressed by digital interventions and the long-standing issues, sites and stakeholders that, over the course of decades, have become a primary focus for social and environmental activism within cities. This is a tension that we capture through establishing a distinction between "the digital as a form of activism" (data-led interventions) and "the digital as it encounters *pre-existing* forms of activism" (situated interventions).

Our empirical focus is on the global South, specifically Brazilian cities, where the unequal provision of urban services and infrastructures, profound asymmetries of power and growing socioeconomic gaps exacerbate the political tensions at the crossover of activism and the environment. We pay special attention to four cities with the greatest concentration of digital interventions on the urban environment: Rio de Janeiro, São Paulo, Recife and Porto Alegre. The analysis looks at the tensions and differences between data-led forms of activism drawn from digital and civic-tech practices and situated interventions involving traditional forms of activism learning to mobilise digital technology for their own political ends. We conclude with a case study of *data_labe*, a Rio de Janeiro-based collective mobilising data and digital tools to challenge urban exclusion and create new narratives about favelas. *data_labe*, we argue, illustrates the coming together of traditional forms of urban activism and digital/data-based interventions, and in doing so it exemplifies a situated intervention where digital practices and epistemologies encounter *pre-existing* forms of activism.

From digital activism to platform capitalism—and back!

Recent academic analyses of civic hacking and what is known within practitioner circles as the civic-tech community have emphasised their role as forms of data activism and advocacy (Schrock, 2016; Schrock and Shaffer, 2017; Luque-Ayala and Marvin, 2020). Civic hacking "has enriched the original meaning and purposes of "hacking", transforming itself into alternative place-making practices [... re-creating] urban governance, community engagement, and the meaning and practice of urban everyday life (Townsend, 2013)" (Perng and Kitchin, 2018: p. 2). Hacking, praised for its desire to challenge and transform the status quo (Wark, 2006), inevitably has implications for the political-economic order—often linked to a transformation in forms of production and the mechanisms for capital accumulation (Luque-Ayala and Marvin, 2020). For some, hackers are a separate social class with the ability to liberate information for the common good, bypassing capitalist forms of production and creating a new societal order (Wark, 2006). For others, a reading of hackers as a counter-class movement is both partial and short of idealistic given their entanglement with entrepreneurial logics and involvement with the corporate and business stakeholders advancing the digital economy (Luque-Ayala and Marvin, 2020).

Urban platforms are clearly connected to activist or civic tech movements. Many popular digital platforms are derived from the work of civic hackers and makers, or from the hackathons and civic tech events and communities often sponsored by well-established platforms and digital economy business such as Amazon, eBay, Facebook and GitHub. Hackathons and civic hack nights are popular with aspiring start-up entrepreneurs as well as venture capitalists and have resulted in commercially successful urban platforms. For example, Roadify and Embark NYC are transit platforms that emerged from the NYC BigApps competition while SpotCrime.com is a security platform that emerged from Baltimore's civic tech scene. When used within a business model, these platforms can capture value within a process of capitalization—"valori[sing] potential for monopoly rents, [and prioritizing] up-scaling and the direct and/or indirect extraction rent from circulations and accompanying data trails" (Langley and Leyshon, 2017: p. 25). The popularity of collaboration between the business world, civic tech communities and entrepreneurial governance substantiates the view of urban platforms as mainly commercial products promoting platform capitalism (cf. Srnicek, 2017).

However, urban platforms can also act as forms of intermediation outside capital accumulation processes. Examining civic tech spaces reveals "platforms in the making"—both for-profit and not-for-profit, as well as both successes and failures. When digital interventions and platforms are appropriated, reassembled and redirected towards socio-environmental issues, they also allow the civic tech community to experiment with social engagement, citizen action and technopolitical agendas. Thus, digital interventions and platforms can also be framed as forms of activism in which civic tech practices allow "ordinary people" to use their skills to "change current political, social, and/or economic circumstances, policies, and values" (Takahashi, Kitchin and Thrift, 2009).

Thinking the city through data activism in four Brazilian cities

The empirical material for this chapter was collected via a web-based desktop review of seven cities in Brazil, where the popularity of the free and open-source software (FOSS) movement has shaped the civic tech community (Richter, Zo and Maruschke, 2009; Shaw, 2011; Leister and Frazier, 2014; Evangelista, 2018). We interrogated the Brazilian version of the Google search engine (google.com.br) against a pre-defined list of descriptors to identify civic tech community activities and digital interventions targeting urban socio-environmental issues and ecological flows. All sources in the review were posted between 2013 and 2018 and resulted in the identification of over 60 digital interventions in the following categories: hackathons, fablabs/maker spaces, academic initiatives, individual/private projects, hacking collectives; smartphone apps, and training workshops. Of the seven cities, four stood out by the number of initiatives or by their characteristics in terms of organised groups, project funding and connections to debates and topics linked to the urban environment: Rio de Janeiro, São Paulo, Recife and Porto Alegre (all state capitals within large metropolitan regions). The following pages describe the emerging landscape of digital interventions in each city, along with a brief description of their socio-environmental conditions.

Rio de Janeiro

Rio has over 6.7 million inhabitants and a unique topography that forces the majority of its population to live in narrow strips of land between its hills and the sea. The city's striking natural landscape makes it a sought-after location for global mega-events, such as the 2016 Olympics, the 2014 World Cup, and large United Nations conferences. Despite the significant investment and recognition brought by these events, Rio is often portrayed as an international exemplar of social and economic inequality that coexists with and reinforces pressing environmental challenges. For example, over 90% of the city's sewage is estimated to be released into the environment without any type of treatment (Instituto Trata Brasil, 2016). At the same time, the nearly 1.5 million people living in favelas face poor access to energy, water and transport infrastructures—a condition of inequality that is compounded by a persistent narrative that associates favelas with violence and crime (Ribeiro and Lago, 2001). In this context, issues of socio-environmental sustainability are often a forgotten topic.

Rio has also hosted a range of hackathons and civic tech events, many of which focus on mobilising data and digital tools to solve urban problems. In 2013 the municipality sponsored the Hackathon 1746, which became the first in a series of city-wide hackathons organised by different types of stakeholders. These hackathons culminated in the 2018 and 2019 Hacking.Rio events, the largest hackathons in Latin America. Common themes included promoting open data logics, advocating for the use of open data sets and open data platforms and encouraging technologies for transparency and accountability (a particular concern given Brazil's history of power and corruption). As is typical of hackathons, these were relatively fleeting events lasting only a few days. Maker spaces and FabLabs (fabrication laboratories), also popular in Rio, account for a set of digital interventions with greater permanence in the city. Several maker spaces and FabLabs have focused on minorities and urban peripheries (e.g. favela dwellers), following a narrative around the advancement of social transformation via technology.

The largest share (46%) of the digital interventions identified in Rio de Janeiro was associated primarily to either urban infrastructure or transparency/accountability topics with a smaller proportion targeting urban mobility, environment or sustainability and gender. Strikingly, digital interventions focused on issues of security, race or sanitation were largely absent in Rio.

São Paulo

São Paulo is Brazil's largest city, with 12 million inhabitants, and also its main centre of economic activity. Like Rio, the city experiences a broad range of environmental problems, including high levels of air pollution, water shortages and poor sanitation infrastructure. In 2013, a combination of severe drought and poor water management resulted in São Paulo's most serious environmental crisis in recent decades—the 2013 drought was considered the worst in recorded history, leaving the state on the verge of water supply collapse (Martirani and Peres, 2016). City authorities are concerned with chronic water shortages, poor air quality linked to high levels of car ownership, river pollution and a shortage of landfill space. São Paulo has the highest prison population in Brazil and is the source of some of the country's most powerful organised crime groups, so violence and security have historically preoccupied city inhabitants and driven public policy.

As one of the most economically active cities in Latin America, São Paulo competes with Rio for the title of the largest technology hub in the region. In 2016, the majority of the city's creative enterprises focused on information technology and data, representing more than a third of the total number of businesses of this type nationally (Observatório de Sâo Paulo - DIEESE, 2016). City authorities have supported the development of FabLabs, and the city hosts a number of hacking collectives that work alongside digital start-ups to drive rapid growth within the city's digital economy. Digital interventions identified in São Paulo focused on cybersecurity, with fewer addressing environmental problems, transport and mobility.

Recife

With 1.6 million inhabitants, Recife is the most dynamic hub of economic activity of the Northeast of Brazil. The city is located at the confluence of the Capibaribe and the Beberibe rivers—a feature of cultural, economic and environmental importance. Known as the "mangrove town" because of the previously abundant ecosystems now replaced by urban settlements, Recife's land-use patterns have created long-standing ecological problems. Historical tensions over the city's delicate ecological configuration persist between real estate markets, planners, environmental activists and politicians. Despite strict legislation protecting valuable ecosystems, mangrove devastation has continued over the past decades as growing informal settlements mount pressure on the natural environment (Sobrinho and Andrade, 2011). Over 40% of informal settlements in Recife are in areas of floodplains, wetlands and mangroves, deteriorating the city's ecology but also affecting issues of drainage, sanitation and environmental comfort (Souza, 2012).

Recife is also well known within Brazil for innovation and creativity, with the *Porto Digital* (Digital Port) housing a considerable group of technology firms in the renovated historical downtown area. This public-private initiative was established in 2000 to transform the historic city centre into a "Brazilian Silicon Valley" hub of innovation and entrepreneurship. In partnership with the municipal informatics institute, the Digital Port supports the growth of civic-tech communities through "digital creative clusters", hacking events and more (Marques and Borba, 2017). Since 2013, the annual *Hacker Cidadão* (Citizen Hacker) event encourages civic hackers to use the municipality's open data portal to identify solutions to urban problems (Prefeitura de Recife, 2017; Gonçalves and Santos da Gama, 2018). The majority of the identified digital interventions focused on mobility and transport, gender, transparency/accountability and health, with environmental issues playing only a minor role.

Porto Alegre

Comparable to Recife in size, Porto Alegre has an international reputation for innovative public management through citizen engagement (via, for example, participatory budgeting; see Célérier and Botey, 2015; Amaral and Carvalho, 2018). This reputation for transformative change was cemented through the city's leadership in establishing the World Social Forum, a counterpart to the World Economic Forum. Porto Alegre has become a symbol of Brazil's re-democratisation following the military dictatorship between 1964 and 1985, and a world reference for participatory democracy. Relative to other Brazilian cities, Porto Alegre has lower levels of social inequality (Sintomer, Röcke and Herzberg, 2016), although its high levels of industrialisation and agro-economy come with environmentally damaging practices such as the intense use of pesticides (Pasquetti et al., 2009).

Porto Alegre also hosts the annual International Free Software Forum (FISL), which is an important event for the global free and open-source software (FOSS) community. Partly as a result of this, the city has also nurtured a local community of civic technologists working on public issues. An early survey of the city's digital activism scene identified a substantive amount of work on urban issues, distributed over two broadly defined camps: those working on environmental issues and those focused on urban renovation, infrastructure and enterprise (Moraes, 2012). In contrast to the previous three cities, the primary focus of digital interventions identified in Porto Alegre addressed issues of social entrepreneurship, with the majority of initiatives being maker spaces or FabLabs (55%). Like in Recife, a large number of initiatives focus on urban mobility, transparency and accountability.

Distant dreams of digital activism

Figure 17.1 shows that nearly 70% of the identified interventions are either one-off events or inactive at the time of the survey, and that almost half (46%) are hackathons. Although ephemeral, the latter are understood to have the potential to generate entrepreneurship via platforms. Previous research on hackathons in global North contexts have identified the critical relevance of outcomes that "address place-specific needs" (Johnson and Robinson, 2014: p. 355); whether hackathons produce such results is inconclusive (ibid). Hackathons in Brazilian cities typically revolve around creating digital interventions to solve sectoral problems with the support of companies, universities and local governments. The typical participant conforms to civic hacker stereotypes: young, affluent, and male data enthusiasts[1], drawing on loose or weak institutional affiliations to work on public issues in the context of dedicated events or commissioned projects.

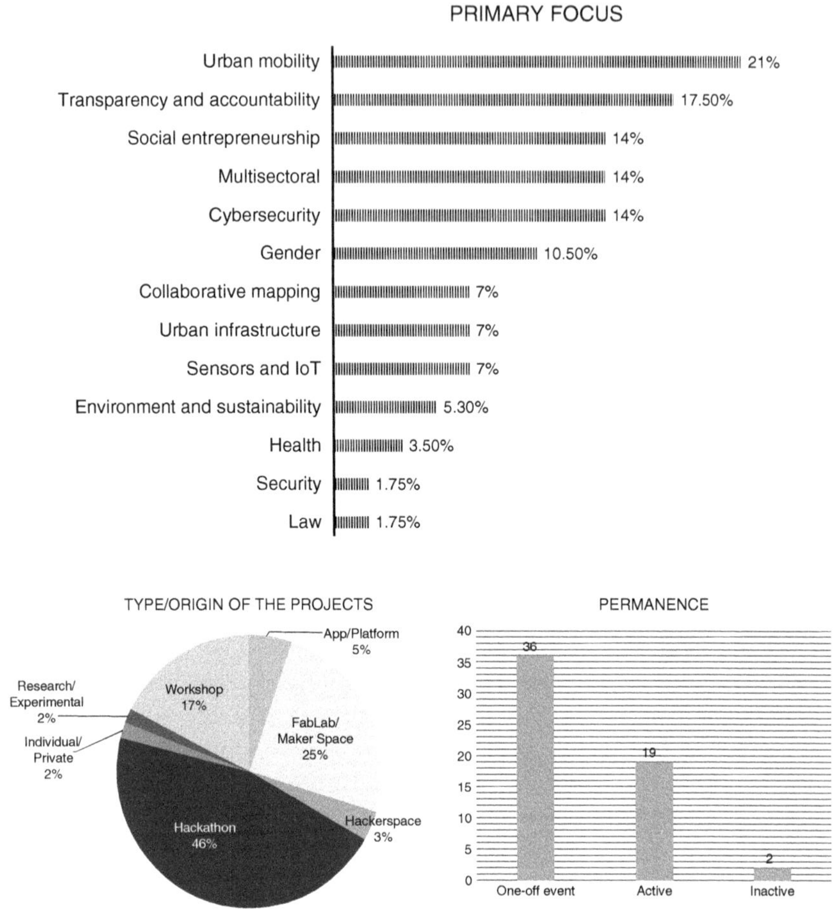

FIGURE 17.1 Combined results for all 57 cases in all 4 cities.
"Primary focus" shows the percentage of initiatives having each theme as one of their key foci. Thus, one single initiative can be counted on more than one topic simultaneously.

The focus of these corporate-driven hackathons is dominated by traditional themes, with urban mobility, transparency/accountability and social entrepreneurship among the most popular. Participant motivations were usually associated with product design, the mobilisation of data, and the development of digital tools to tackle narrowly defined problematics via short-term solutions. This favours an entrepreneurial approach to urban management (Barns, 2016) relying on a collective ability to perform data-based calculations and visualisations (Luque-Ayala and Marvin, 2020). The combined results of all four cities suggest a disconnect between the civic tech communities or digital interventions and the profound, challenging and historically grounded socio-environmental problematics within the conurbations. The actions of civic hackers and the civic tech community, including data activism and advocacy for the open data movement, show "the digital as a form of activism" or what we term *data-led interventions.*

Digital encounters with pre-existing forms of activism

The data-led interventions above stand in contrast to "the digital as it encounters pre-existing forms of activism", or what we call *situated interventions.* The final section showcases the work of *data_labe* (www.datalabe.org)—a collective based, in the Northern area of Rio de Janeiro. *data_labe* illustrates the appropriation of digital technologies towards situated and pre-existing forms of urban activism. *data_labe* has arguably emerged as a situated urban platform that both challenges and transcends traditional epistemologies and political orientations associated with civic-tech domains.

Established in 2016, *data_labe* has three aims: technical and political training of favela residents in journalism and data science; the generation and distribution of content on several social media and digital platforms; and "citizen-generated data" or the primary collection of data for local issues. Its operations focus exclusively on the *Complexo da Maré,* a conglomerate of 16 favelas with more than 140,000 residents.[2] Maré has over 40,000 households in a variety of dwelling types, from informal settlements to social housing estates, illustrating the unevenly distributed public services and socio-economic factors characteristic of large Brazilian cities. Schools, healthcare, social services, basic sanitation and cultural facilities are present, but in numbers and quality far below those of other sectors of Rio. Like many other favelas in Rio, Maré is seen by the state as an unruly, lawless, needy and violent territory.

data_labe sees itself as "a laboratory of data and narratives". Its members are data-journalists and young favela inhabitants who experiment with ways "to make data visible, in order to safeguard new narratives that allow the development of communities and their rights".[3] The city's imaginaries, pervasive centre-periphery dichotomies (both in terms of subjects and territories) and the agency of its inhabitants all play a central role within *data_labe's* debates and projects. As an activist movement rooted within digital culture logics (Wright and McCarthy, 2004; Gere, 2009), *data_labe* uses conceptual and aesthetic references that promote encounters between technology, class consciousness, racial empowerment, public policy, human rights, horizontality and entrepreneurship. Projects developed by the organisation focus on diverse locally relevant topics ranging from gender, race and sexuality to sanitation, territory and technology. One recent project compares the 2014 and 2018 military occupations of Maré and looks at the differentiated role of state and federal governments—as seen through locally generated data on how the military interventions affected residents' perception of security. Other recent projects mobilise both data and narrative to report' on the everyday stigma and racism experienced by migrant Angolan communities, or on the experiences of young black HIV-positive residents of Rio—both of which challenge existing narratives by showing alternative data in support of appropriate public policies.[4]

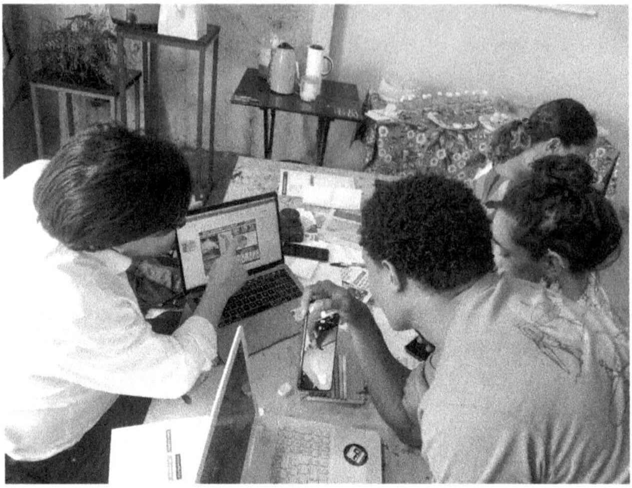

FIGURE 17.2 CocôZap Hackathon at data_labe's workshop (December 2019), aimed at imagining ways of using data as a form of infrastructural activism around sanitation.

Source: The Authors.

One project with the potential to intervene in the environment and ecological flows of the city is *CocôZap*, a citizen-led database of sanitation problems in three favelas at Maré. *Cocô* is an informal Portuguese word for faeces and *Zap* is Brazilian slang for the WhatsApp messaging service. *CocôZap* is a reporting mechanism that operates via smartphones equipped with WhatsApp (Figure 17.2), allowing residents to provide details (including images and video) of recurrent sanitation issues like waste accumulation or sewage problems (Figure 17.3). The project allows *data_labe* to evidence, record and map unsanitary conditions in order to exert pressure on local authorities and to challenge official claims that 100% of Rio de Janeiro is served by waste collection services (SNIS, 2014). The data assembled in the *cocozap.datalabe.org* platform includes a digital map, photos, a Google Sheets spreadsheet and JavaScript APIs for anybody to download and use. Monthly meetings with residents, students, teachers, healthcare professionals and resident associations examine and discuss the data to promote debate on health and environmental issues. By coupling citizen data, WhatsApp, web-based platforms and social media, *data_labe* and partners[5] have opened channels for reporting problems, debating conditions, and proposing ideas around basic sanitation infrastructure in Maré.

CocôZap is one of several data-interventions developed in collaboration with communities within Maré—a move described by *data_labe* as a "peripheral data revolution". This recognises the emerging possibilities for engagement afforded by digital technologies and their assembly into urban platforms for those peripheral to political debates. Here data becomes a vehicle for disrupting and

FIGURE 17.3 Poor sanitation conditions in Maré, under a graffiti forbidding littering.
Source: The Authors.

problematising the common narratives that dominate how favelas and peripheral areas of Rio de Janeiro are imagined and the ways in which the state responds. Within *data_labe*'s operations, juxtaposing the notion of "narrative" with that of "data" is important as it challenges the calculative epistemology of data. Meaningful information capable of affecting change, thus, can take a multiplicity of forms, both numeric and textual *data_labe* understands the production of data, the act of telling stories, and the practice of disputing access to data as political and cultural matters; by mobilising citizens to tell their own stories, as well as providing access to data as much as creating it, *data_labe* situates digital activism within the political and cultural agendas of poor communities in Rio.

Conclusions

Our survey revealed an active community of digital activists in Brazilian cities. From hackathons sponsored by municipal authorities and FabLabs working on digitally enabled devices to hacker collectives and academic research projects, many of these digital interventions use the city as a primary site for and topic of digital experimentation. The thematic focus of these interventions is broad and varied, although the extent to which they meaningfully target "place-specific needs" (cf. Johnson and Robinson, 2014) is inconclusive. In keeping with long-standing concerns within the civic-tech community worldwide, issues of transparency and accountability take a prominent role, followed by transport and mobility and by social entrepreneurship. Problems associated with urban ecological flows and natural environments are much less commonly tackled.

Our preliminary analysis has identified multiple disconnects between the issues addressed by civic tech community-driven and data-led digital interventions and the issues and sites of greatest concern for social and environmental activists and suggests a twofold explanation. First, there is a significant geographic and demographic distance between those that participate in civic tech communities (university students, upper middle class professionals, predominantly white and male, living in middle-class neighbourhoods) and those most directly impacted by the problems targeted by traditional socio-environmental activism (often working class, low income, racially mixed, gender diverse and inhabiting the peripheral or informal urban spaces). Second, the activism envisioned within the civic tech community frames the urban environment in narrow and occasionally exclusionary ways by operating strictly through the data and digitalisation of urban processes (cf. (Luque-Ayala and Marvin, 2020). Effectively, these urban digital interventions can only access and act upon concerns through numeric data and processes of calculation.

Our analysis highlights the need to differentiate between digital activism that originates from within the city's civic-tech community and digital activism geographically and socio-politically grounded within traditional forms of urban activism. These "situated interventions" are llustrated through the *data_labe* collective in Rio de Janeiro, an organization that uses data and digital technologies as the primary tools for locally grounded political activism on issues of gender, race and the environment. Within *data_labe*, data reveals not only injustice and inequality, but also the state inaction in response to local concerns. Yet, data does not play a hegemonic role in imagining and advocating for a different city; *data_labe*'s operations always establish a—both productive and conflicting—dialogue between data and narrative; between objective evidence and a subjective truth. While both data-led and situated digital interventions generate urban politics, the politics they produce differ significantly. The former, exemplified by civic hackathons and their primary modus operandi via open data logics, data and digital process, seeks to remake the urban world through place-detached calculative epistemologies. In the latter, exemplified by urban activists asking for particular rights and improvements in living conditions associated with a specific territory, digital interventions are subservient to pre-existing socio-political struggles. These promote an engagement with digital technologies where the primacy of data and data-based epistemologies is challenged. Research leading to this chapter was supported by the British Academy project "Hacking the urban environment: smart cities and the role of civic hackers in remaking the city" (NAF2R\170051).

Notes

1 It is not surprising that most participants in hacker collectives and hackathons are white, young and upper-class males. Between 2000 and 2013 in Brazil, only 17% of computer science graduates were women (Maia, 2016) while most computer enthusiasts and those in the free and open-source-software movement (FOSS; a particularly active community of practice in Brazil; see Richter, Zo and Maruschke, 2009; Shaw, 2011; Takhteyev, 2012; Leister and Frazier, 2014) were men under 30 years old (Gilboa, 1996; Jordan and Taylor, 1998; Schell and Holt, 2009). Paz (2013)

and (Natansohn, 2013) argue that the 'digital gender division' is reinforced by a low female participation in such activities and the relatively few women leading software and digital system companies.

2 Maré is one of Rio de Janeiro's largest favela complexes. It was formally recognised as a neighbourhood in 1988 but dates back to the 1940s settlements associated with the construction of Avenida Brasil (Brazil Avenue), an important city thoroughfare.

3 From the original in Portuguese: "Como tornar os dados visíveis a ponto de garantir novas narrativas que permitam o desenvolvimento de comunidades plenas de direitos?" (http://datalabe.org/#narrativas).

4 The report celebrates the 2018 AIDS awareness month and was prepared by an HIV-positive young black resident following an invitation from *data_labe*.

5 *CocôZap* is supported by the Casa Fluminense, the Fundo Socioambiental Casa, Redes da Maré and other partners, including Durham University (UK) and the Pontifícia Universidade Católica do Paraná (Brazil).

References

Amaral, C. T. G. do and Carvalho, F. T. de (2018) 'Democracia e deliberação: a escolha popular das políticas públicas locais-o caso do orçamento participativo de Porto Alegre (OPPOA)', *Revista de Direito da Cidade*, 10(1), pp. 463–485.

Barns, S. (2016) 'Mine your data: Open data, digital strategies and entrepreneurial governance by code', *Urban geography*, 37(4), pp. 554–571.

Célérier, L. and Botey, L. E. C. (2015) 'Participatory budgeting at a community level in Porto Alegre: A Bourdieusian interpretation', *Accounting, Auditing & Accountability*, 28(5), pp. 739–772.

Evangelista, R. (2018) *Beyond Machines of Loving Grace: Hacker Culture, Cybernetics and Democracy*. São Paulo: Edições SESC/SP.

Gere, C. (2009) *Digital Culture*. London: Reaktion Books.

Gilboa, W. (1996) 'Elites, lamers, narcs, and whores: Exploring the computer underground.', in Cherny, L. and Weise, E. R. (eds) *Wired Women: Gender and New Realities in Cyberspace*. Seattle, WA: Seal Press.

Gonçalves, B. A. and Santos da Gama, K. (2018) 'Transparência e dados abertos do Recife: Uma estratégia bem sucedida de publicação', *Cadernos BAD*, 1, pp. 157–164. Available at: https://www.bad.pt/publicacoes/index.php/cadernos/article/viewFile/1901/pdf.

Instituto Trata Brasil (2016) *Pesquisa Saneamento Básico em àreas irregulars - Relatório Brasil*. São Paulo. Available at: http://www.tratabrasil.org.br/datafiles/estudos/areas-irregulares/volume2/relatorio-completo-areas-irregulares.pdf.

Johnson, P. and Robinson, P. (2014) 'Civic hackathons: Innovation, procurement, or civic engagement?', *Review of policy research*, 31(4), pp. 349–357.

Jordan, T. and Taylor, P. (1998) 'A sociology of hackers', *The Sociological Review*, 46(4), pp. 757–780.

Langley, P. and Leyshon, A. (2017) 'Platform capitalism: The intermediation and capitalisation of digital economic circulation', *Finance and Society*, 3(1), pp. 11–31.

Leister, V. and Frazier, M. (2014) 'Brazil's role in emerging global open-source initiatives', in *12th Congress of the Brazilian Studies Association (BRASA XII)*. London.

Luque-Ayala, A. and Marvin, S. (2020) *Urban Operating Systems: Producing the Computational City*. Cambridge, MA: MIT University Press.

Maia, M. M. (2016) 'Limites de gênero e presença feminina nos cursos superiores brasileiros do campo da computação', *Cadernos Pagu (Universidade Estadual de Campinas)*, 46, pp. 223–244.

Marques, L. and Borba, C. (2017) 'Co-creating the city: Digital technology and creative tourism', *Tourism Management Perspectives*, 24, pp. 86–93.
Martirani, L. A. and Peres, I. K. (2016) 'Water crisis in São Paulo: News coverage, public perception and the right to information', *Ambiente & Sociedade*, 19(1), pp. 1–20.
Moraes, F. D. de (2012) *A "Cidade digital" de Porto Alegre (RS): um estudo sobre espaço urbano e tecnologias de informação e comunicação a partir da apropriação do estado e de grupos (CIBER) ativistas, Programa de Pós-Graduação em Geografia*. Porto Alegre: Federal University of Rio Grande do Sul.
Natansohn, G. (2013) 'Introdução: Qué têm a ver as tecnologias digitais com o gênero?', in Natansohn, G. (ed.) *Internet em código feminino: teorias e práticas*. Buenos Aires: La Crujía, pp. 15–38.
Observatório de Sâo Paulo - DIEESE (2016) *Indicadores*. Sâo Paulo. Available at: http://saopaulo.dieese.org.br/indicadores.php.
Pasquetti, G. G. et al. (2009) 'Meio Ambiente e Desenvolvimento na Região Metropolitana de Porto Alegre: notas introdutórias baseadas no espaço rural metropolitano', in *X Salão de Iniciação Científica da PUCRS*. Pontifical Catholic University of Rio Grande do Sul, Porto Alegre. Available at: http://www.pucrs.br/edipucrs/XSalaoIC/Ciencias_Sociais_Aplicadas/Economia/70943-GIORGIA_GONCALVES_PASQUETTI.pdf.
Paz, M. de S. D. (2013) 'A divisão digital de gênero no movimento software livre do Brasil', in Natansohn, G. (ed.) *Internet em código feminino: teorias e práticas*. Buenos Aires: La Crujía, pp. 151–162.
Perng, S.-Y. and Kitchin, R. (2018) 'Solutions and frictions in civic hacking: Collaboratively designing and building wait time predictions for an immigration office', *Social & Cultural Geography*. Routledge, 19(1), pp. 1–20. doi: 10.1080/14649365.2016.1247193.
Prefeitura de Recife (2017) *Hacker Cidadão 5.0*. Recife. Available at: http://www2.recife.pe.gov.br/etiquetas/hacker-cidadao-50.
Ribeiro, L. C. de Q. and Lago, L. C. (2001) 'A oposição favela-bairro no espaço social do Rio de Janeiro', *São Paulo em Perspectiva*, 15(1), pp. 144–154.
Richter, D., Zo, H. and Maruschke, M. (2009) 'A comparative analysis of open source software usage in Germany, Brazil, and India', in *Fourth International Conference on Computer Sciences and Convergence Information Technology*. Seoul, South Korea: IEEE, pp. 1403–1410.
Schell, B. H. and Holt, T. J. (2009) 'A profile of the demographics, psychological predispositions, and social/behavioral patterns of computer hacker insiders and outsiders', in Chen, K. and Fadlalla, A. (eds) *Online Consumer Protection: Theories of Human Relativism*. London: IGI Global, pp. 190–213.
Schrock, A. R. (2016) 'Civic hacking as data activism and advocacy: A history from publicity to open government data', *New Media & Society*, 18(4), pp. 581–599.
Schrock, A. and Shaffer, G. (2017) 'Data ideologies of an interested public: A study of grassroots open government data intermediaries', *Big Data & Society*, 4(1), p. 2053951717690750.
Shaw, A. (2011) 'Insurgent expertise: The politics of free/livre and open source software in Brazil', *Journal of Information Technology & Politics*, 8(3), pp. 253–272.
Sintomer, Y., Röcke, A. and Herzberg, C. (2016) *Participatory Budgeting in Europe: Democracy and Public Governance*. New York: Routledge.
SNIS – Sistema Nacional de Informações sobre Saneamento (2014) *Diagnóstico do Manejo de Resíduos Sólidos Urbanos*. Brasilia: Governo Federal (Brasil). Available at: http://www.snis.gov.br/diagnostico-residuos-solidos/diagnostico-rs-2014.

Sobrinho, M. A. da M. and Andrade, A. C. (2011) 'O desafio da conservação de manguezais em áreas urbanas: identificação e análise de conflitos socioambientais no Manguezal do Pina–Recife–PE–Brasil', *Unimontes Científica*, 11(1/2), pp. 8–16.

Souza, C. P. (2012) *Políticas públicas ambientais e gestão do ecossistema manguezal da Bacia do Pina – Recife/PE: análise do licenciamento ambiental do Sistema Viário*. Universidade Federal de Pernambuco, Recife. Available at: https://repositorio.ufpe.br/handle/123456789/10513.

Srnicek, N. (2017) *Platform Capitalism*. Cambridge: Polity.

Takahashi, L. M., Kitchin, R. and Thrift, N. (2009) 'Activism', in *International Encyclopedia of Human Geography*. Oxford: Elsevier, pp. 7–14.

Takhteyev, Y. (2012) *Coding Places: Software Practice in a South American City*. Cambridge, MA: MIT Press.

Townsend, A. M. (2013) *Smart Cities: Big Data, Civic Hackers, and the Quest for a New Utopia*. New York and London: W. W. Norton & Company.

Wark, M. (2006) 'Hackers', *Theory, Culture & Society*, 23(2–3), pp. 320–322.

Wright, P. and McCarthy, J. (2004) *Technology as Experience*. Cambridge, MA: MIT Press.

INDEX

Note: **Bold** page numbers refer to tables; *italic* page numbers refer to figures and page numbers followed by "n" denote endnotes.

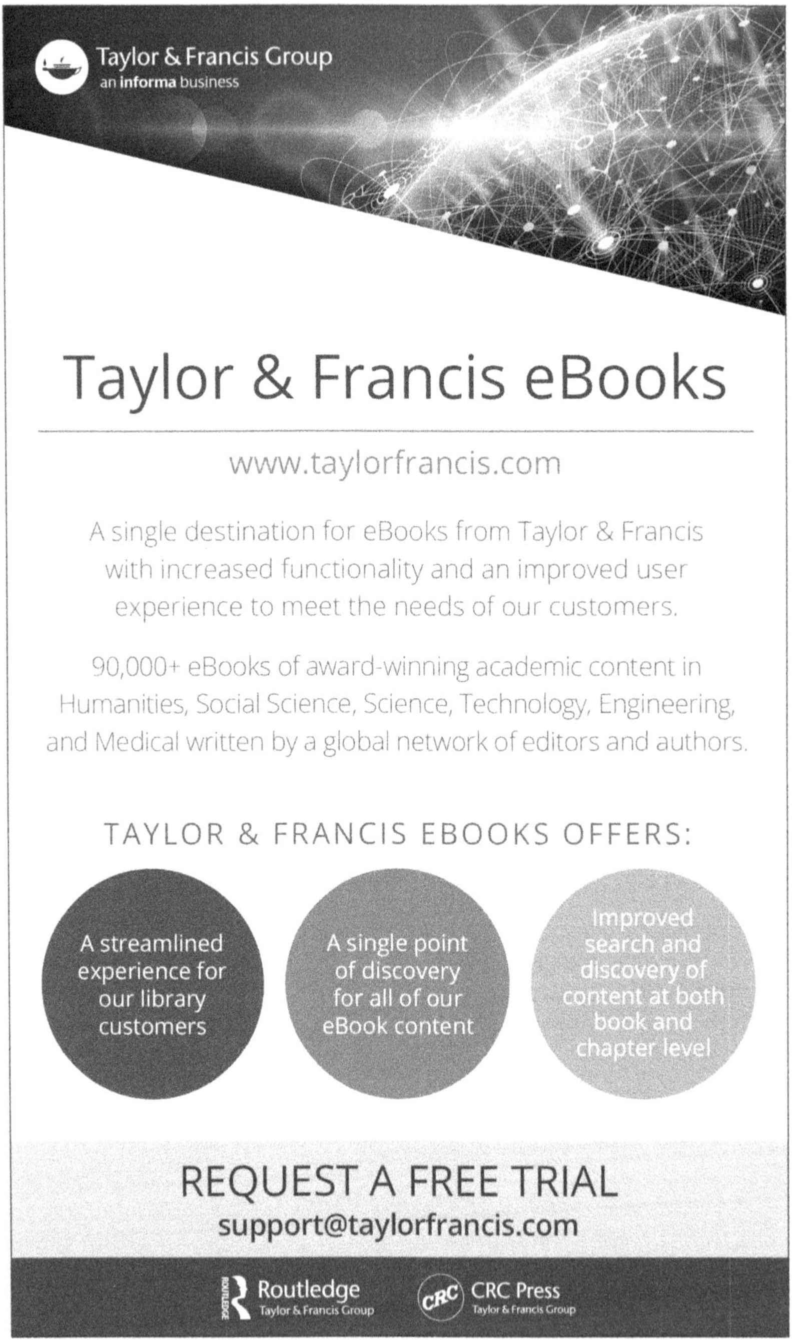
Taylor & Francis Group
an informa business
Taylor & Francis eBooks
www.taylorfrancis.com
A single destination for eBooks from Taylor & Francis with increased functionality and an improved user experience to meet the needs of our customers.
90,000+ eBooks of award-winning academic content in Humanities, Social Science, Science, Technology, Engineering, and Medical written by a global network of editors and authors.
TAYLOR & FRANCIS EBOOKS OFFERS:
A streamlined experience for our library customers
A single point of discovery for all of our eBook content
Improved search and discovery of content at both book and chapter level
REQUEST A FREE TRIAL
support@taylorfrancis.com
Routledge
Taylor & Francis Group
CRC Press
Taylor & Francis Group

For Product Safety Concerns and Information please contact our EU
representative GPSR@taylorandfrancis.com
Taylor & Francis Verlag GmbH, Kaufingerstraße 24, 80331 München, Germany

www.ingramcontent.com/pod-product-compliance
Ingram Content Group UK Ltd.
Pitfield, Milton Keynes, MK11 3LW, UK
UKHW022027061126
477531UK00002B/72
* 9 7 8 0 3 6 7 3 3 4 1 8 5 *